Supplements to the 2nd Edition of

RODD'S CHEMISTRY OF CARBON COMPOUNDS

ELSEVIER SCIENCE B.V.
Sara Burgerhartstraat 25
P.O. Box 211, 1000 AE Amsterdam, The Netherlands

Library of Congress Cataloging-in-Publication Data

(Revised for vol. IV (pts. C-D)

Supplements to the 2nd edition (editor S. Coffey) of Rodds's Chemistry of carbon compounds.

Cover title: Supplements to the 2nd edition of Rodd's Chemistry of carbon compounds.
Includes bibliographical references and indexes.
Contents: v. 1. Aliphatic compounds. pt. A. Hydrocarbons; halogen derivatives. pt. B.
Monohydric alcohols, their ethers and esters; sulphur analogues; nitrogen derivatives;
organometallic compounds.--pt. C. Monocarbonyl derivatives of aliphatic hydrocarbons, their
analogues and derivatives. pt. D. Dihydric alcohols, their oxidation products and derivatives.--
pt. E. Unsaturated acyclic hydrocarbons, trihydric alcohols, their oxidation products and
derivatives.--[etc.] v.IV pt. C. Five-membered heterocyclic compounds with two hetero-atoms in
the ring from groups V and/or VI of the periodic table.--pt. D. Five-membered heterocyclic
compounds with more than two hetero-atoms in the ring.--[etc.].
1. Chemistry, Organic.   2. Carbon compounds.   I. Coffey, S. (Samuel).   II. Ansell, Martin F.
(Martin Frederick).   III. Rodd's Chemistry of carbon compounds.   IV. Supplements to the
2nd edition of Rodd's Chemistry of carbon compounds.
QD251.R6 1964 Suppl 547   77-375408
ISBN 0-444-89932-4

ISBN 0-444-89932-4
© 1994 ELSEVIER SCIENCE B.V. All rights reserved.

This book is printed on acid-free paper.

Printed in The Netherlands

Supplements to the 2nd Edition of

RODD'S CHEMISTRY OF CARBON COMPOUNDS

VOLUME I

ALIPHATIC COMPOUNDS
★

VOLUME II

ALICYCLIC COMPOUNDS
★

VOLUME III

AROMATIC COMPOUNDS
★

VOLUME IV

HETEROCYCLIC COMPOUNDS
★

Supplements to the 2nd Edition (Editor S. Coffey) of

RODD'S CHEMISTRY OF CARBON COMPOUNDS

A modern comprehensive treatise

Edited by
MARTIN F. ANSELL
Ph.D., D.Sc.(London) F.R.S.C. C. Chem.
Reader Emeritus, Department of Chemistry,
Queen Mary College, University of London, Great Britain

Supplement to

VOLUME IV HETEROCYCLIC COMPOUNDS

PART C:

Five-Membered Heterocyclic Compounds with
Two Hetero-Atoms in the Ring from Groups V and/or VI
of the Periodic Table

PART D:

Five-Membered Heterocyclic Compounds with More than
Two Hetero-Atoms in the Ring

ELSEVIER
Amsterdam – London – New York – Tokyo   1994

CONTRIBUTORS TO THIS VOLUME

John D. Hepworth, B.Sc., Ph.D., C.Chem., F.R.S.C.
Department of Chemistry
The University of Central Lancashire, Preston, PR1 2HE, England

John H. Little, B.Sc., Ph.D.
Department of Chemistry
Sheffield Hallam University, Sheffield, S1 1WB, England

Stephen T. Mullins, B.Sc., Ph.D., C.Chem, M.R.S.C.
Department of Chemistry
Brunel University, Uxbridge, Middlesex, UB8 3PH, England

David J. Rowe, B.Sc., Ph.D.
Department of Chemistry
University of Sunderland, Sunderland, SR2 7EE, England

Mark Wainwright, B.Sc., Ph.D., C.Chem., M.R.S.C.
Department of Chemistry
University of Central Lancashire, Preston, PR1 2HE, England

Malcolm Sainsbury, Ph.D., D.Sc., C.Chem., F.R.S.C.
Department of Chemistry
The University of Bath, Bath, BA2 7AY, England
(Index)

PREFACE TO SUPPLEMENT IV CD

Rodd's Chemistry of Carbon Compounds continues to maintain its position as the most comprehensive text on Organic Chemistry by the publication of Supplements to keep it up to date. Although each chapter in this supplement stands on its own, it is intended to be read in conjunction with the parent chapter in the second edition.

At a time when there are many specialist reviews, monographs and reports available there is still in my view an important place for a book such as *Rodd,* which gives a broader coverage of organic chemistry. One aspect of the value of this work is that it allows the specialist in one field to find out quickly what is happening in other fields of organic chemistry. On the other hand a chemist looking for information in a particular field of study will find in *Rodd* an outline of the important aspects of that field of chemistry together with leading references to other works where more detailed information may be found.

The appearance of this volume nearly completes the supplementation of the second edition. Only the publication of the supplement to Volume IV IJ, which is now in preparation, is required.

As editor I have been fortunate in my team of contributors who have maintained the high standard set in previous supplements and produced concise and readable reviews of the individual topics. I am grateful to them as I know that each of them has had to fit in their writing with the many other demands upon their time. My thanks also goes to my contributors' secretarial assistants who have produced the very clear manuscripts. Again I have pleasure in thanking Dr. Malcolm Sainsbury for the preparation of the extensive and detailed index which adds considerably to the value of this supplement. Finally I wish to thank the production staff at Elsevier for transforming the authors' manuscripts into the published work.

Martin F. Ansell   July 1993

CONTENTS

VOLUME IV C and D

Heterocyclic Compounds: Five-membered Heterocyclic Compounds with Two Hetero-atoms in the Ring from Groups V and/or VI of the Periodic Table

Five-membered Heterocyclic Compounds with More than Two Hetero-atoms in the Ring

Chapter 16. Five-membered Heterocyclic Compounds with
Two Nitrogen Atoms in the Ring
by S.T. MULLINS

Chapter 17. Five-membered Heterocyclic Compounds with
Two Different Hetero-atoms in the Ring
by J.D. HEPWORTH and M. WAINWRIGHT

Chapter 18. Five-membered Heterocyclic Compounds with Three Hetero-atoms in the Ring
by D.J. ROWE

Chapter 19. Five-membered Heterocyclic Compounds with
Four Hetero-atoms in the Ring
by J.H. LITTLE

OFFICIAL PUBLICATIONS

B.P.   British (United Kingdom) Patent
F.P.   French Patent
G.P.   German Patent
Sw.P.   Swiss Patent
U.S.P.   United States Patent
U.S.S.R.P.   Russian Patent
B.I.O.S.   British Intelligence Objectives Sub-Committee
                Reports
F.I.A.T.   Field Information Agency, Technical Reports
                of U.S. Group Control Council for Germany
B.S.   British Standards Specification
A.S.T.M.   American Society for Testing and Materials
A.P.I.   American Petroleum Institute Projects
C.I.   Colour Index Number of Dyestuffs and Pigments

SCIENTIFIC JOURNALS AND PERIODICALS

With few obvious and self-explanatory modifications the
abbreviations used in references to journals and periodicals
comprising the extensive literature on organic chemistry,
are those used in the World List of Scientific Periodicals.

LIST OF COMMON ABBREVIATIONS AND
SYMBOLS USED

A	acid
$\overset{\circ}{\text{A}}$	Ångström units
Ac	acetyl
a	axial; antarafacial
as, $asymm.$	asymmetrical
at	atmosphere
B	base
Bu	butyl
b.p.	boiling point
C, mC and μC	curie, millicurie and microcurie
c, C	concentration
C.D.	circular dichroism
conc.	concentrated
crit.	critical
D	Debye unit, 1×10^{-18} e.s.u.
D	dissociation energy
D	dextro-rotatory; dextro configuration
DL	optically inactive (externally compensated)
d	density
dec. or decomp.	with decomposition
deriv.	derivative
E	energy; extinction; electromeric effect; Entgegen (opposite) configuration
E1, E2	uni- and bi-molecular elimination mechanisms
E1cB	unimolecular elimination in conjugate base
e.s.r.	electron spin resonance
Et	ethyl
e	nuclear charge; equatorial
f	oscillator strength
f.p.	freezing point
G	free energy
g.l.c.	gas liquid chromatography
g	spectroscopic splitting factor, 2.0023
H	applied magnetic field; heat content
h	Planck's constant
Hz	hertz
I	spin quantum number; intensity; inductive effect
i.r.	infrared
J	coupling constant in n.m.r. spectra; joule
K	dissociation constant
kJ	kilojoule

k	Boltzmann constant; velocity constant
kcal	kilocalories
L	laevorotatory; laevo configuration
M	molecular weight; molar; mesomeric effect
Me	methyl
m	mass; mole; molecule; *meta-*
ml	millilitre
m.p.	melting point
Ms	mesyl (methanesulphonyl)
[M]	molecular rotation
N	Avogadro number; normal
nm	nanometre (10^{-9} metre)
n.m.r.	nuclear magnetic resonance
n	normal; refractive index; principal quantum number
o	*ortho-*
o.r.d.	optical rotatory dispersion
P	polarisation, probability; orbital state
Pr	propyl
Ph	phenyl
p	*para-*; orbital
p.m.r.	proton magnetic resonance
R	clockwise configuration
S	counterclockwise config.; entropy; net spin of incompleted electronic shells; orbital state
S_N1, S_N2	uni- and bi-molecular nucleophilic substitution mechanisms
S_Ni	internal nucleophilic substitution mechanisms
s	symmetrical; orbital; suprafacial
sec	secondary
soln.	solution
symm.	symmetrical
T	absolute temperature
Tosyl	p-toluenesulphonyl
Trityl	triphenylmethyl
t	time
temp.	temperature (in degrees centigrade)
tert.	tertiary
U	potential energy
u.v.	ultraviolet
v	velocity
Z	zusammen (together) configuration

α	optical rotation (in water unless otherwise stated)
$[\alpha]$	specific optical rotation
α_A	atomic susceptibility
α_E	electronic susceptibility
ε	dielectric constant; extinction coefficient
μ	microns (10^{-4} cm); dipole moment; magnetic moment
μ_B	Bohr magneton
μg	microgram (10^{-6} g)
λ	wavelength
ν	frequency; wave number
χ, χ_d, χ_μ	magnetic, diamagnetic and paramagnetic susceptibilities
$\sim$	about
$(+)$	dextrorotatory
$(-)$	laevorotatory
$(\pm)$	racemic
$\ominus$	negative charge
$\oplus$	positive charge

Chapter 16

FIVE-MEMBERED HETEROCYCLIC COMPOUNDS WITH TWO NITROGEN ATOMS IN THE RING

Stephen T. Mullins

1. Pyrazole

Several reviews are available. The synthesis of pyrazole derivatives up to 1985 has been covered by M.H. Elnagdi *et al.* (Heterocycles, 1985, 23, 3121), and the chemistry of 3*H* and 4*H* pyrazoles has been reviewed by M.P. Sammes and A.R. Katritzky in two articles (Advances in Heterocyclic Chemistry, 1983, 44, 1 and 53).

(a) Preparation

(b) Cyclisation

Syntheses, in which the final step is a cyclisation, are still the most widely used routes to pyrazoles. Few new methods have been reported but many new compounds have been prepared.

Type 1: -C-C-C- + -N-N- ⟶

The reaction of hydrazine with ß-diketones, a widely used method for pyrazole synthesis, has been used to prepare derivatives with substituted phenolic groups at the 5-position of the pyrazole ring (V.N. Ingle, J.Indian chem.Soc., 1988, 65, 852). The method simply involves heating a solution of the diketone and hydrazine hydrate in ethanol for 1h. The products (1) are obtained in moderate to high yield.

(1)

4-Benzoyl-3-aryl-5-(1'-hydroxynaphthyl)pyrazoles (**2**) are prepared by the reaction of 3-aroyl-2-arylchromones and hydrazine hydrate. The substituted chromone is the ß-diketone and ring opens during the cycloaddition (J.T. Makode and V.N. Ingle, J.Indian chem Soc., 1990, <u>67</u>, 176).

(2)

A similar ring opening of a chromone occurs when the 2-pyrone (**3**) is heated with phenylhydrazine in either ethanol or a mixture of acetic acid and acetic anhydride to yield the four substituted-pyrazoles (**4-7**). A plausible mechanism is given in Scheme 1 (A.B. Saranovic, B.K. Razem, and I. Susnik, Heterocycles, 1989, <u>29</u>, 1559).

The above cyclisations normally occur under neutral conditions. ß-Diketones with hydrazines under basic conditions give reduced pyrazoles (M. Brie, I.A. Silberg, and N. Palibroda, Rev. Roumaine de Chimie., 1989, <u>34</u>, 945). The o-nitrobenzoyl diketone (**8**) with hydrazine in the presence of a catalytic amount of 3% aqueous sodium hydroxide solution yields 3-methyl-5-(2-nitrophenyl)-1*H*-pyrazoline (**9**). The proposed mechanism for this reaction is given in Scheme **2**. Excess hydrazine allows the formation of a bis-hydrazone which loses nitrogen. Cyclisation then gives the pyrazoline (**9**).

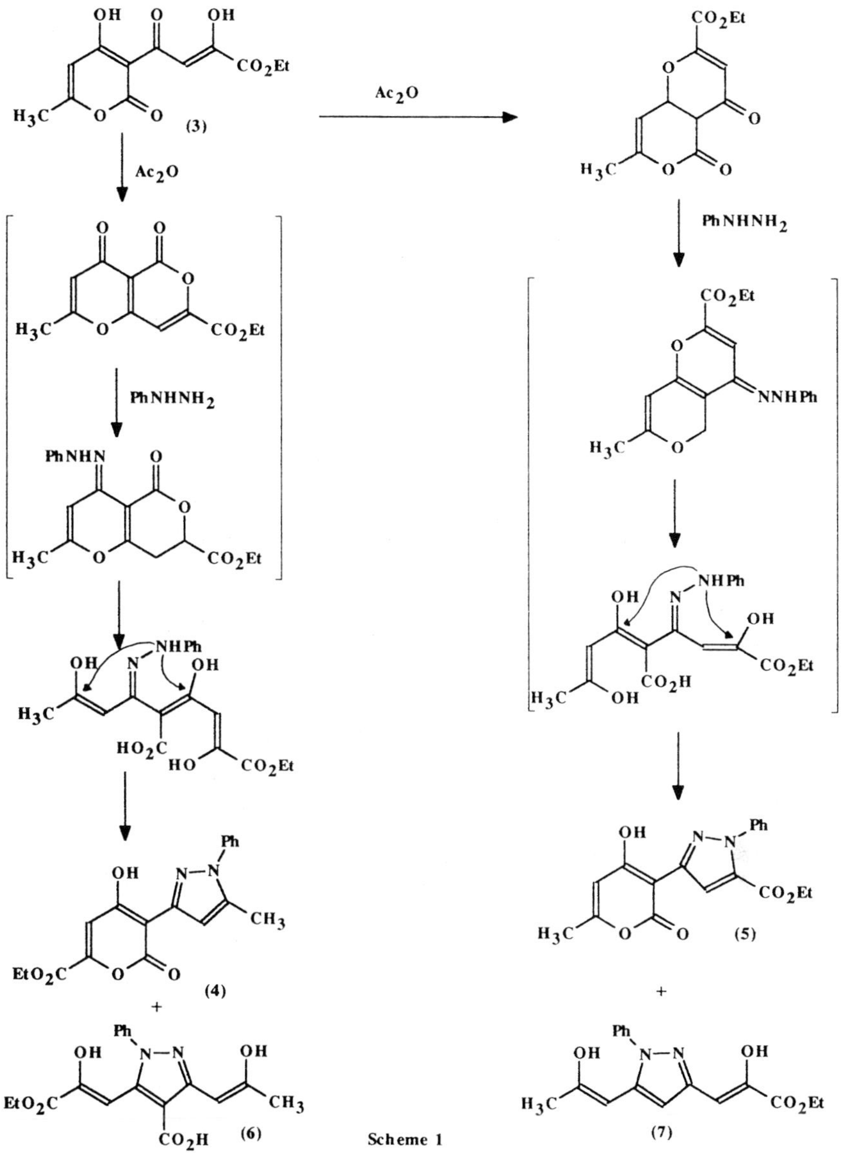

Scheme 1

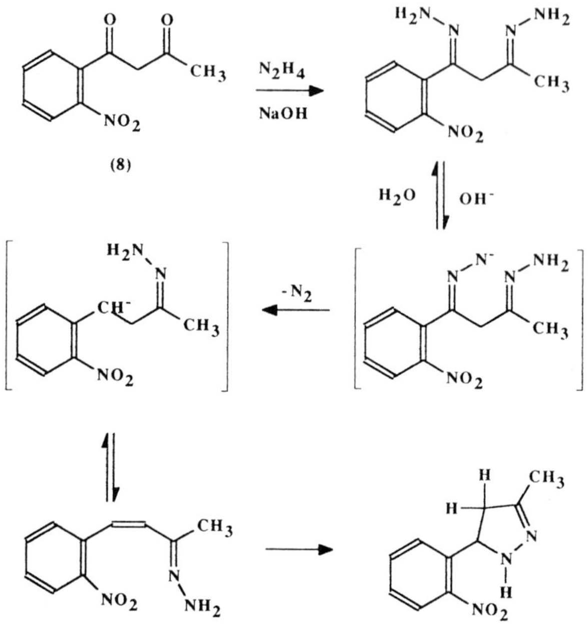

4

Scheme 2   (9)

Polarized alkenes having both electron donating (amino or methylthio) and electron accepting (cyano, carbamoyl, carbomethoxy) substituents react readily with hydrazines to give pyrazole derivatives (Y. Tominaga *et al.*, J.heterocyclic Chem., 1990, 27, 647). Thus the substituted alkenes (10) with hydrazine hydrate give the 3-substituted-5-aminopyrazole-4-carbonitriles (11) in good yield. The alkenes (10) do not react with phenyl-hydrazine under these conditions, the morpholino group must be replaced by a methylthio group for reaction to occur.

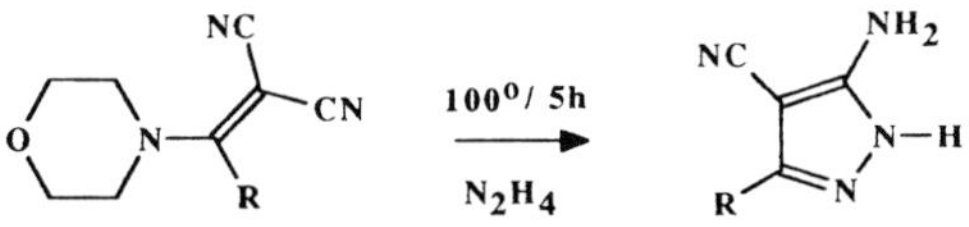

(10)   (11)

Pyrazole derivatives containing fluorinated substituents have been synthesised from protected trifluoromethyl-ß-diketones (J.W. Lyga and

R.M. Patera, J.heterocyclic Chem., 1990, <u>27</u>, 919). The ß-diketone is first treated with pyrrolidine at 0°C to give (12). Addition of phenylhydrazine to (12) at room temperature gives the reduced pyrazole (13) which is readily dehydrated.

A group of pyrazole derivatives of quinazolinone, which show anti-inflamatory activity, has been synthesised by condensation of hydrazine derivatives with 3-aryl-2-(3-aryl-3-oxopropenyl)-4(3H)-quinazolinones (14) (A.M. Farghaly, et al., Arch.Pharm., 1990, <u>323</u>, 311).

Type 2: -C-C-C-N-N- ⟶

A new route to pyrazoles is the cyclisation of N-allyl-N-nitrosomides (T. Momose, et al., Heterocycles, 1990, <u>30</u>, 789). Heating N-crotyl-N-

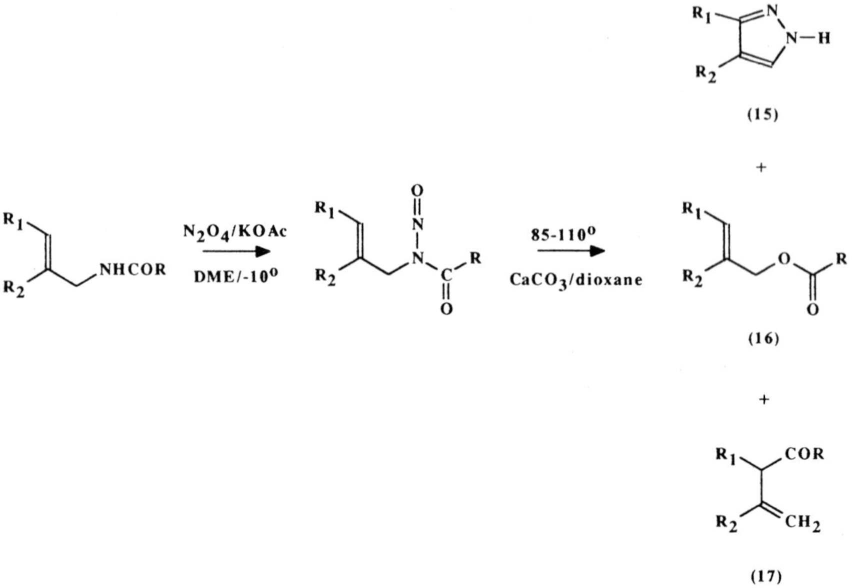
R1
R2
NHCOR
N2O4/KOAc
DME/-10°
O
N
N
C
R
O
85-110°
CaCO3/dioxane
N
N—H
(15)
+
O
R
O
(16)
+
COR
CH2
(17)

Table 1

Amide			% Yield	
R	R_1	R_2	(15)	(16)+(17)
Me	Me	H	21	-
Bn	Me	H	22	54
c-hexyl	Me	H	21	27
Me	Me	H	58	-
Bn	H	Me	54	29
Me	$-(CH_2)_3-$		11	47
Me	$-(CH_2)_4-$		30	40
Bn	$-(CH_2)_4-$		14	53
Me	$-CH_2CH{=}CHCH_2-$		28	46
Bn	$-CH_2CH{=}CHCH_2-$		13	64
Me	*p*-MeC_6H_4	H	-	29

nitrosocyclohexanecarboxamide in dioxane with calcium carbonate would be expected to yield crotyl-cyclohexanecarboxylate and 3-buten-2-yl-cyclohexane-carboxylate *via* a Huisgen-White rearrangement. The major product, however, is 2-methylpyrazole. The reaction is quite general and a series of substituted pyrazoles has been prepared form *N*-(substituted allyl)-*N*-nitrosamides, Table 1.

The postulated mechanism (Scheme 3) involves the nucleophilic addition of the olefin to the γ- position of the allylic π-system. Dehydration of the oxime intermediate then leads to the substituted pyrazole.

Scheme 3

Cyclisation of a suitable *gem*-dithiol yields a pyrazole. Thus 1-(4-biphenyl)-3,3-bis(methylthio)-1-phenylhydrazonoyl-propene yields 3-(4-biphenyl)-1-phenyl-5-methylthio pyrazole in 90% yield (S.E. Zayed, *et al.*, *Arch.Pharm.*, 1989, 322, 841).

Ar = 4-Ph-C$_6$H$_4$

Acylhydrazones from 1,3-diketoesters undergo cyclisation, in acidified ethanol, to the corresponding pyrazole but only in 30-40% yield (H.M.Mokhtar, J.chem.Soc.Pak., 1988, 10, 414).

Type 3: -C-C- + -C-N-N- ⟶

Classic examples of the C-N-N synthon are diazo compounds. Diazo compounds undergo 1,3-dipolar cycloadditions with a variety of two carbon units to give pyrazoles. For example acetylene derivatives readily react with diazo compounds (F. Fariña, Heterocycles, 1989, _29_, 967). Cycloaddition of diazomethane (the dipole) to an acetylenic ester (the dipolarophile) affords a pyrazole derivative in high yield under mild conditions. A typical procedure is to add a solution of diazomethane in ether to an ethereal solution of the acetylenic ester at 0°C. The reaction temperature is kept at 0°C for between 0.5 and 72h after which time the solvent is removed. Cycloadditions with ethyl diazoacetate require a higher temperature (20°C).

Dipolarophile Z	Dipole R	Products [A] : [B]	% Yield [A] + [B]
$CH(OMe)_2$	H	100 : 0	92
CH_2OH	H	100 : 0	90
$CH(OMe)_2$	CO_2Et	100 : 0	95
CHO	CO_2Et	35 : 65	95
CH_2OH	CO_2Et	95 : 5	89

Phenylacetylene undergoes cycloaddition reactions with diazo compounds derived from heterocyclic aldehydes (K. Sato, Heterocycles, 1989, _29_, 1537). Thus the sodium salt of thiophen-2(or 3)-carbaldehyde tosylhydrazone reacts with phenylacetylene to afford the expected thiophene substituted pyrazoles. The reaction conditions (anhydrous diglyme at 130°C) are more forcing than those required for cycloaddition of acetylenic esters. Yields of the resulting pyrazole derivatives are rather low, the highest yield recorded is 60%, but typically they are about 15%.

3-Substituted pyrazoles are formed by the cycloaddition of polyfluorinated diazoalkanes with electron deficient alkenes. Thus 3,3,4,4,4-pentafluoro-2-pentafluoroethyl-2-trifluoromethyldiazobutane (18) at room temperature in the dark, with methyl and ethyl propenoate gives the methyl and ethyl ester of pyrazole-3-carboxylic acid respectively (P.L. Coe and M.I. Cook, J.fluoro.Chem., 1989, 45, 323). The reaction is unusual as electron deficient diazoalkanes do not normally undergo cycloaddition reactions. It is believed that the reaction initially follows the mechanism expected for addition of a diazoalkane to an olefin and intermediate (19) is formed. When hydrocarbon diazo compounds are used the reaction stops at this stage i.e. with the formation of a pyrazoline. The perfluorinated alkyl side chain, however, is a good leaving group and so elimination of $(C_2F_5)_2CF_3CH$ occurs yielding (20). The latter undergoes a proton shift to give the 3-substituted pyrazole.

Intramolecular addition reactions produce fused pyrazole systems. Benzopyrano[4,3-c]pyrazoles are formed by such an addition, between nitrile imine species and acetylene substituents (D. Janietz and W.D. Rudorf, Tetrahedron, 1989, 45, 1661). 2-Alkynyloxybenzaldehydes are converted into their corresponding hydrazones by treatment with toluene-*p*-sulphonylhydrazine. The hydrazone with lead (IV) acetate results in the pyrazole (21). The reaction is useful in that it shows how nitrile imines are able to react with unactivated triple bonds under mild conditions and it provides a shorter synthetic route to benzopyrano[4,3-c]pyrazoles than the one previously available.

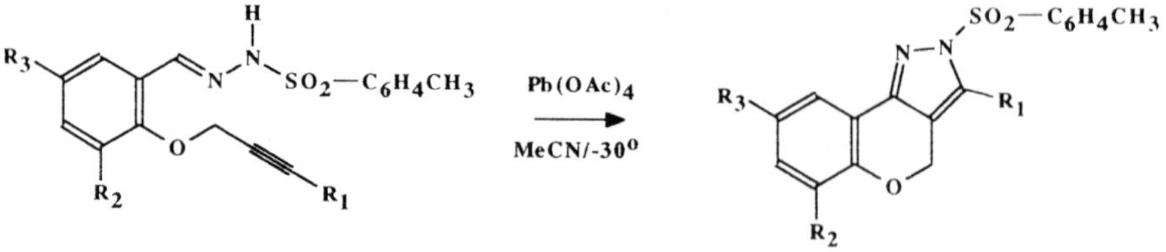

$R_1 = R_2 = H$, $R_3 = Br$

(21)

$R_1 = CH_2OH$, $R_2 = H$, $R_3 = Cl$

Nitrile imines react intermolecularly with α,β-disubstituted acrylonitriles to give substituted pyrazoles (H.M. Hassaneen *et al.*, J.pract.Chem., 1988, 330, 558).

Type 4: -C- + -C-N-N-C- ⟶

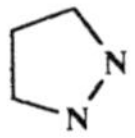

N-Alkylpyrazoles (22) have been synthesised using the Vilsmeier-Haack reaction. A phenylhydrazone and one equivalent of the Vilsmeier-Haack reagent (DMF/POCl₃) are stirred together at room temperature for 6h and the reaction mixture then neutralised to give the pyrazole derivative. A possible mechanism for the reaction is given in Scheme 4 (R.A. Pawar and A.P. Borse, J.Indian.chem.Soc., 1989, 66, 203).

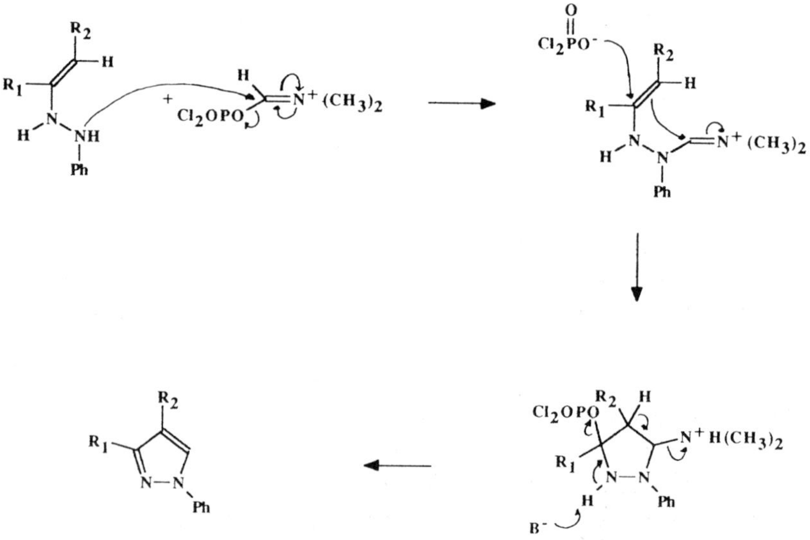

(22)

Scheme 4

(c) From Other Heterocycles

It is possible to prepare pyrazoles from other heterocycles by ring contraction. Thiadiazines, with hydrazine, readily undergo ring contraction to pyrazoles in good yield. The reaction is useful as it provides an efficient route to N-methylsulphamide a key intermediate in the synthesis of N-SO$_2$-N containing heterocycles. 2,3,5-Trimethyl-1,2,6-thiadiazine-1,1-dioxide with hydrazine gives N-methylsulphamide and 3,5-dimethylpyrazole (J. Elguero, P. Goya, and A. Martínez, Heterocycles, 1989, 29, 245).

(d) Pyrazole Derivatives

(i) N-Oxides

The structure, synthesis and chemistry of pyrazole N-oxides has been reviewed (A. Kotali and P.G. Tsoungas, Heterocycles, 1989, 29, 1615).

(ii) N-Alkyl and N-Aryl Derivatives

N-Alkylation of an unsymmetrically substituted pyrazole usually results in a mixture of isomeric products and so any approach which allows the regioselective alkylation of such a pyrazole is valuable. Alkylation of (23) with isobutylene can be achieved essentially exclusively either at N-1 to give (24a) or N-2 to give (24b) by the choice of catalyst (E.V.P. Tao et al., J.heterocyclic Chem., 1988, 25, 1293).

Regioselective alkylation of pyrazoles by tertiary carbocations is catalysed by sulphuric acid, under quite mild conditions, or by boron trifluoride (J.R. Beck et al., J.heterocyclic Chem., 1989, 26, 3). Thus treatment of ethyl-5-cyano-1H-pyrazole-4-carboxylate with 2-methyl-1-pentene in acetonitrile and a catalytic quantity of sulphuric acid gave ethyl-N-(1,1-dimethylbutyl)-

5-cyano-pyrazole-4-carboxylate in 50% yield, no trace of the corresponding 3-cyano ester was detected though unreacted starting material was recovered.

It is interesting to note that the use of strong Lewis acids, e.g. BF_3 reverses the regioselectivity. Ethyl-5-cyano-1H-pyrazole-4-carboxylate reacts with 2-ethyl-1-butene and boron trifluoride (two equivalents) to give ethyl-N-(1-ethyl-1-methylpropyl)-5-cyano-1H-pyrazole-4-carboxylate in 83% yield. In this case no 5-cyano ester is formed. These alkylations are only effected by olefins capable of forming tertiary carbocations.

Electroreductive *N*-alkylation results in good yields of *N*-alkylpyrazoles under very mild conditions.

Various spectroscopic studies have been carried out on pyrazoles derivatives. In (**25**) the methyl signal is observed at 2.4 ppm whereas in isomer (**26**) it is at 2.8 ppm. This difference is striking in view of the fact that such a difference between the chemical shift of a 3-methyl group and that of a 5-methyl group is not observed in other substituted pyrazoles (S.P. Singh, S. Sehgal, and L.S. Tarar, Indian J.Chem., 1989, 288, 27). ^{13}C-NMR data have been reported for a number of *N*-alkylpyrazoles (G. Heinisch and W. Holzer, Heterocycles, 1988, 27, 2443).

(25)   (26)

A series of 3-methyl-5-arylpyrazoles has been prepared and the members characterized by infra-red spectroscopy (M. Brie and I.A. Silberg, Rev.Roumaine de Chemie, 1989, 34, 733).

(iii) N-Acyl Derivatives

N-Acylation of an unsymmetrically substituted pyrazole results, as is the case with *N*-alkylation, in the formation of two isomers. A reaction has been reported, however, which results in the isolation of one isomeric product (J.T. Drummond and G. Johnson, J.heterocyclic.Chem., 1988, 25, 1123). The substituted pyrazole (**27**) with phenylisocyanate yields the isomeric products (**28**) and (**29**). The minor component, (**28**), can be extracted from the product mixture with boiling ethyl acetate, while the major isomer (**29**) remains as an insoluble crystalline solid. The minor product may be converted into (**29**) by heating it in boiling ethyl acetate or by heating the solid to its melting point and allowing it to cool. This conversion is almost quantitative.

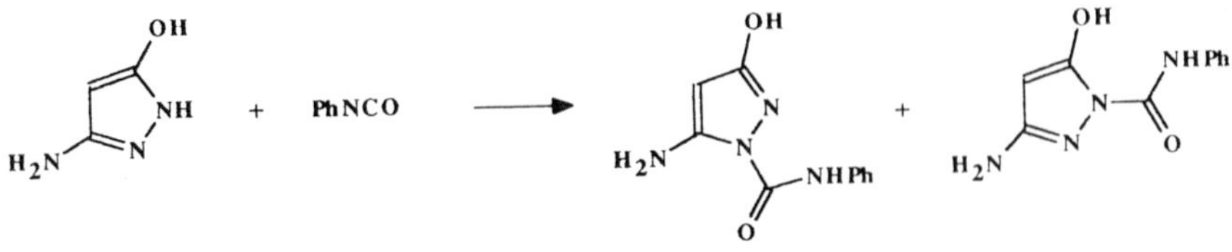

(27)   (28)   (29)

Scheme 5

More complex compounds are synthesised by cyclisation reactions rather than direct acylation. Indol-2-yl pyrazol-1-yl ketones readily synthesised by the condensation of indole-2-carbonylhydrazides with acetylacetone or ethyl acetoacetate (S.P. Hiremath *et al.*, Indian J.Chem., 1988, 27B, 758). A similar synthetic route can be used to prepare 1-arylpropanoyl-3,5-dimethylpyrazoles (e.g. **30**) (C.S. Andotra and S.K. Sharma, Proc.Nat.Acad.Sci.India, 1988, 58, 215).

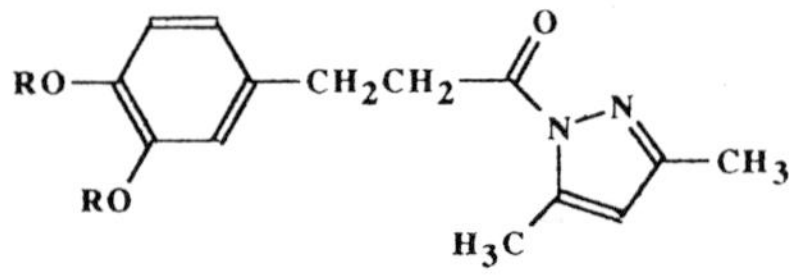

(30)

1-Methoxycarbonylpyrazole with trimethylchlorosilane, HMPA, and magnesium affords pyrazole in 47% yield (K. Saito *et al.*, Heterocycles, 1989, 29, 1545). The elimination is believed to occur *via* formation of radical anions by electron transfer from magnesium to pyrazole. Thus reaction of 1-methoxycarbonylpyrazole with six-molar equivalents of trimethylchlorosilane in HMPA in the presence of magnesium at 85°C for 24h afforded pyrazole after aqueous work up. The reaction mechanism is shown in Scheme **6**.

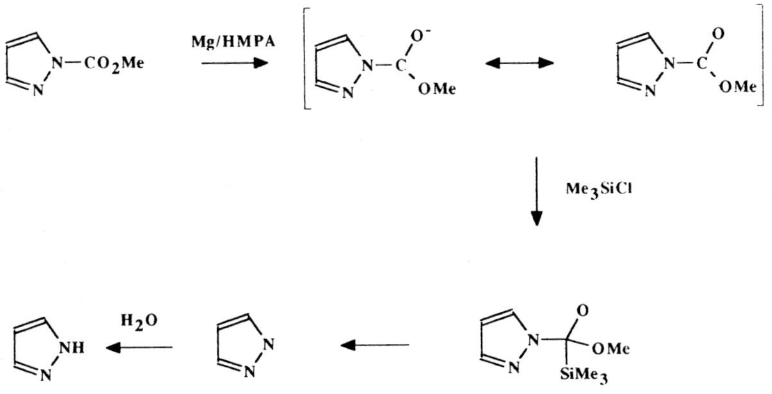

Scheme 6

N-Acylpyrazoles with molybdenum hexacarbonyl undergo ring expansion to yield pyrimidines (M. Nitta, T. Hamamatsu, and H. Miyano, Bull.chem.Soc.Jpn., 1988, 61. 4473). The mechanism involves N-N bond cleavage and carbon insertion (Scheme 7).

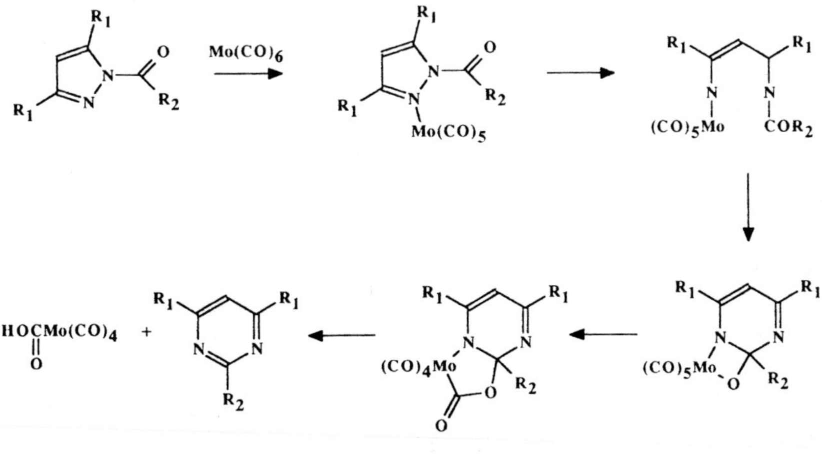

Scheme 7

(iv) C-Alkyl and C-Aryl Derivatives

Methods for C-alkylation of pyrazole normally require initially N-protection. A useful methodology would be one where the N-protection and subsequent deprotection can be effected without the isolation of intermediates, and one in which the N-protecting group activates other positions on the pyrazole ring to substitution. Such a method using formaldehyde for N-protection is available (A.R. Katritzky, P.Lue and K. Akutagawa, Tetrahedron, 1989, 45, 4253). The reaction sequence (Scheme 8) involves three steps: N-protection of the pyrazole by aqueous formaldehyde in tetrahydrofuran at 20°C; lithiation using n-butyl lithium or lithium diisopropylamide at 20°C to give a dilithiated product; treatment with an electrophile; deprotection by either acid catalysed hydrolysis or silica assisted fission.

Scheme 8

The advantages of this one-pot hemiaminal method for substitution at the 5-position are: the protecting group is readily introduced; no purification step is required; a variety of lithiating agents may be used; deprotection occurs under mild conditions, and a good overall yield of substituted product is obtained.

1,3-Disubstituted pyrazoles may be prepared by condensation of an eneamine with hydrazines (D.Sedzik-Hibner and W. Czuba, Pharm., 1986, 38, 516), and the synthesis of 4-arylidenepyrazoles (31) from condensation of hydrazines with chloro-substituted-2-aroylacrylophenones (32) has been reported (V.S. Dubey and V.N. Ingle, J.Indian chem.Soc., 1989, 66, 174).

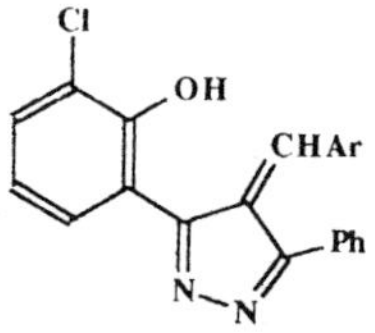

(31) 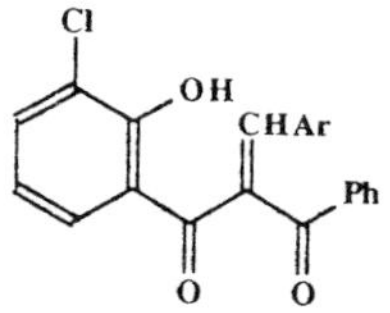   (32)

C-Substituted pyrazoles may be used as precursors of other heterocyclic systems. Ring expansion of pyrazoles, effected by trichloromethyl radicals, leads to the formation of pyrimidine derivatives in moderate to high yields. The effects of pyrazole substituents on the yields of pyrimidines has been studied (M. Ehsan, Pak.J.sci.ind.Res., 1989, 32, 377). Increasing the number of electron donating groups in the pyrazole nucleus increases the yield of pyrimidine. Porphyrin analogues with pyrazole groups replacing pyrrole rings have been prepared as shown in Scheme 9. These compounds are able to extract alkali metal cations from an aqueous phase into an organic phase. The ion most readily complexed is Na$^+$ with a percentage extraction limit of 16% (G. Tarrago *et al.*, J.org.Chem., 1990, 55, 421).

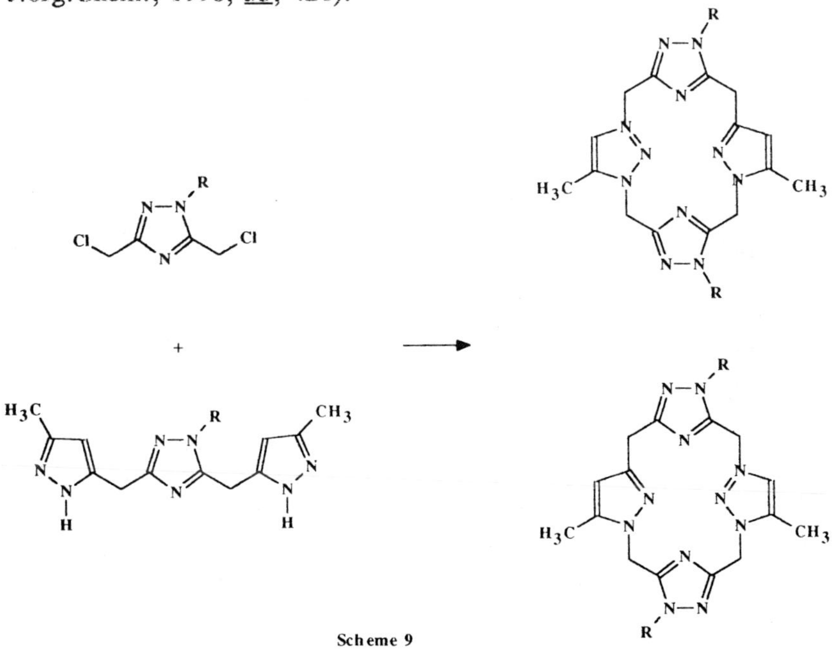

Scheme 9

(e) Fused Systems

There are two general classes of fused pyrazole derivatives; those in which pyrazole is fused to another heterocyclic ring and those in which it is fused to a hydrocarbon system. The first class is the more widely studied.
Pyrano- and thiopyrano- pyrazoles are synthesised by condensation of a hydrazine with a 1,3-dicarbonyl compound. Thus the reaction of the ketosulphide **(33)** with a substituted hydrazine results in tetrahydrothiopyrano[3,2-c]pyrazole **(34)**. If the hydrazine derivative is deactivated then the arylhydrazone **(35)** is formed. The latter compound does *not* cyclise to the pyrazole under any of the conditions reported (S. Fatutta *et al.*, J.heterocyclic Chem., 1989, 26, 183).

Similarly a pyrano[2,3-c]pyrazole is formed, in 65-70% yield, by the reaction of hydrazine hydrate with a 2-amino-3-cyanopyran (B.Y. Riad and S.M. Hassan, Sulphur Letters, 1989, 10, 1). An alternative route is to form the pyran ring by cyclisation of a suitably substituted pyrazole. Thus the 1,5-diketone **(36)** with polyphosphoric acid results in the formation of the corresponding pyrano[3,2-c]pyrazole.

Replacing polyphosphoric acid by P_2S_5 in dry xylene produces thiopyranopyrazoles (M.I. Younes and A.M. Nour, Bull.Fac.Sci.Assuit Univ., 1988, 17, 1).

Crystal structures of pyrazolopyrazoles and functionalised derivatives have been studied. Compound (37) exists as a zwitterion in the solid state with intermolecular hydrogen bonding between the amino groups in the cationic part and the anionic oxygens thus forming a polymeric layer with alternating orientation of the ions (38) (G. Zvilichovsky, J.heterocyclic.Chem., 1988, 25, 1301).

(37)

(38)

Cinnamonitriles may be used as precursors to pyrazolopyrimidines as well as pyranopyrazoles. Pyrazolopyrimidines are active as adenosine cyclic monophosphate phosphodiesterase inhibitors (T. Novinson et al., J.med.Chem., 1975, 18, 460), and antischistosomal agents (K.U. Sadek et al., Synthesis, 1983, 739). Aminopyrazolo[1,5-a]pyrimidines are prepared via the reaction of an aminopyrazole with a cinnamonitrile (Y.R. Elnagdi, J.heterocyclic Chem., 1983, 20, 667). The ring nitrogen of the pyrazole adds to the activated double bond system of the cinnamonitrile and the resulting Michael adduct cyclises to give the 5-amino-6,7-dihydropyrazolo[1,5-a]pyrimidine derivative (M.H. Elnagdi et al., Coll.Czech.chem.Comm., 1989, 54, 1082 and J.L. Soto et al., Synthesis, 1987, 33). A new synthesis of pyrazolopyrimidines (A.O. Abdelhamid, A.M. Negm and I.M. Abbas, Egypt.J.pharm.Sci., 1989, 30, 103. and J.prakt.Chem., 1989, 331, 31) involves the intermediate cyanoaminopyrazole pyrazole derivative (39).

(39)

The latter reacts with formic acid, formamide, guanidine, and hydrazine hydrate to give pyrazolo[3,4-a]pyrimidinones and pyrazolo[3,4-d]pyrimidines (Scheme **10**).

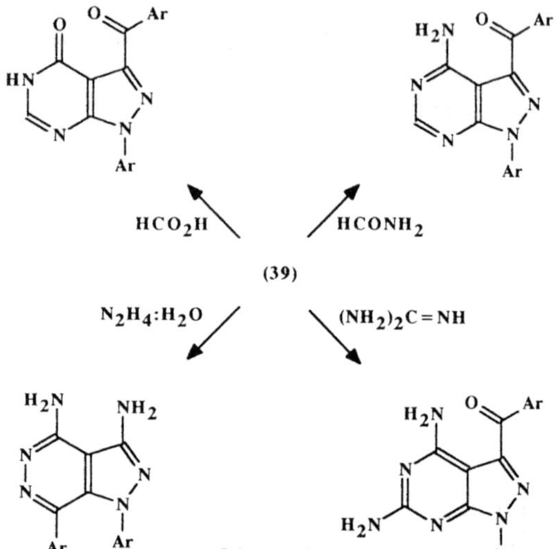

Scheme 10

Diazomethylcoumarins may be used as precursors to benzo-pyranopyrazoles. Thus 4-diazomethylcoumarin in boiling toluene yields benzopyrano[3,4-c]-pyrazol-4(3H)-one (85%). A variety of coumarin analogues and related heterocycles have been thermolysed in this way to give fused pyrazoles (K.Ito and J. Maruyama, J.heterocyclic Chem., 1988, 25, 1681).

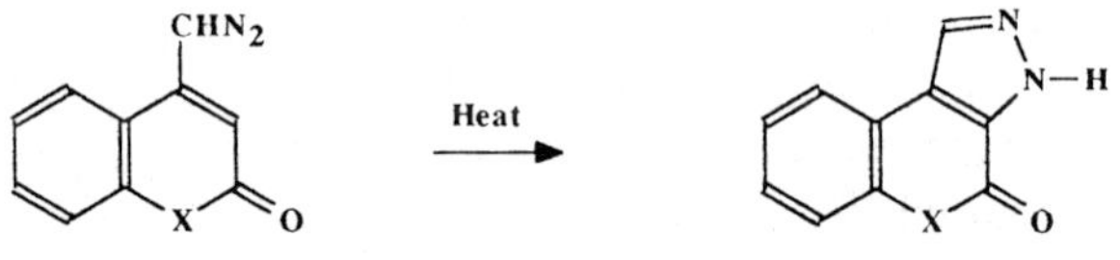

X = O, NCH₃, S

Other fused pyrazole derivatives which have been prepared include benzoxepino[5,4-c]pyrazoles (A.K.Saxena et al., Indian.J.Chem., 1989, 28B, 37), imidazo[4,5-c]pyrazoles via cyclisation of 4-nitroso-5-alkylaminopyrazoles (C.B. Vincentini et al., Tetrahedron, 1988, 47,

6171) and pyrazolo[1,5-a]pyridines by the reaction of cinnamonitriles with 5-amino-4-cyano-3-cyanomethylpyrazoles(A.H.H. Elghandour and M.R.H. Elmoghayar, J.prakt.Chem., 1988, 330, 657).

Pyrazoles fused to a non-heterocyclic ring have not been a widely studied. The fused naphthoquinopyrazoles are prepared by the addition of diazomethanes or diazoacetic esters to naphthoquinones (V.K. Tandon *et al.*, Arch.Pharm., 1990, 323, 383).

ca 90%

Substituted 2,3,3a,4,5,6-hexahydrobenzo[6,7]cyclohepta[1,2-c]pyrazoles have been synthesised by condensing hydrazines with 2-arylidene-1-benzosuberones in methanol, followed by *N*-bromination (using bromine) and dehydrobromination (N.R. El-Rayyes and N.H. Bahtiti, J.heterocyclic Chem., 1989, 26, 209).

(f) Reduced Systems

Methods for the preparation of 2-pyrazolines are related to those used for pyrazoles. The cycloaddition of a *C*-phenylamino-carbonyl-*N*-arylformohydrazidoyl chloride (40) to an α,β-unsaturated ketone results in the formation of a 1,4-diaryl-3-phenylaminocarbonyl-5-aroyl-2-pyrazoline (41). The isomeric product (42) is not formed under the reaction conditions reported (H.M. Hassaneen *et al.*, Org.prep.Proc.Intern., 1989, 21, 119). The regioselectivity of this reaction can be interpreted in terms of frontier orbital energies and was found to be independent of the solvent used. The interaction of (40) with electron deficient dipolarophiles, for example α,β-disubstituted acrylonitriles and benzylacetophenones, is controlled by HOMO (dipole)-LUMO (dipolarophile) interaction (G. Bianci *et al.*, J.chem.Soc.Perkin 1, 1973, 1, 1148). In contrast the reaction of (40) with electron rich dipolarophiles, such as phenacyl cyanide, is controlled by LUMO (dipole)-HOMO (dipolarophile) interaction.

$$\underset{(40)}{PhNH\underset{O}{\overset{Cl}{\underset{\|}{C}}}C=NNC_6H_4X} \quad + \quad Ph\underset{O}{\overset{\|}{C}}CH=CHC_6H_4Y$$

(41)   (42)

Organometallic reagents have been used to prepare 2-pyrazolines from 4*H*-pyrazoles. Organolithiums and Grignard reagents react smoothly with 4,4-dimethyl-3,5-disubstituted-4*H*-pyrazoles to give 4,4-dimethyl-3,4-dihydro-3,4,5-trisubstituted-2*H*-pyrazoles (2-pyrazolines) in high yield (A.L. Baumstark *et al.*, J.heterocyclic.Chem., 1990, 27, 291). Some regioselectivity is shown by the reaction in that 4,4-dimethyl-3-alkyl-5-phenyl-4*H*-pyrazoles undergo addition exclusively at the 3-alkyl position due to stabilisation of the negative charge by the 5-phenyl ring. This is consistent with the observation that the organometallics react with phenyl substituted 2*H*-pyrazoles rather than with alkyl substituted 2*H*-pyrazoles.

2-Pyrazolines are intermediates in the synthesis of pyrazoles by cyclocondensation of a hydrazine with a ß-diketone. The 5-hydroxy-2-pyrazoline intermediates are normally too unstable to isolate and only observed by NMR spectroscopy. However 5-hydroxy-2-pyrazolines which contain a powerful electron withdrawing group may be isolated. For example in the synthesis of mono- and di-(trifluoromethyl) substituted pyrazoles from trifluoro- and hexafluoro- acetylacetone respectively the trifluoromethylhydroxypyrazoline has been isolated. It is postulated that the trifluoromethyl groups hinder the heterolytic removal of the 5-hydroxyl group and so the intermediates are stabilised (J. Elguero and G.I. Yranzo, J.chem.Res., 1990, 120).

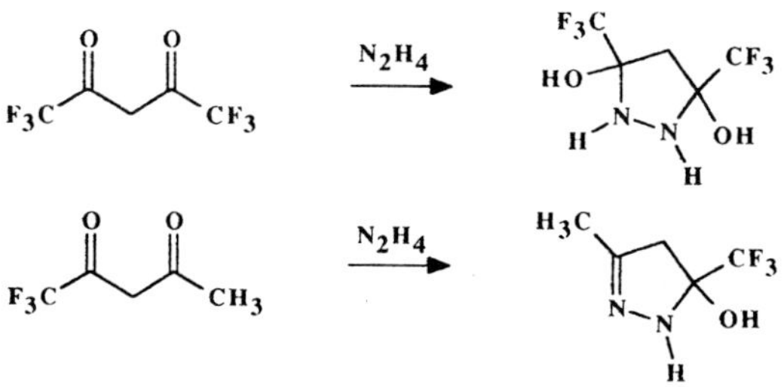

α,β-Unsaturated ketones with hydrazines give stable 2-pyrazolines. Thus the 2-naphthyl ketone (**43**) yields the 2-pyrazoline derivative (**44**). These N-aryl-2-pyrazolines are stable, crystalline materials regardless of the nature of the aryl group (H.A.A. Regaila, N. Latif, and I.H. Ibrahim, Egypt.J.pharm.Sci., 1989, <u>30</u>, 1).

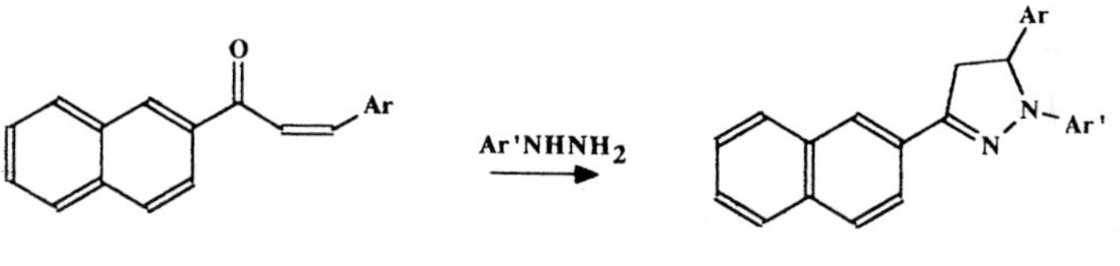

(43)   (44)

2-Pyrazolines undergo addition reactions with ketenes. It might be expected that treatment of a 2-pyrazoline derivative containing an exomethylenic group would result in a fused four membered ring *via* a [2+2] cycloaddition as in the case of the reactions of methylenecycloalkanes. Cycloaddition of the ketene to the pyrazoline does not occur and the product is a stable, crystalline pyrazole benzoate. For example 3,4,4-trimethyl-1-benzoyl-5-methylene-2-pyrazoline (**45**) reacts with diphenylketene to give the 3-alkyl-4,4-dimethyl-5-methylene 1-(diphenylacetyl)2-pyrazoline (**46**). The proposed mechanism for this reaction is shown below (J.S. Stephanatou *et al.*, J.org.Chem., 1990, <u>55</u>, 4732).

24

(45)

(46)

2. Indazoles

(a) Preparation

The three classical routes to indazoles are cyclisation of aromatic diazo compounds, the reaction of o-chloroaromatic ketones with aryl hydrazines and the cyclisation of N-nitroso-2-methyl anilines.   The majority of preparative routes involve the cyclisation of a benzenoid derivative.

(b) From Benzoid Precursors

Type 1:

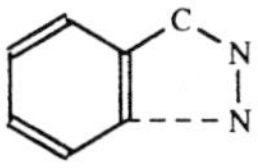

Arylazoacetates, obtained by oxidation of arylhydrazones, cyclise to indazoles in the presence of Lewis acids.   In the case of arylhydrazones derived from 2-benzoylpyridine the yield of indazole is low, however higher yields have been obtained from phenylhydrazone derivatives of 3-acylazoles.   Oxidation of the phenylhydrazone of 3-benzoyl-5-phenyl-1,2,4-oxadiazole (47) with lead (IV) acetate (LTA) gives the aza-acetate (48) which with aluminium trichloride gives the indazole (49) (N. Vivona et al., J.heterocyclic Chem., 1985, 22, 29).

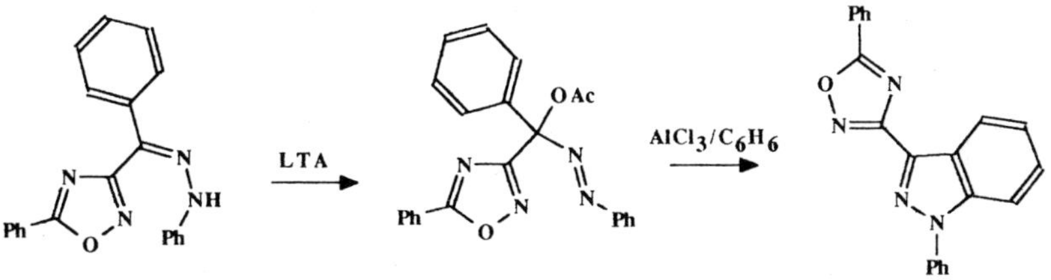

(47)   (48)   (49)

Similarly 3-benzoyl-5-phenylisoxazole and 3-benzoyl-4-methyl-1,2,5-oxadiazole, form the corresponding indazoles in 94% and 97% yield respectively.

Cyclisation by displacement of a halogen from the 2-position of a benzoic hydrazide leads to an indazole.   Thus thermal cyclisation of the N,N'-disubstituted-2-chlorobenzoic hydrazide (50) gives the 1,2-pentamethylene-

5-nitro-1,2-dihydro-3*H*-indazol-3-one **(53)**. The reaction is believed to proceed *via* the salt **(51)**, which undergoes ring expansion *via* the substituted indazole **(52)** to give the indazolediazepine product **(53)** (C. Foces-Foces, F.H. Cano, and M. Martinez-Ripoll, J.heterocyclic Chem., 1985, <u>22</u>, 1743).

(50)   (51)

(53)   -HCl   (52)

Type 2:

N-N Bond formation leading to indazoles has been achieved in a variety of ways. For example thermolysis of *o*-azidobenzimidate **(54)** and *o*-azidobenzamidine **(55)** leads to 3-ethoxy and 3-amino-1*H*-indazole respectively. The reaction involves a 1,5 proton shift to generate the 1*H* indazole (M.A. Ardakani, R.K. Smalley, and R.H. Smith, J.chem.Soc.Perkin 1, 1983, 2501).

Substituted indazoles with substituents on both the benzene and the heterocyclic ring, have been prepared by treating *o*-azidobenzanilides with thionyl chloride. The expected product is an imidoyl chloride (H. Ulrich, 'The Chemistry of Imidoyl Halides', Plenum Press, New York, 1968,

X = OEt or NH$_2$

p.55) but the chloroindazole is formed directly, probably by an assisted loss of nitrogen (see below).

A cyclisation reaction involving *N-N*-bond formation occurs when a 2,4-diaryl-4-(2,4-dinitroaryl)-5(4*H*)-oxazolone rearranges to give a 1-hydroxy-1*H*-indazole. The reaction is catalysed by toluene-*p*-sulphonic acid in methanol and the proposed mechanism involves nucleophilic attack of the oxazolone nitrogen on the protonated nitro group of intermediate (56) followed by a migration of the *C*-2 hydroxyl group to give a cyclic intermediate which loses carbon dioxide and rearranges to give (57). Compound (57) is readily converted into either the 1-hydroxyindazole by hydrolysis or the 1-(acyloxy)indazole by acylmigration (M. D'Anello, *et al.*, Ber., 1988, 121, 67).

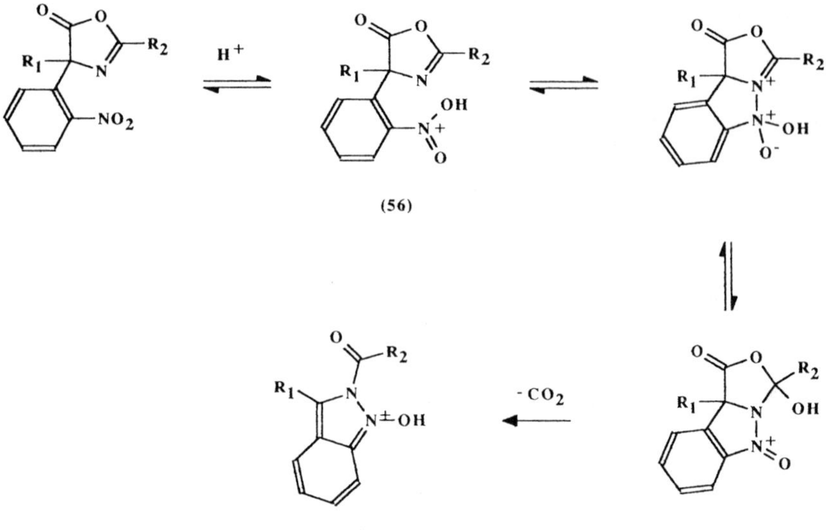

(56)

(57)

Type 3:

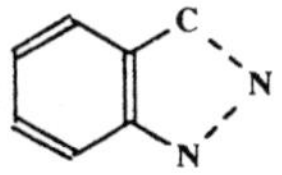

Cyclisation of o-methyl and o-ethylbenzenediazonium tetrafluoroborates (58) with potassium acetate in chloroform, in the presence of 18-crown-6, gives indazoles in high yield (R.A. Bartsch and I.W. Yang, J.heterocyclic Chem., 1984, 21, 1063).

(58)

X = NO_2, Cl, Me, MeO   R = Me, Et

This method is reasonably general and indazoles bearing both electron withdrawing (nitro, chloro) and donating (methyl, methoxy) substituents have been prepared. No indazole is produced in the absence of the crown ether. It carries the insoluble tetrafluoroborate and the potassium acetate into the chloroform phase where the reaction takes place.

Thermal rearrangements have also been used as routes to indazoles, for example, when 2,5,6-trifluoro-4-(2,4,6-trimethylphenylazo)pyrimidine (**59**) is heated under reflux in xylenes for 2h, 5,7-dimethyl-2-(2,5,6-trifluoropyrimidin-4-yl)-2*H*-indazole (**60**) is formed in 23% yield together with the expected diazepine derivative (**61**) (A.C. Alty and R.E. Banks, J.fluorine Chem., 1988, 41, 439).

| | (59) | (60) | + | (61) |

Thermal cylcisation of 2,6-dimethylphenylazo compounds also yields indazoles. For example heating ethyl [(2,6-diethylphenyl)imino][(2,6-dimethylphenyl)azo]acetate (**62**) in refluxing xylene with a catalytic amount of DABCO results in the formation of the indazole (**63**) and the benzotriazepine (**64**). The mechanism proposed for this reaction involves tautomerization of the starting material to a methylenecyclohexadienone imine which can cyclise to a dihydroindazole. Oxidation, possibly by another molecule of (**62**) then results in (**63**). The reduced form of the starting material (**62**) is observed in the product mixture. If the positions of the methyl and ethyl groups in (**62**) are reversed the triazepine is not formed and a good yield of indazole and reduced starting material are obtained (R. Fusco, A. Marchesini, and F. Sannicolo, J.heterocyclic Chem., 1987, 24, 773). This is a new route to indazoles.

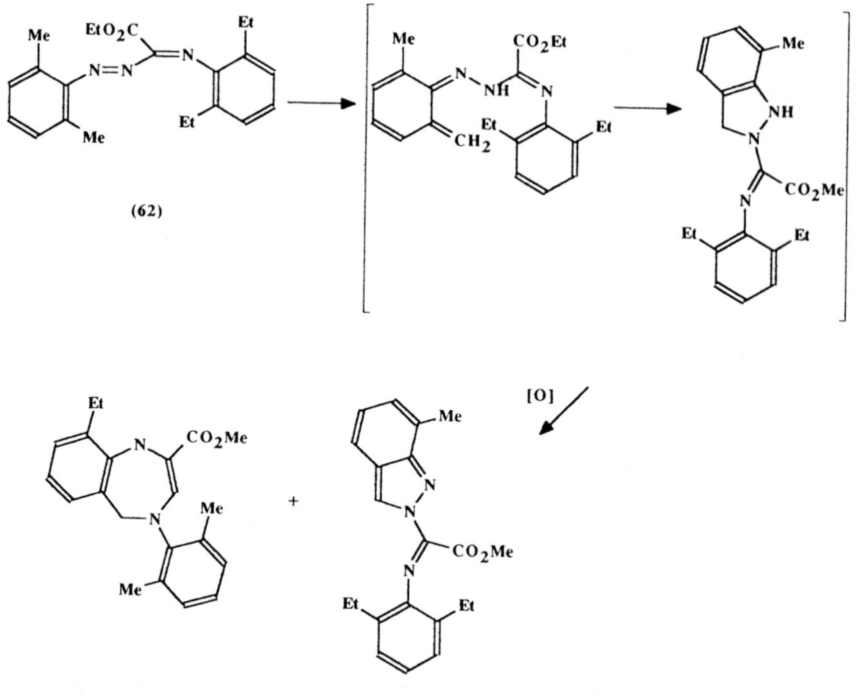

(62)

(64) + (63)

[O]

Diazonium salts can be precursors of indazoles (G.M. Shutske *et al.*, J.med.Chem., 1983, <u>26</u>, 1307).   Thus 2-aminobenzophenones on diazotisation in strongly acidic media, followed by reduction with sodium dithionite, give 3-hydroxy-3*H*-indazoles.   Reduction of the latter with tin (II) chloride gives indazoles (**65**).

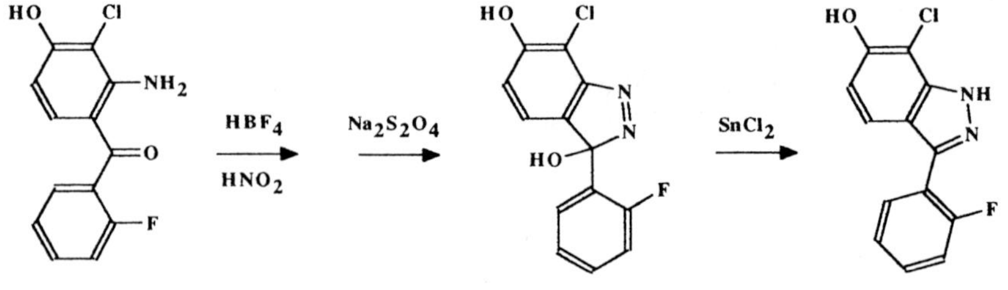

(65)

Azobenzenes also readily react with carbenes to generate indazoles. A variety of *p*-substituted azobenzenes **(66)** have been treated with isopropylidenecarbene generated form the corresponding triflate. Yields range from 18 to 43%.

(66)

X and Y = various H, Me, MeO, Cl, F

The reaction may involve either a carbene-nitrogen zwitterionic ylid **(67)** as an intermediate or a concerted 1,4 addition *via* the transition state **(68)**. The study of substitution effects on this cyclisation and molecular orbital calculations on the addition of singlet carbenes to 1,3-dienes suggest that the more probable mechanism is that involving the ylid intermediate **(67)** (K. Krageloh, G.H. Anderson, and P.J. Stang, J.Amer.chem.Soc., 1984, 106, 6015).

(67)   **(68)**

(c) Miscellaneous

The cyclisation, by hot polyphosphoric acid, of hydrazidoyl chlorides (**69** and **70**) to indazoles is not an efficient nor general route. Only the compounds (**69a** and **b**) give the expected products in low (6-7%) yield (L. Baiocch and M. Giannangeli, J.heterocyclic Chem., 1983, 20, 225).

(69)

(70)

(69a) R = Me, (69b) R = Ph

A successful, general and mild route to indazoles is based on the cycloaddition of a hydrazine derivative to 2-acylcyclohexane-1,3-diones. This addition results in tetrahydroindazolinones which are readily reduced, by sodium borohydride, to the hydroxy derivatives (**71**). Subsequent dehydration/aromatization, effected by 10%-palladium on charcoal, gives the indazole derivatives (**72**; 70% to 95%) (P.D. Croce and C.L. Rosa, Synthesis, 1984, 982).

(71)

(72)

Tropone tosylhydrazone sodium salt (**73**), in the presence of silver chromate, rearranges to 2-tosyl-2*H*-indazole (K. Saito, T. Toda and T. Mukai, Bull.Soc.Chem.Jpn., 1984, 57, 1567).

Highly substituted indazoles may be prepared from substituted pyrazol-3-ones (S.Matsugo et al., Chem.pharm.Bull.,1984, 32, 2146) and by the reaction of dialkyl acetylenedicarboxylates with substituted 1,6-dihydropyrano[2,3-c]pyrazoles (S. Matsugo and A. Takamizawa,

Synthesis, 1983, 852).

(73)

(d) Indazole Derivatives

(i) Alkyl and Aryl Derivatives

N-Alkyl indazoles may be prepared by the reaction of an N-H indazole with a suitable alkylating agent, for example 3-phenylindazole with ethyl bromoacetate and sodium hydride in DMF yields 1-(ethoxycarbonyl)methyl-3-phenylindazole in high yield, which gives N-substituted indazole (76) on treatment with methylamine. Compound (76) can also be prepared by the rearrangement of 3,4-dihydro-1-methyl-6-phenyl-1,4,5-benzotriazocin-2(1H)-one (74) via a spiro intermediate (75). The rearrangement occurs smoothly at room temperature in chloroform solution. The 8-chloro-, 8-bromo-, and 8-fluoro- derivatives of (74) also undergo the reaction. Yields of the N-alkylindazole products range from 84% to 90% (Y. Fujimura et al., Chem.pharm.Bull., 1984, 32, 3252).

(74)   (75)   (76)

Few routes to *C*-alkylated indazoles, particulary where the alkyl group contains a functional group in the ⍵-position, are known. A simple method involves the selective *ortho*-acylation of anilines by treatment with an α-cyano-⍵-chloroalkane in the presence of boron trichloride and aluminium trichloride. Cyclisation of the 2-(⍵-chloro-alkanoyl)aniline *via* diazotisation and reduction leads to the substituted indazole in *ca.* 85% yield (K.Sasakura, A. Kawasaki, and T.Sugasawa, Synth.Comm., 1987, 741; 1988, 259). By replacement of the ⍵-chlorine functionalised *C*-alkylated indazoles can be obtained.

(ii) Carboxylic Acid Derivatives and Indazolediones

The synthesis and the biological properties of indazole carboxylic acids have been reviewed (B. Silvestrini, Chemotherapy, 1981, <u>27</u>, 9). Indazole-3-carboxylic acid (**77**) is a useful intermediate in the synthesis of biologically active indazoles, some of which have applications as anti-arthritic (G. Alunni, Farmaco, Ed.Sci., 1981, <u>36</u>, 315) and antifertility agents (F. Coulston, Chemotherapy, 1981, <u>27</u>, 98).

Phenylhydrazine with chloral hydrate and hydroxylamine hydrochloride in aqueous acid yields *N*-acetylaminoisonitrosoacetanilide (76%). The latter with concentrated sulphuric acid gives indazole-3-carboxylic acid (**77**, 77%), *via* the intermediate (**78**). This new synthesis is superior, with respect to cost and yield, to those previously described (M. Sisti *et al.*, J.heterocyclic Chem., 1989, <u>26</u>, 531).

(77)   (78)

2*H*-Indazole-4,7-diones are prepared by 1,3-dipolar cycloadditions of *p*-toluquinone with 3-phenylsydnone. The reaction shows no regioselectivity and both possible isomeric indazoles are formed in equimolar amounts (S. Nan'ya, K. Katsuraya, and E. Maekawa, J.heterocyclic Chem., 1987, 24, 971).

Interest in the antitumour activity of indazole derivatives prompted the synthesis of a number of 5,6-disubstituted-1(2)*H*-indazole-4,7-diones (G.A. Conway, L.J. Loeffler, and I.H. Hall, J.med.Chem., 1983, 26, 876). Indazole-4,7-diones are formed by the 1,3-dipolar cycloaddition of diazomethane to substituted *p*-benzoquinones, a method particulary suited to symmetrically- substituted benzoquinones. In 5,6-dichloro-1*H*-indazole-4,7-dione (79), the 5-chloro group is readily replaced by ethyleneimine and ammonia to give the 5-aziridinyl- and the 5-amino- derivative respectively. Calculations indicate that for (**79**) the electron density at the 4-carbonyl position is higher than that at the 7-position. This will effect a difference in the degree of positive charge at *C*-5 and *C*-6 and may account for the difference in the reactivities towards nucleophilic displacement.

(iii) Aminoindazoles

Aminoindazoles are readily prepared by the catalytic reduction (H_2/Pd/C) of nitroindazoles (B. Beyer *et al.*, Acta Cientifica Venezolana, 1984, 35, 103). One route to aminoindazoles involves the rearrangement of a 3-(2-aminophenyl)-5-aryl-1,2,4-oxadiazole (**80**) to 3-acylaminoindazoles (**81**). This rearrangement is promoted by an electron withdrawing group (R) at the amino nitrogen of (**80**) (D. Korbonits, I. Kanzel-Szvoboda and K. Horvath, J.chem.Soc., Perkin 1, 1982, 759).

(80)   (81)

(iv) Halogenated Indazoles

The position of electrophilic substitution in indazoles is very much dependant upon the reaction conditions. The situation is further complicated in 2-phenylindazoles as, in principle, electrophilic substitution can also occur in the phenyl ring. Bromination of 2-phenyl-2*H*-indazole takes place exclusively in the indazole rings under all the reaction conditions studied. In acetic acid at room temperature one equivalent of bromine causes substitution at the 3-position almost exclusively. 2-Phenylindazole at elevated temperatures with 2.4 equivalents of bromine yields 3-bromo (major product), 3,5-dibromo- and 4,7-dibromo-indazoles. The various products may be separated by column chromatography (P.S. Waalwijk, P.Cohen-Fernandes, and C.L. Habraken, J.org.Chem., 1984, 49, 3401).

Indazoles bearing trifluoromethyl substituents have been prepared *via* the cycloaddition reaction of *N,N*-dimethyl-2,4-bis(trifluoroacetyl)-1-naphthylamine with a hydrazine. Unsymmetrical hydrazines yield the isomeric 1*H*- and 2*H*- substituted indazoles (M. Hojo, R. Masuda, and E. Okada, Synthesis, 1990, 481).

(e) Fused Systems

(i) Pyrazoloindazoles

Pyrazolo[1,2-a]indazole (**82**) is susceptible to aerial oxidation but can be prepared (and used in solution) as indicated below. The route is similar to that used for pyrazolo[1,2-a]pyrazole.

In DMSO at 0°C pyrazolo[1,2-a]imidazole (82) reacts with one molecule of dimethyl acetylene dicarboxylate (DMAD) to yield (83: 43%) (Y. Fukimara *et al.*, Heterocycles, 1987, 26, 133). At room temperature two molecules of DMAD are involved and the products are (84) and (85). In protic solvents the main product (86) arises by the addition of 2 molecules of DMAD followed by a proton shift. Compound (83) is an intermediate in the formation of (84), (85), and (86) (G. Bettinetti *et al.*, Heterocycles, 1988, 27, 1207).

Pyrazoloindazoles readily undergo a Hofmann degradation. Thus 2,3-dihydro-7-methyl-9-phenyl-1*H*,9*H*-pyrazolo[1,2-a]indazole with ethyl iodide gives a mixture of the inseperable indazolium iodides (87) and (88). The reaction of this mixture with potassium hydroxide yields the tetrahydrobenzodiazocine (89; 65%) and the substituted benzophenone (90; 19%) (Y. Fujimara and M. Hamana, Heterocycles, 1986, 24, 3187).

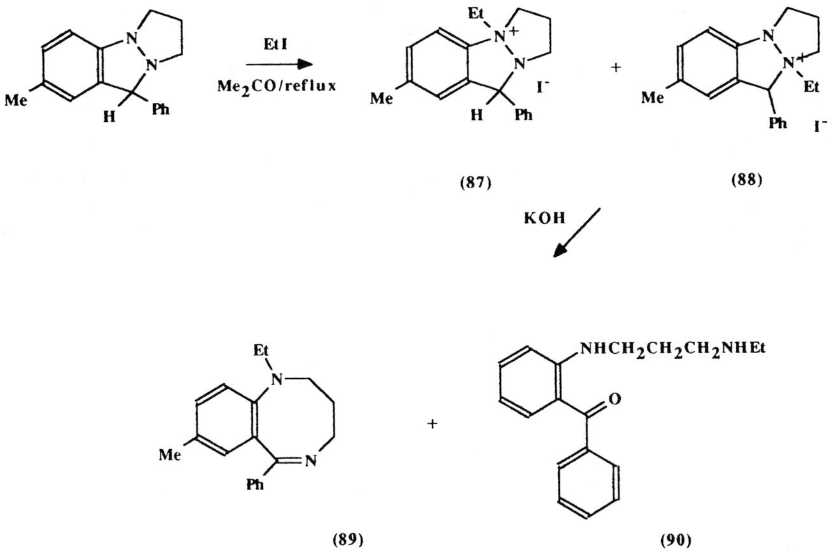

(87)   (88)

(89)   (90)

(ii) Triazoloindazoles

The 1,3-addition reaction of a nitrile imine to a 3-indazolinone results in

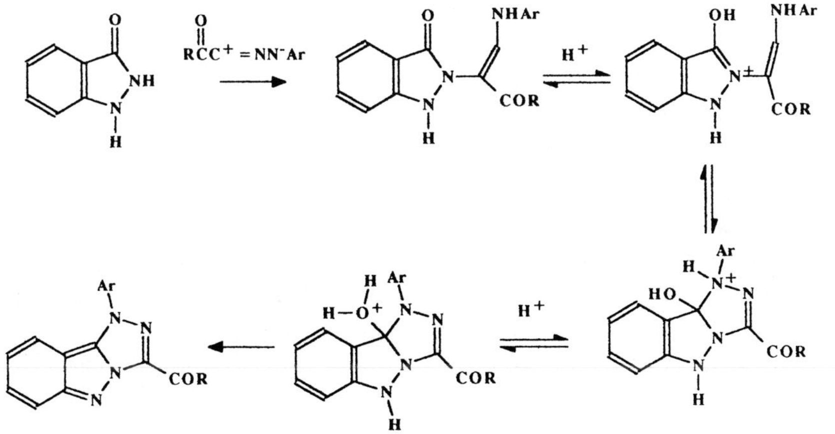

Scheme 11

the formation of a 2-acylindazole-arylhydrazone. Acid-induced electrocyclic ring-closure of the latter yields a 1*H*-1,2,4-triazolo[4,3-b]indazole (scheme 11) (G. Cusmano *et al*, Heterocycles, 1989, <u>29</u>, 2149).

The phosphazide (91) synthesised, from *o*-azidobenzaldehyde as shown below, reacts with an isothiocyanate in refluxing toluene to yield the triazoloindazole derivative (92). Cyclisation of phosphazide (91) with carbon dioxide in toluene in sealed tube at 120°C results in the 2-oxo-2,3-dihydroindazole (93), and if carbon disulphide is used the thio analogue (94) is obtained in 60% yield (P. Molina, A.A. Arques, and M.V. Vinader, J.org.Chem., 1990, <u>55</u>, 4727).

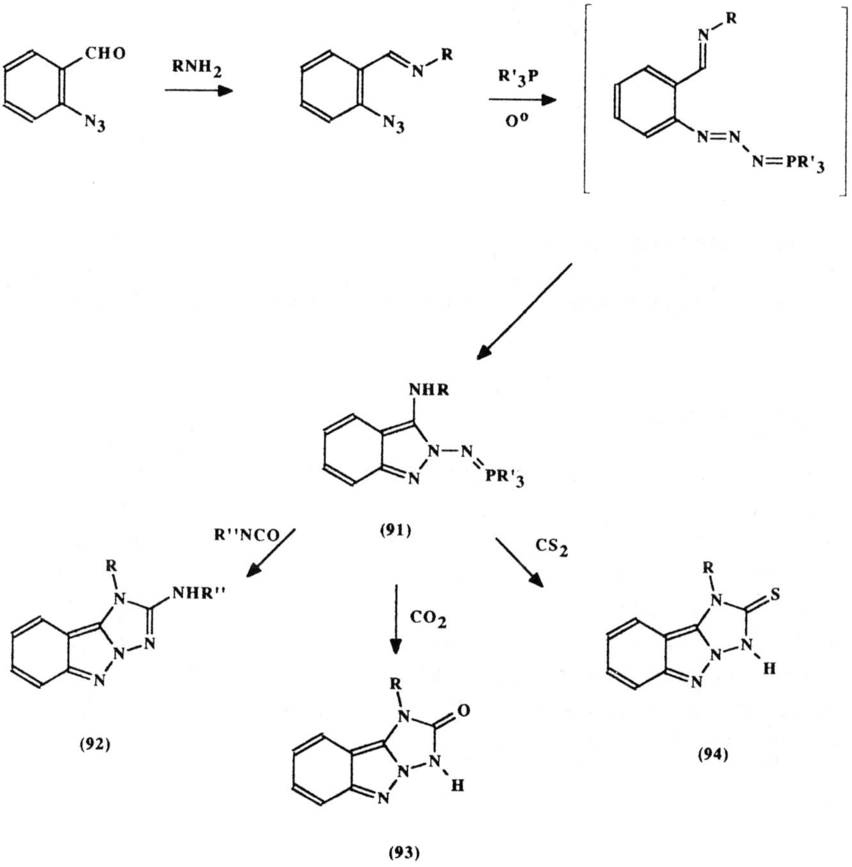

Other fused indazoles which have been reported include pyrano[2,3-e]indazoles (L. Mosti, G. Menozzi, and P. Schenone, J.heterocyclic Chem., 1984, 21, 361), formed by the reaction of dichloroketenes and N,N-disubstituted-α-aminomethyleneketones. Reaction of benzyne with pyridinium-N-imines results in pyrido[1,2-b]indazoles *via* a 1,3-dipolar cycloaddition rection (M. Masumura *et al.*, Chem.Lett., 1980, 1133). Cyclocondensation of 6-aminoindazole with benzoin occurs in refluxing glacial acetic acid to give the pyrrolo[2,3-g]indazole (95; 85%) (S. Sequeria and S. Seshardi, Indian J.Chem., 1987, 26B, 437).

(95)

Indazole-4,7-diones are useful precursors for the synthesis of fused systems, for example condensation of 5-methyl- or 6-methyl-2-phenyl-2H-indazole-4,7-diones with 4-substituted-2-aminophenols in pyridine results in 9-substituted-5-methyl-2-phenyl-4H-pyrazolophenoxazin-4-ones (S. Nan'ya, J.heterocyclic Chem., 1988, 25, 1373).

(iii) Benzindazoles

Benz[g]indazoles are readily prepared by the reaction of a 2-arylidene-1-tetralone with a hydrazine. Yields are high (>85%), and the reaction conditions (boiling ethanol) mild (N.R. El-Rayyes and A. Al-Jawhary, J.heterocyclic Chem., 1986, 23, 135). Benz[c,d]indazole is an aza analogue of acenaphthylene and potential precursor of 1,8-didehydronaphthalene. The heterocycle has not yet been unambiguously synthesised but has been detected, by low-temperature NMR spectroscopy, in the photolysis product of 1,8-diazidonaphthalene (H. Nakanishi, A.Yabe, and K. Honda, Chem.Comm., 1982, 86).

(f) Reduced Indazoles

The condensation reaction of cyclic ß-diketones with hydrazines has been used to synthesise some novel 1,4,5,6,7-pentahydro-1H-indazoles. Thus the cyclohexanone (**96**) with hydrazine hydrate in bioling ethanol yields a racemic mixture of 5-benzoyl-4,6-diaryl-3,7-diphenyl-1,4,5,6,7-pentahydro-1H-indazole (**97**) (F.H. Al-Hajjar and H.S. Hamoud, J.heterocyclic Chem., 1981, <u>18</u>, 591).

(96)   (97)

Ar = Ph, p-CH$_3$C$_6$H$_4$, Cl, p-CH$_3$OC$_6$H$_4$

Reduced indazoles containing a fused cyclooctane ring (**99**) are formed in fair yield by the action of hydrazines on 1-chlorodiisophor-2,7-en-3-ones (**98**) (F. Kurzer and S.S. Langer, J.heterocyclic Chem., 1990, <u>27</u>, 871).

(98)   (99)

R = H, Me   Y = Cl, OH   Ar = Ph, o-CH$_3$C$_6$H$_4$, p-O$_2$NC$_6$H$_4$

3. Imidazoles

More has been published on the chemistry and properties of the imidazole ring system, and its derivatives, than on any other azole. Several books and numerous reviews have dealt with the vast number of imidazole compounds and their increasing importance as pharmaceutical products. Reviews concerning the biochemistry of imidazoles include an overview of their antifugal properties (R.A. Fromtling, Drugs of Today, 1984, 20, 325), detailed studies of their action as fungicides (M. Borgers, Reviews of Infectious Diseases, 1980, 2, 520 and W.H. Beggs, F.A. Andrews, and G.A. Sarosi, Life Sciences, 1981, 28, 111) and a review on their effects on gastric secretion (W.S. Rehm, G. Carrasquer, and M. Schwartz, Membrane Biophysics, pp. 229-246, 1981, Alan R. Liss, Inc., NY). The involvement of imidazoles in the aging of proteins has been the subject of three reviews (G.N. Somero, Transport Processes, Iono- and Osmoregulation, ed. R. Gilles and M. Gilles-Baillen, Springer-Verlag, Berlin, 1985, pp 454-468, P. Ulrich et al., Modification of Proteins During Aging, 1985, 7, 83, and D. Berg et al., Mykosen, 1986, 29, 221).

The chemistry of imidazole derivatives which are inhibitors of sterol biosynthesis has been reviewed (P.A. Worthington, Sterol Biosynthesis Inhibitors, ed. D. Berg and M. Plempel, Ellis Horwood, Chichester, 1988, pp.19-55). This review also covers simple nitrogen heterocycles, piperizines, pyridines, triazines and morpholines, all of which show fungicidal activity and inhibition of sterol biosynthesis. The chemistry of 2H-imidazoles has been reviewed by M.P. Sammes and A.R. Katritzky (Advances in Heterocyclic Chemistry, 1984, 35, 375).

(a) Preparation

Two main routes to the imidazole ring system are cyclisation or cycloaddition reactions, and ring transformation reactions of other heterocyclic compounds. The majority of methods involve the cyclisation of a suitable precursor.

(b) Cyclisation

Type 1: C-C + N + C + N ⟶ (imidazoline ring structure)

The reaction of glyoxal with an aldehyde and two equivalents of hydroxylamine hydrochloride results in the formation of a 1-hydroxy-1H-imidazole-3-oxide (**100**) in moderate to high yield, which on reduction affords the 2-substituted-1-hydroxy-1H-imidazole (**101**) (G. Lause, J. Stadlwieser, and W. Klotzer, Synthesis, 1989, 773). This route has been used for the synthesis of 1-aryloxy- and 1-arylalkyloxy-1H-imidazoles and for the first synthesis of 1-hydroxy-1H-imidazole.

(100)   (101)

R = H, 68%; R = Me, 74%; R = Et, 70%.

Several new imidazole derivatives have been synthesised, in moderate to good yields, by the reaction of a diketone with ammonium acetate and an aldehyde (I. Isikdag, U. Ucucu, and B. Cakir, J.fac.pharm.Gazi., 1989, 6, 49).

Novel imdazoles where the imidazole ring is attached to a quaternary carbon centre have been prepared. The key compound in this synthesis is the nitrile (**102**) as the imidazole ring is synthesised by elaboration of the nitrile substituent. The latter reacts with methylmagnesium bromide to give a methyl ketone. Elaboration of the imidazole ring is achieved by bromination the methyl ketone and cyclocondensation of the product with formamide (G.B. Gregory, A.L. Johnson, and W.C. Ripka, J.org.Chem., 1990, 55, 1479).

(102)

Type 2: C-C-N + C-N ⟶ [imidazole ring structure]

The combination of a C-C-N unit and a C-N unit has been used to prepare a series of bis(carbamoylmethyl) derivatives of imidazoles which exhibit antineoplastic activity in the murine P388 lymphocytic leukemia assay (W.K. Anderson, D. Bhattacharjee, and D.M. Houston, J.med.chem., 1989, 32, 119). Diethyl 1-phenylimidazole-4,5-dicarboxylate (103) is prepared by treatment of the ethyl ester of N-phenyl-N-formylglycine (104) with diethyl oxalate and sodium ethoxide to give the intermediate (105), which on treatment with acid and potassium thiocyanate yields the mercaptoimidazole (106) in 48% yield. Desulphurization is achieved with sodium nitrite and nitric acid in 93% yield. The diester is converted to the biscarbamate (107) by hydride reduction of the diester followed by condensation of the resulting diol, with methylisocyanate.

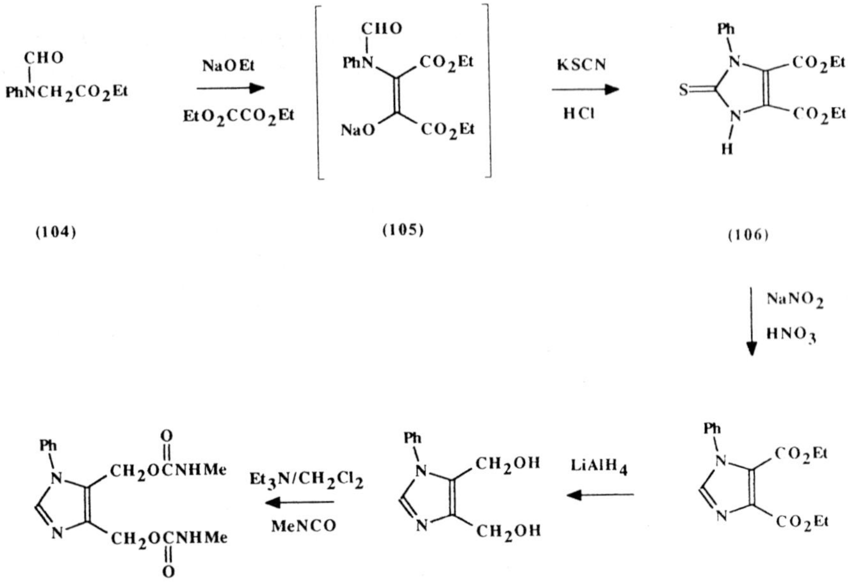

(104)　　　　　　　　　　　(105)　　　　　　　　　　　(106)

(107)　　　　　　　　　　　　　　　　　　　　　　　　(103)

Two other cyclisation routes to imidazoles are reported in this paper;　the cyclisation of the diaminomaleonitrile (**108**), prepared by reaction of diaminomaleonitrile with trimethylorthoformate, and reaction of tartaric acid dinitrate (**109**) with ammonium hydroxide and a suitable aldehyde.

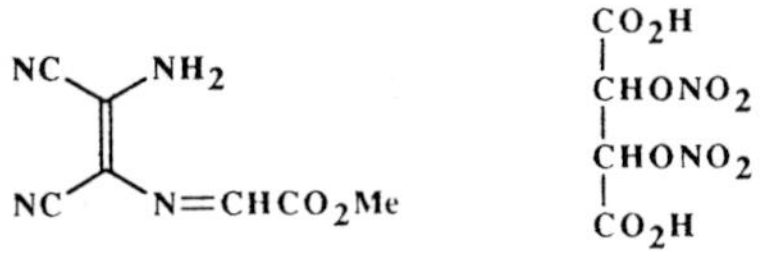

(108)　　　　　　　　　　　(109)

Type 3: C-N-C-C-N　　⟶　

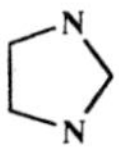

(*E,Z*)-1-Amino-2-benzyloximinoethane　with　carbon　disulphide　and dicyclohexyl carbodiimide leads to the mercaptoimidazole (**110**) *via* the isothiocyanate intermediate (**111**) (H. Hauser *et al.*, Sci. pharm., 1988,

56, 235). The mercapto imidazole can be converted into a 1*H*-imidazole by *S*-alkylation with dimethyl sulphate, oxidation to the disulphide or, vigorous oxidation to give 1-benzyloxy-1*H*-imidazole.

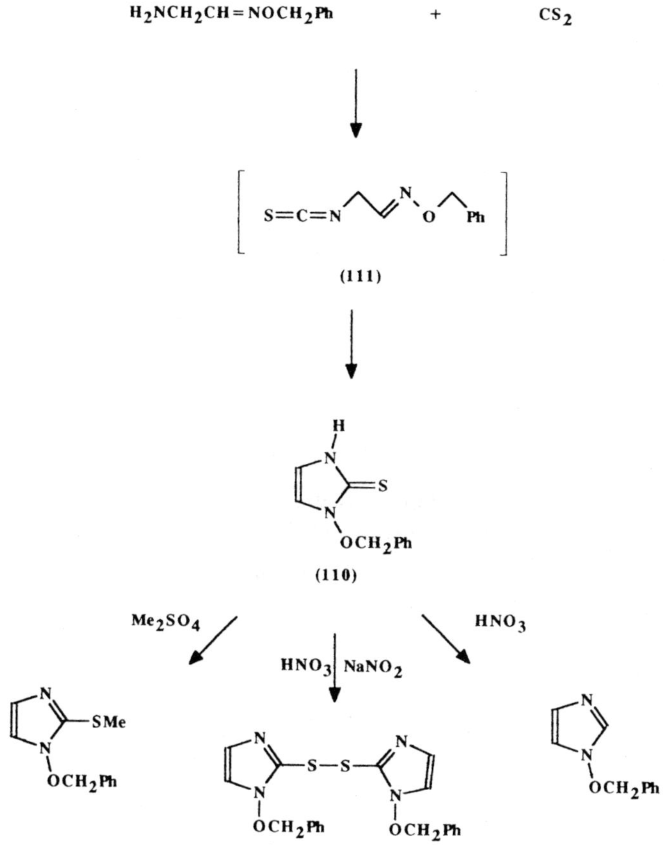
H2NCH2CH=NOCH2Ph
+
CS2
S=C=N
N
O
Ph
(111)
H
N
S
N
OCH2Ph
(110)
Me2SO4
HNO3
HNO3
NaNO2
N
SMe
N
OCH2Ph
N
S—S
N
N
N
OCH2Ph
OCH2Ph
N
N
OCH2Ph

Type 4: C-N-C-N-C ⟶ 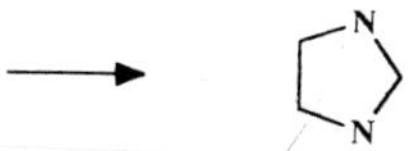

Cyclisation of the *N*-acylated compound (**112**) by phosphorus oxychloride gives the fused imidazole (**113**) (J. Liebscher and K. Feist, J.pract.chem., 1988, 330, 175).

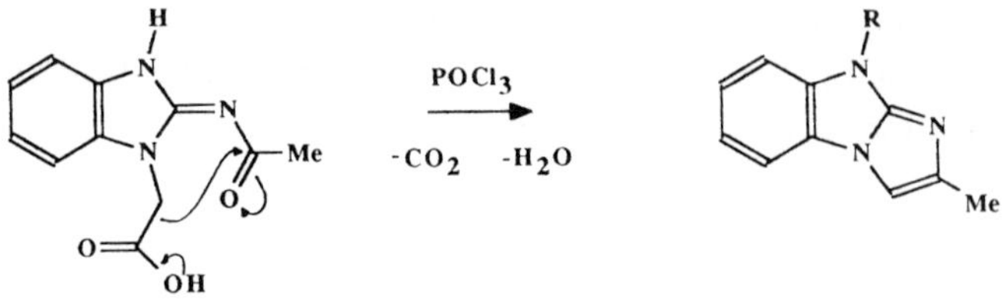

(112)   (113)

Non-fused imidazole systems are also prepared by this route (scheme 12). The *N*-aryl-*N'*-(4-nitrobenzyl)benzamidine (114) reacts with acetic anhydride to give intermediate (115), which cyclises upon treatment with base losing water to give the substituted imidazole (116). The electron withdrawing *p*-nitrobenzyl group renders the CH_2 group very susceptible to attack by base.

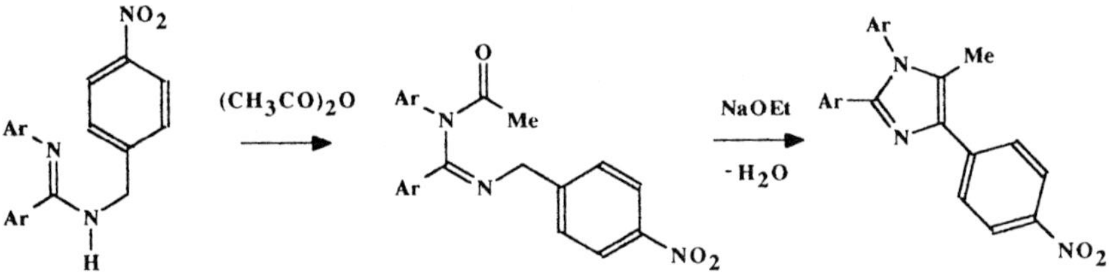

(114)   (115)   (116)

Scheme 12

Type 5: C-N-C + C-N ⟶

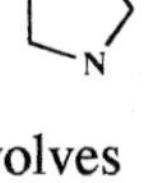

A novel route to imidazoles involves the reaction of a *N*-alkylisocyanoacetamide (117) with an arylsulphenyl chloride (118) to give an arylcarbon-imidochlorodithioate (119). Cyclisation of the latter to an imidazole-3-oxide is effected by triethylamine (R. Bossio *et al.*, Synthesis, 1989, 641).

(117)   (118)   (119)

The condensation of tosylmethyl isocyanide with various imines has been used to prepare fluorinated analogues of the insect growth-regulator KK-42 (**120**) (A. Messeguer *et al.*, Heterocycles, 1990, <u>31</u>, 67).

R_1/R_2 = F/Me; H/ ... ; H/ ...

(**120**)

(c) Other Cyclisation Reactions

Protonated 1,3-diaza-4,4-diphenyl-2-(methylthio)butadienes (**121**) undergo [1+4] cycloaddition reactions with isocyanides (**122**) to give imidazoles (**123**) and (**124**). The latter is formed by the loss of isobutene from the salt of (**123**) when R=Bu^t. Numerous examples are given (G.Morel, E. Marchand, and A. Foucaud, J.org.chem., 1989, <u>54</u>, 1185). Rearranged imidazoles (**125**) and (**126**) are some times formed in low yield.

Under acylating conditions 1,4-diaza-1,3-dienes give imidazoles (D. Armesto *et al.*, Tetrahedron Letters, 1987, **28**, 4605). Thus the bis-imine (**127**) with base gives the anion (**128**), which cyclises *via* a radical process, to an imidazole when treated with an acid chloride.

Cycloaddition of KS^{14}CN with ethyl 2-aminoacetoacetate hydrochloride gives the ^{14}C-labelled thiol (**129**). Desulphurization is effected by activated Raney nickel to give the imidazole (**130**) in 68% yield (A.J. Villani *et al.*, J.labelled Compds and Radiopharm., 1989, <u>27</u>, 1395).

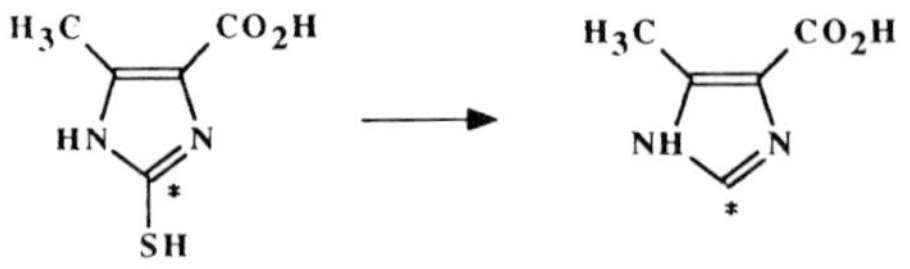

(129) (130)

(d) From Other Heterocycles

1-Benzyloxy-3-(2,2-diethoxyethyl)thiourea (**131**) with dilute hydrochloric acid results in the formation of 2-benzyloximino-5-ethoxythiazolidine (**132**) in 63% yield (H.Hauser *et al.*, Sci.Pharm., 1988, <u>56</u>, 235). This sulphur containing heterocycle is readily converted into the mercaptoimidazole (**133**).

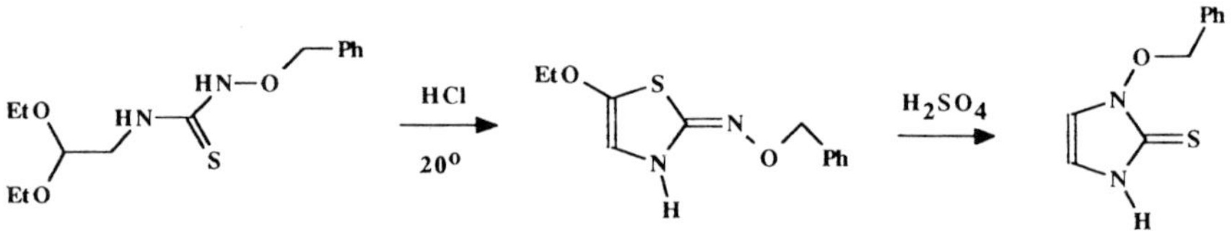

(131) (132) (133)

1,3-Disubstituted-5-acylamino-6-methyluracils (**134**) undergo ring contraction when treated with sodium hydroxide in aqueous ethanol (J. Sakakibara *et al.*, Tetrahedron Letters, 1988, <u>29</u>, 4607).

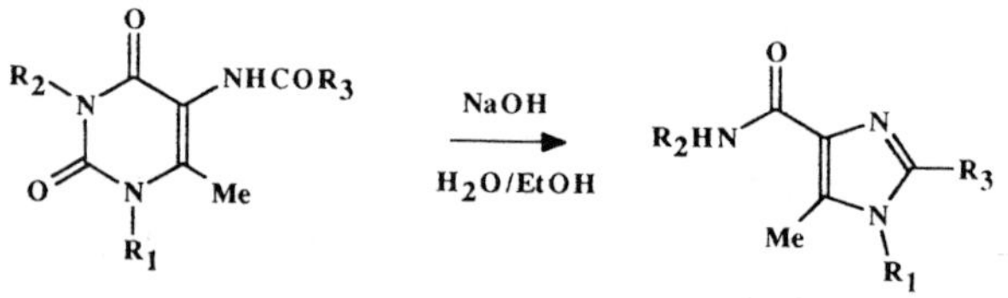

(134)

R_1, R_2, R_3, = combination of Me and Ph

(e) Imidazole Derivatives

(i) N-Substituted Derivatives

N-Substituted imidazoles may be prepared either from non-heterocyclic precursors or *via* substitution of suitable N-H imidazoles. The latter method may lead to a mixture if a 4(5)-substituted imidazole is used due to the ready 1,3 tautomerisation of N-H imidazoles.

Imidazole has been alkylated by 3-bromopropyl acetate and ethylene carbonate to give (135) and (136) respectively (A. Banif *et al.*, J.heterocyclic Chem., 1990, 27, 251) and by 2(α-bromoacetyl)naphthalene to yield (137) (V. Calis, S. Dalkara, and M. Ertan, Arch.Pharm., 1988, 321, 841).

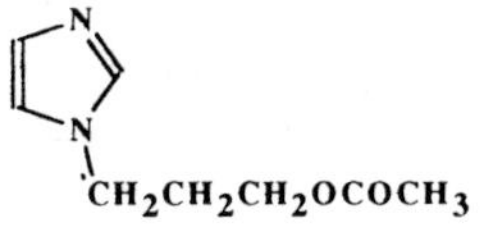

(135) 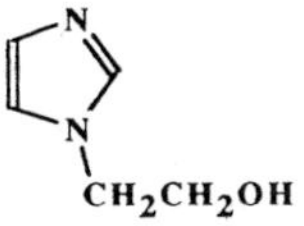　　　　　(136) 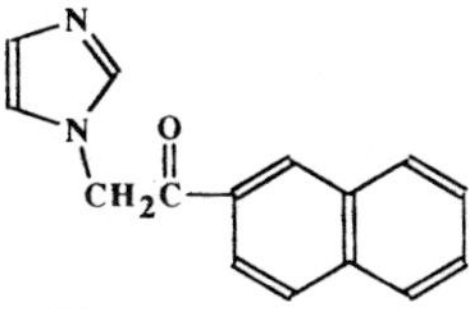　　　　　(137)

Imidazole derivatives (138) are readily synthesised by arylation of imidazole with the fluorobenzophenone (139) (G. Stefanich *et al.*, Arch.Pharm., 1990, 323, 273).

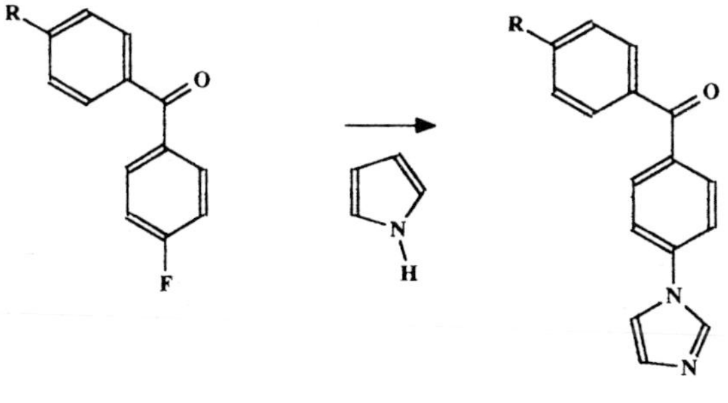

(139)

(138)

The ring contraction of a benzothiopyranol (**140**) in the presence of imidazole leads to *N*-substituted imidazole (**141**). The *cis*- stereochemistry of this product is determined by the reaction mechanism (D.F. Rane *et al.*, Tetrahedron, 1988, <u>44</u>, 2397).

(99% cis; 1% trans)

(140)   (141)

N,N-disubstitution of imidazoles leads to quaternary salts. 1-(Alkoxymethyl)-2-[(hydroxyimino)methyl]-3-methylimidazolium halides have been prepared as agents which reactivate organophosphorus inhibited *acetylcholine esterases* (C.D. Bedford *et al.*, J.med.Chem., 1989, <u>32</u>, 493, 504). Synaptic *acetylcholine esterase* is inhibited very effectively by organophosphorus pesticides and by phosphorus agents used in chemical warfare. One particularly toxic compound is 3,3-dimethyl-2-butylmethylphosphono fluoridate (**142**). Drugs based on pyridinium oximes, which are normally effective at reversing the symptoms of poisoning by nerve gasses or pesticides, are not effective against (**142**), however, a class of bis(pyridinium) dimethyl ether derivatives is. In order to mimic their activity several imidazolium halides (**143**) have been prepared (C.D.Bedford *et al.*, J.med.Chem., 1989, <u>32</u>, 504).

(143)

(ii) C-Alkyl- and C-Arylimidazoles

Substitution at a carbon in the imidazole ring is easily achieved by preparing a metallated imidazole which is then quenched with a suitable electrophile. Metallation of 1,2-dimethyl imidazole can yield 1,2-dimethylimidazole-5-yl-lithium and 1-methyl-2-(lithiomethyl)imidazole. The site of metallation is dependent on the metallating agent, the solvent and the reaction conditions used (B. Iddon and B.L. Lim, J.chem.Soc. Perkin I, 1983, 271, 279). 1,2-Dimethylimidazole is metallated exclusively at the 2-methyl group when treated with *n*-butyllithium in diethyl ether at -110°C. The 5-lithio derivative is formed by a trans-metallation reaction and is more stable than the lithiomethyl compound which decomposes almost entirely at 5°C. Thus the position of substitution may be controlled by careful monitoring of the temperature at which the lithiated imidazole is quenched. 1-Methylimidazole metallates preferentially at the 2-position. It is possible to dimetallate *N*-protected imidazoles thus allowing 1,2,5-trisubstituted imidazoles (e.g. **144**) to be readily prepared (D.J. Chadwick and R.I. Ngochindo, J.chem.Soc. Perkin I, 1984, 481).

(144)

2-Vinylimidazoles are precursors of imidazole polymers (A.S. Rothberg *et al.*, U.S.P. 4410706, 1983; Chem.Abs., 1984, 100, 34542). Compounds of type (**145**) are anti-eubacterial agents, and so simple routes to these compounds and their analogues are of considerable interest.

(145)

2-Vinyl-1*H*-imidazoles (**146**) are readily prepared in high yield by dehydration of hydroxyalkylimidazoles (**147**) (S. Ohata *et al.*, Synthesis, 1990, 78) as shown below.

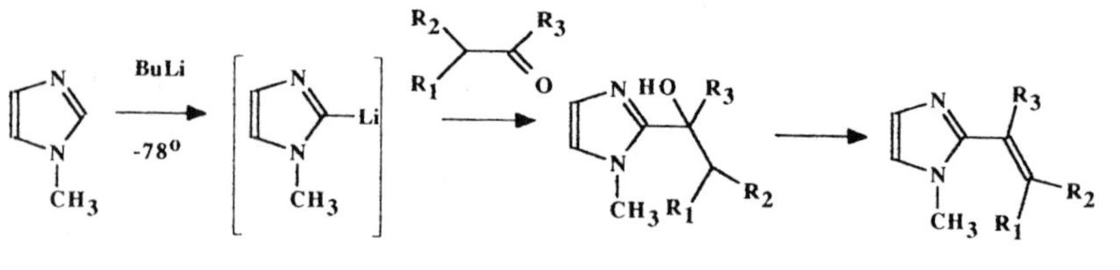

(147)   (146)

R_1, R_2 = H or alkyl

R_3 = Allyl, Aryl, or Hetero

Reagents other than organolithiums have been used in the synthesis of *C*-substituted imidazoles. Pyridine substituted imidazoles may have commercial applications as selective metal ion extracting agents (D. Undsay, *et al.*, Chem. Comm., 1987, 1270). The usual route to these imidazole derivatives is *via* ring closure of a suitably functionalised pyridine precursor and few routes involving aryl ring coupling have been reported. 2-Lithioimidazoles undergo trans-metallation with anhydrous zinc chloride, addition of tetrakis(triphenylphosphine)palladium, 2-bromopyridine and excess zinc chloride then leads to aɪ *N*-substituted-2-(2-pyridinyl)imidazole (**148**) in high yield. The coupling reaction does not occur in the absence of zinc chloride and it is considered that an intermolecular imidazolylzinc intermediate is formed which is unreactive towards heteroarylhalides. Coupling only occurs when excess zinc chloride is added to break down this complex so that palladium catalysed cross coupling can occur (A.S. Bell, D.A. Roberts, and K.S. Ruddock, Tetrahedron Letters, 1988, 29, 5013).

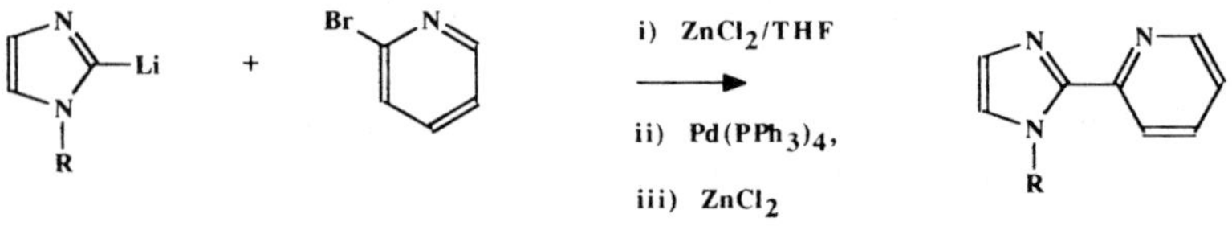

i) $ZnCl_2$/THF

ii) $Pd(PPh_3)_4$,

iii) $ZnCl_2$

(148)

R = Me, 93%; CH_2Et, 93%; SO_2NMe_2, 60%

It has been reported that soft electrophiles such as thienyl bromides, benzyl chlorides and benzyl bromides undergo direct substitution reactions with 2-substituted-NH-imidazoles to give 4(5)-substituted imidazoles (J.P. Whitten *et al.*, J.heterocyclic Chem., 1988, 25, 1845). Modest yields of 4(5)-alkylated products together with N-alkylated derivatives are formed. Thus 2-phenylimidazoles reacts with 3-thienylmethyl bromide and sodium hydroxide to give 4-(3-thienylmethyl)-2-phenylimidazole in 45% yield, 4,5-bis(3-thienylmethyl)-2-phenylimidazole in 26% yield and only a trace of N-(3-thienyl)-2-phenylimidazole. The nature of the electrophile strongly influences the product distribution, the harder the electrophile the more N-substitution takes place.

The synthetic utility of *o*-xylylenes has lead to interest in their heterocyclic analogues. Thus imidazole analogue (149) has been prepared by flash pyrolysis of a *C*-alkylimidazole. 1-Methylimidazole-4,5-xylylene condenses with thiophenol to give the isomeric adducts (150) and (151) in 65% yield (R.C. Storr *et al.*, Tetrahedron Letters, 1990, 31, 1487).

(149)   (150)   (151)

The biological properties of *C*-alkylimidazoles have been widely studied and in particular the imidazole alkaloids. A detailed review of their natural occurrence, isolation, synthesis and properties has been published (L. Maat and H.C. Beyermen, The Alkaloids, 1983, 22, 281). J.R. Lewis has published a series of papers reviewing the literature from 1982 (Natural Product Reports, 1984, 387; 1985, 245; 1986, 586; 1988, 351). The immunoreppressive activity of substituted imidazoles has also been covered in a very detailed review (J.J. Miller and J.R. Salaman, Curr.Status.Mod.Ther., 1981, 7, 111).

Water soluble polymers based on *C*-alkylated imidazoles are active as catalysts in esterolytic reactions. Imidazole polymers have been prepared as models for biologically active macromolecules and have a high level of catalytic activity. The imidazole polymers are studied since imidazole groups are involved in the catalytic activity of most hydrolases (J.A. Pavlisko and C.G. Overberger, "Biomedical and Dental Applications of Biopolymers", Polymer Science and Technology, Vol 14, ed. C.G. Gebelein and F.F. Koblitz, Plenum Press, New York, 1981, p. 257 and C.G. Overberger and R. Tomko, ACS Symposium Ser., 1983, 212, 13).

C-Alkylated imidazoles undergo an interesting ring cleavage reaction to give vicinal diamines (J. Altman and M. Wilchek, Ann., 1989, 493). For example scheme 13 shows how the reaction is used in the preparation of 5,6-diaminohexanoic acid. Considering the usual difficulties assiciated with the synthesis of vicinal diamines this provides a useful route which can be used in the preparation of metal amine complexes.

R = $CH_2CH_2CH(CO_2Et)_2$

Scheme 13

(iii) Halogenated Derivatives

Halogens are readily introduced into the imidazole ring by either direct electrophilic halogenation or by cleavage of a metal carbon bond. 2-Lithio-N-methylimidazole with bromine leads to 2-bromo-N-methylimidazole in 80% yield (M.El Borai et al., Polish J. Chem., 1981, 55, 1659), further bromination at the 5-position is effected by N-bromosuccinimide.

The introduction of fluorine into imidazoles is important due to the modified physiological properties which often result from the introduction of fluorine into a biologically important molecule. Normally electrophilic fluorinating agents, including elemental fluorine, cannot be used to fluorinate imidazoles due to reaction at the nitrogen and this problem of selective fluorination of imidazole has not yet been fully resolved. Organotin derivatives of imidazole have been used to mediate the reaction between elemental fluorine and imidazole, and to introduce selectivity into the reaction (M.R. Bryce, R.D. Chambers, S.T. Mullins, and A. Parkin, Bull:Soc.chim.France, 1986, 6, 930). The trimethyltin group is introduced via an organolithium derivative.

Trifluoromethyl groups are readily introduced into the imidazole nucleus. The reaction of imidazole-2-carboxylic acid with sulphur tetrafluoride affords 2-trifluoromethylimidazole in 61% yield (D. Owen, R.G. Plevey, and J.C. Tatlow, J.fluorine Chem., 1981, 17, 179).

Photochemical methods are also used to trifluoromethylate imidazole. The trifluoromethyl radical, generated photochemically from trifluoromethyl iodide, reacts with N-acylhistidine and N-acylhistidine esters to give, after ester hydrolysis, a mixture of 2-(152) and 4-(153)-trifluoromethylated imidazoles (H. Kimoto, S. Fujii, and L.A. Cohen, J.org.Chem., 1985, 49, 1060). Yields of trifluoromethylated products are rather low ranging from 18 to 27%.

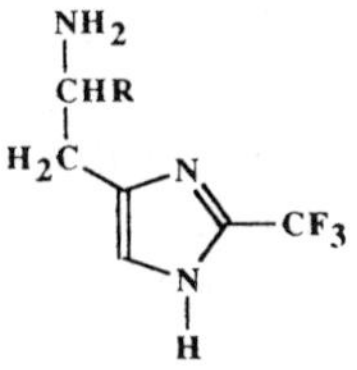

60

(152) 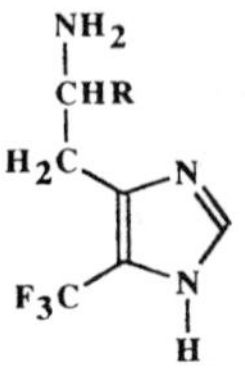   (153)

R = H or CO₂Me

Thermal cyclisation of the dimethylhydrazones (154) leads to 1-methyl-4-aryl-5-trifluoromethylimidazoles (155) in up to 80% yield. If the same reaction is carried out in the presence of silica gel then the regioisomer (156) is formed as the major product. The suggested mechanism for the cyclisation of (154) to (156) is shown below. The hydrazino group migrates from the azomethine carbon to the trifluoroacetyl carbonyl carbon of (154) on the silica gel surface to give the intermediate (157), which affords (156) upon cyclisation. (M. Hojo et al., J.heterocyclic Chem., 1990, 27, 487).

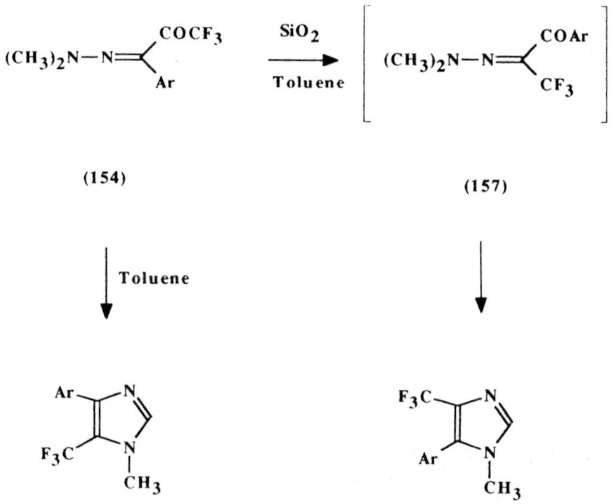

(154)   (157)

Toluene

(155)   (156)

Trichloroethylene can be used to introduce a chlorinated side chain into the imidazole ring. Dropwise addition of trichloroethylene to a mixture of 50% aqueous sodium hydroxide, diethyl ether, dimethylsulphoxide and the substituted imidazole results in the formation of the 1-(1,2-dichlorovinyl)imidazole derivative (158) in about 60% yield (J. Pielichowski and D. Bogdal, Stud.org.Chem., 1988, 35, 473).

The stereoselectivity of the reaction is good and the E-isomer is almost exclusively formed. The reaction is a multistage process as shown below. Formation of (159) occurs via the chlorovinyl anion (160). The former attacks the imidazole derivatives to afford the 1-(1,2-dichlorovinyl)-imidazole anion (161) which is quenched by water to give the product (158) (J. Pielichowski and D. Bogdal, Bull.Pol.Acad.Sci., 1989, 37, 123).

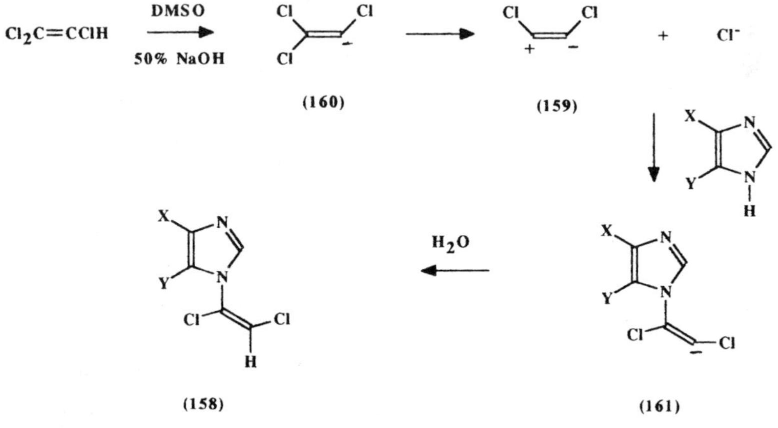

X/Y = H/H; Me/H; Ph/H; H/Ph.

Imidazoles bearing halogenated side chains can be used in the preparation of vinylimidazoles (J. Altman and M. Wilchek, J.heterocyclic Chem., 1988, 25, 915). 1-Triphenylmethyl-4-(2-bromoethyl)imidazole undergoes base catalysed elimination of hydrogen bromide to give 4(5)-vinylimidazole. Imidazoles which are N-unsubstituted do not undergo the base catalysed elimination due to formation of an imidazolium anion by abstraction of the N-H proton. This prevents removal of a proton from the α-position of the side chain.

(iv) Nitrogen Derivatives

Simple nitroimidazole derivatives have been studied because of their potential intestinal antiprotozoal activity (V.P. Arya, K. Nagarajan, and S.J. Shenoy, Indian J.Chem., 1982, 21B, 1115). The preparation of 5-substituted-1-methyl-4-nitroimidazoles involves nitration of 5-chloro-1-methylimidazole at the 4-position. The 5-chloro substitutent can be used to introduce other functional groups by nucleophilic displacement. Nitration is readily effected by treating the imidazole derivative with potassium nitrate in concentrated sulphuric acid (A.P. Bhaduri et al., Indian J.Chem., 1989, 28B, 391).

N-Substituted-nitroimidazoles may also be prepared by alkylation of a nitrated precursor. Such N-substituted compounds are of great use as chemotherapeutic agents. Tinidazole and metronidazole, which are both 5-nitroimidazoles, have been widely used in the treatment of protozoal infections such as trichmonoiasis (C.E. Nord, J.antimicob.Chemotherapy, 1982, 10, supp A35). 4-Nitroimidazoles are gaining importance as immunosuppressants, aldehyde-dehydrogenase inhibitors and potential radio-sensitizers. Alkylation of a 4(5)-nitroimidazole generally leads to a mixture of N-alkylated-4-nitroimidazoles and N-alkylated-5-nitroimidazoles. The problem of forming mixtures of products upon alkylation has been overcome by using ethyl vinyl sulphone as the alkylating agent. It is then possible to form only the 4-nitroimidazole derivative (A.K. Rao et al., J.chem.Soc.Perkin 1, 1989, 1352). The reaction of 4-nitro-2-methylimidazole (162) with Michael acceptors (163) leads to a variety of 4-nitro-N-substituted imidazoles (164) (A.K.S.B. Rao, C.G. Rao, and B.B. Singh, J.org.Chem., 1990, 55, 3702).

(162)   (163)   (164)

R = SO_2Et, CN, CO_2H, CO_2Et, COMe.

The biological activity of nitroimidazoles has been associated with reductive metabolism (D.I. Edwards, "Antibiotics : Modes and Mechanism of Microbial Growth Inhibitors", Vol 6, ed. F.E. Hahn, Springer Verlag, Berlin, 1983). It has been suggested that the mechanism of their biological action involves reduction to a radical anion (P. Goldman and J.D. Wuest, J.Amer.chem.Soc., 1981, 103, 6224), a nitrosoimidazole (W.J. Ehlhardt, B.B. Beauliue, and P. Goldman, J.med.Chem., 1988, 31, 323) or hydroxyimidazole. These reduced imidazoles then react with biological targets such as cellular DNA.

This hypothesis has been tested by studying the reactions of nitroimidazoles with thiols under reducing conditions. Thus metronidazole (165) reacts with an excess of the thiol in water to give 4-[(2-aminoethyl)thio]-2-methylimidazole-1-ethanol (166) in 98% yield. This reaction involves first a one electron transfer from the thiol to the nitroimidazole and subsequent coupling of the imidazole anion with the thiyl radical to give a Meisenheimer complex, which undergoes protonation and nitrous acid elimination to give the substituted product (166) (M.P. Crozet et al., Heterocycles, 1989, 28, 849).

N-Alkyl-2-substituted imidazoles have been used as radio sensitisers for the detection of solid tumours. In general increased effectiveness as a radio sensitiser *in vivo* is paralleled by an increase in chronic toxicity (J. Parrick and M. Moazzam, J.chem.Soc.Pakistan, 1988, 10, 357).

(v) Sulphur Derivatives

Interest in thioimidazoles stems from the discovery that the type 1 copper protein, Populus Nigeria Plastocyanin contains a copper ion in a distorted tetrahedral environment formed by two sulphur atoms from methionine and cysteine and two imidazole nitrogens (P.M. Colman *et al.*, Nature, 1978, 272, 319). This copper complex constitutes the active site of the enzyme. Thus attempts have been made to prepare ligands which contain both the imidazole group and sulphur ligands. The general method for the synthesis of thioether-imidazoles involves the condensation of a thiol with a hydroxymethylimidazole in acetic acid (E. Bouwman and W.L. Driessen, Synth.Comm., 1988, 18, 1581).

(iv) Metalloimidazoles

The metallation of imidazole and its derivatives has been referred to in previous sections. Lithio imidazoles are, by far, the most widely studied metalloimidazoles due to their utility as precursors of imidazole derivatives. The position of metallation depends upon the substitution pattern in the imidazole ring. If the 2-position is unsubstituted then lithiation occurs here first, but only if the *N-H* group is protected. The second site for lithiation is the 5-position, followed by the 4-position. A detailed review covers the metallation of and metal-halogen exchange reactions of imidazoles (B. Iddon, Heterocycles, 1985, 23, 417). Lithio imidazoles may be used to prepare other metalloimidazoles, such as silylimidazoles, mercury imidazoles and trialkylstannyl imidazoles (M.R. Bryce, R.D. Chambers, S.T. Mullins, and A. Parkin, Bull.Soc.chim.France, 1986, 930). C-Lithiation of imidazoles takes place when the N-H group is protected and many types of protecting groups are known (T.S. Manoharan and R.S. Brown, J.org.Chem., 1988, 83, 1107). There have been far fewer reports dealing with the protection of the 2-position of the imidazole ring, thus allowing quantitative 5-lithiation. The use of the phenylthio group was reported in 1978 (C.C. Tang et al., J.Amer.chem.Soc., 1978, 100, 3918). This functionality survives the normal rigorous work up procedures following lithiation, but all other groups used for 2-protection do not.

It has now been reported that the presence of either a tertiary amide group or the t-butyldimethylsilyl group at C-2 allows quantitative lithiation at C-5. The methods for the introduction of these groups are shown below. The amide group may be removed by alkaline hydrolysis and the silyl group by reaction with a nucleophile (e.g. Bu₄NF) (R.I. Ngochindo, J.chem.Soc. Perkin I, 1990, 1645).

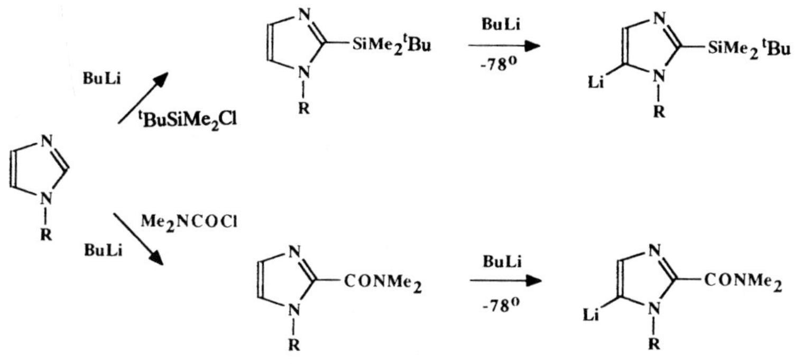

The use of the amide protecting group is restricted to substituted imidazoles which are resistant to the rather forcing conditions used for the deprotection, but these two procedures do represent very useful synthetic methods.

Cobalt imidazole complexes have been studied in attempts to reproduce the mechanism of vitamin B_{12} catalysed biochemical processes (K.C. Dash, Proc.Indian Natn.Sci.Acad., 1989, 55, 212). A review detailing the chemistry of copper (I) and copper (II) imidazole complexes has also been published (P.J.W.L. Birker and J. Reedijk, Copper Coordination Chemistry: Biochemical and Inorganic Perspectives, ed. K.D. Karlin and J. Zubieta, Adenine Press, New York).

(f) Miscellaneous

The imidazole moiety of histidine residues often forms part of the metal binding-site in metallo enzymes, for example zinc is bound to three imidazole rings in carbonic anhydrase (W. Tagaki and K. Ogino, Top.curr.Chem., 1985, 128, 143) and enzymes containing imidazole coordinated to Cu^{II}, Co^{II}, Fe^{II}, and Fe^{III} are known (R.G. Wilkins and P.C. Harrington, Adv.inorg.Chem., 1983, 5, 51). Methods for linking imidazole rings together *via* the 2-position are well known, however in the naturally occurring ligands the imidazole groups are linked *via* the 4-position. A new general route to 4-substituted imidazoles where there is a single carbon link between the rings involves a metal-halogen exchange reaction at the 4-position (A.R. Katritzky *et al.*, J.chem.Soc.Perkin I, 1989, 1139).

This represents a convenient route to 4-substituted imidazoles and does not require protection of the 1 and 2-positions.

New paracyclophanes (**167**) with four imidazole rings are obtained in low yield (5%) by coupling 1,4-bis(2-isocyano-2-tosylethyl)benzene with 1,4-bis(4-halogenophenyl)iminomethyl benzenes (H. Sasaki and T. Kitayawa, Chem.pharm.Bull., 1988, 36, 3646).

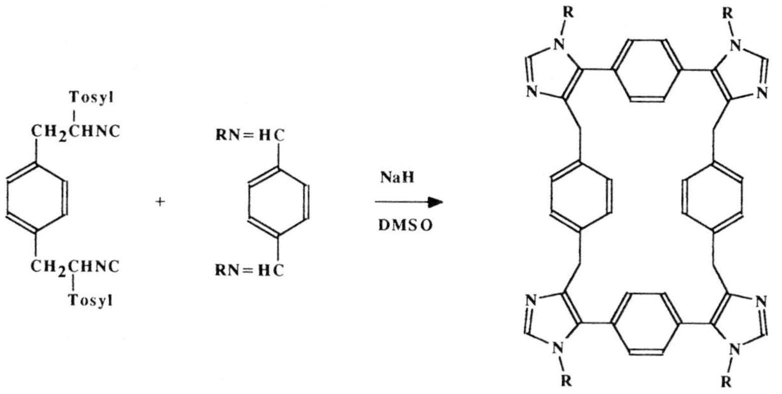

(167)

Crown ethers such as (168) are prepared by coupling the diol (169) with 1,3-bis(bromomethyl)benzene under high dilution conditions (S.C. Zimmermann, K.D. Cramer, and A.A. Galan, J.org.Chem., 1989, 54, 1256).

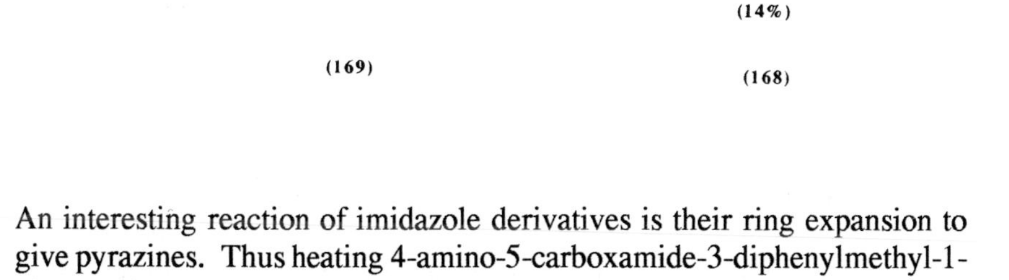

An interesting reaction of imidazole derivatives is their ring expansion to give pyrazines. Thus heating 4-amino-5-carboxamide-3-diphenylmethyl-1-phenacylimidazolium bromide under reflux in methanolic alkali gives the pyrazine derivative (170) in 65% yield. The mechanism is shown below (G. Chattopadhyay et al., Indian J.Chem., 1990, 29B, 1).

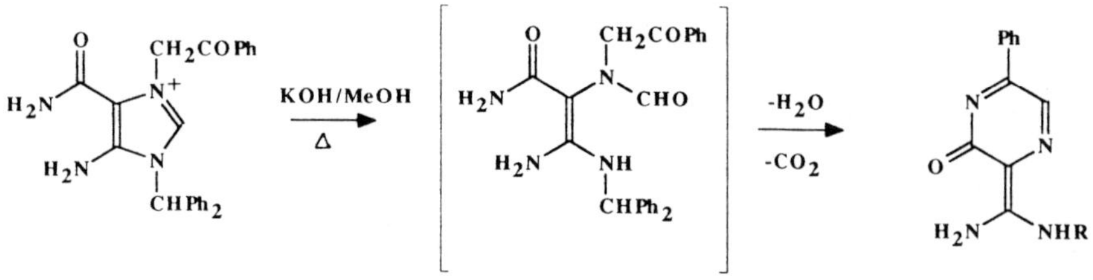

CH2COPh
H2N
N+
KOH/MeOH
Δ
CHO
NH
CHPh2
-H2O
-CO2
Ph
NHR

(170)

(g) Fused Imidazoles

1*H*-Pyrrolo[1,2-c]imidazole derivatives (**171a-d**) are readily prepared by condensation of 2-formylpyrroles (**172a-d**) with ammonium acetate (B. Musicki, M.F. Malley, and J.Z. Gougoutas, Heterocylces, 1989, 29, 1137).

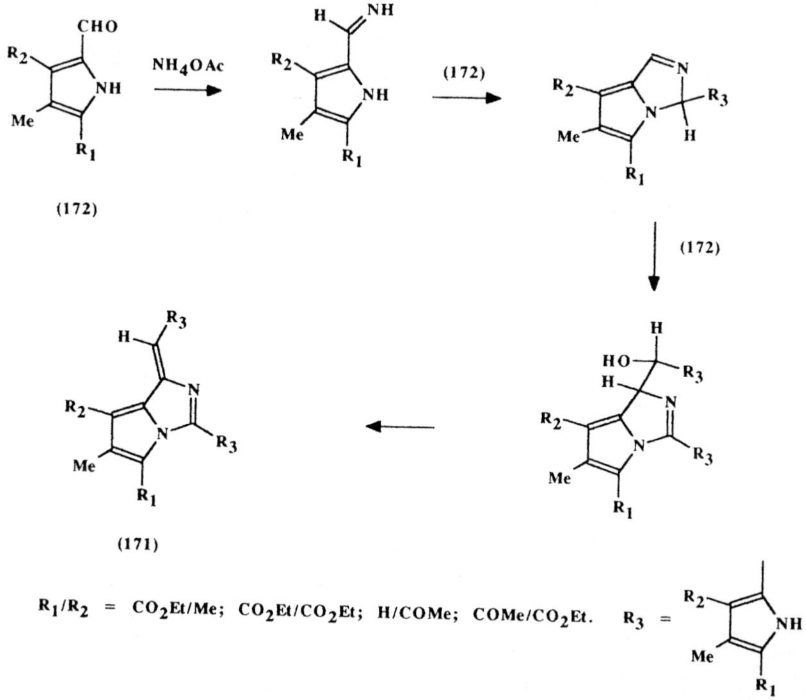

CHO
NH4OAc
(172)
(172)
(172)
(171)
R1/R2 = CO2Et/Me; CO2Et/CO2Et; H/COMe; COMe/CO2Et. R3 =

This is the first reported route to the parent, fused pyrrolo[1,2-c]imidazole, though several routes are known to saturated systems (P.N. Preston, "Condensed Imidazoles 5-5 Ring Systems", 1st edn., John Wiley & Sons, Inc., New York, NY, 1986, pp 42-62).

Activated nitriles may be used in a novel synthesis of pyrrolo[1,2-c]imidazole and also in the preparation of pyrano[2,3-d]imidazoles (M.A. Abdelaziz *et al.*, Phosphorus, Sulphur and Silicon, 1990, 48, 269). The reaction of the activated nitrile, a 3-(2-furanyl)- or 3-(2-thienyl)-acrylonitrile derivative (173a-c) with 2-thiohydantonin (174) gives the pyrrolo[1,2-c]imidazoles (175a-c).

2,3-Diaryl-6,7-dihydro-5H-pyrrolo[1,2-a]imidazoles (176) possess anti-inflammatory activity. Methods for functionalising the C-7 position of these compounds have been developed in order to investigate their structure-activity relationships. The C-7 position is functionalised by first activating the C-7 protons by quaternising the ring nitrogen, followed by condensation with a suitable aldehyde to give (177) (T.F. Gallagher and J.L. Adams, Tetrahedron Letters, 1989, 30, 6599). Thus compound (178) reacts with p-nitrobenzaldehyde in refluxing ethanol with triethylamine to give the C-7 substituted product (179) in 42% overall yield. Removal of the N-alkyl group is effected by treating (178) in DMF at 100°C for 1h.

C-7 Functionalised pyrrolo[1,2-a]-7H-imidazoles are also obtained by direct reaction of 1-acylimidazoles with dimethyl acetylene-dicarboxylate (DMAD), in a novel double transacylation reaction (H.J. Knolker and R. Boese, J.chem.Soc.Perkin I, 1990, 1821). Compound (180) was obtained in 26-40% yield.

Pyrrolo[1,2-c]imidazoles undergo 1,3-dipolar cycloaddition reactions with acetylenic dipolarophiles (B. Musicki, J.org.Chem., 1990, 55, 910). The pyrrolo[1,2-c]imidazole mesomeric betaines (181) are prepared by condensation of an imine with excess of a 2-formylpyrrole. These compounds (181) are analogues of pentalenyl dianion and are of interest from the point of view of their electronic structure and their 1,3-dipolar cycloaddition reactions. They show a strong UV/visible absorption and a strong fluorescence at 500 nm. They undergo a wide variety of 1,3-

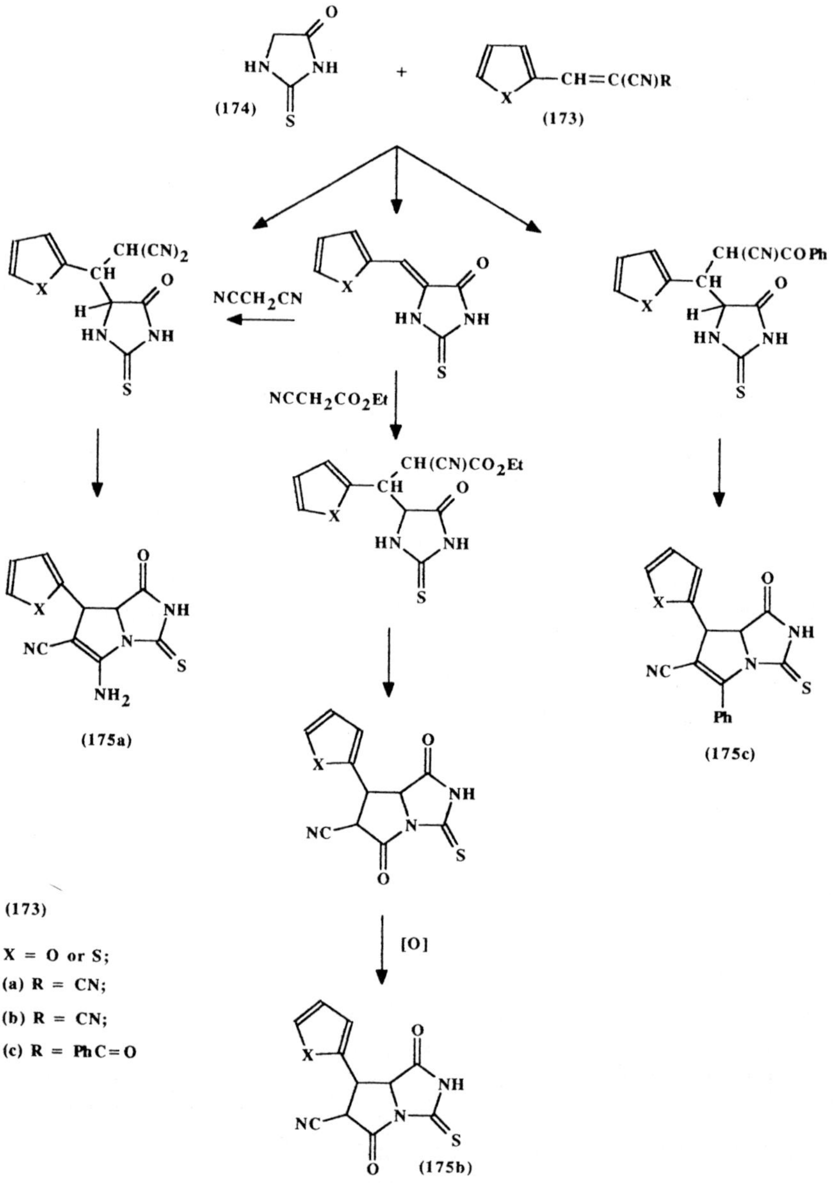

(174) + (173)

(175a)

(175b)

(175c)

(173)

X = O or S;
(a) R = CN;
(b) R = CN;
(c) R = PhC=O

$R_1/R_2 = C_6H_4F/C_6H_4F; \quad C_6H_4OMe/pyridyl; \quad C_6H_4OMe/H \qquad R = p\text{-}NO_2C_6H_4$

$E = CO_2Me$

$R = Me \text{ or } CH_2CH_2C\equiv CH$

dipolar cycloadditions which lead to unique monocyclic and ring-annulated heterocycles (K.T. Potts in, "1,3-Dipolar Cycloaddition Chemistry", ed. A. Podwa, John Wiley & Sons, New York, 1984, 2, pp 1-82).

Typical substituents

$R = CO_2R_1$; R_1 = alkyl; R_3 = aryl

In theory the two azomethine ylide forms of compound (181); basically (182) and (183), both undergo 1,3-dipolar cycloaddition to give (184) and (185) respectively.

When an acetylenic dipolarophile is added to a solution of (181) only the product from the addition of (182) is formed. The isolated products are 2,2'-bipyrroles (186) (only a typical product is illustrated). The reaction occurs smoothly within 2 hours.

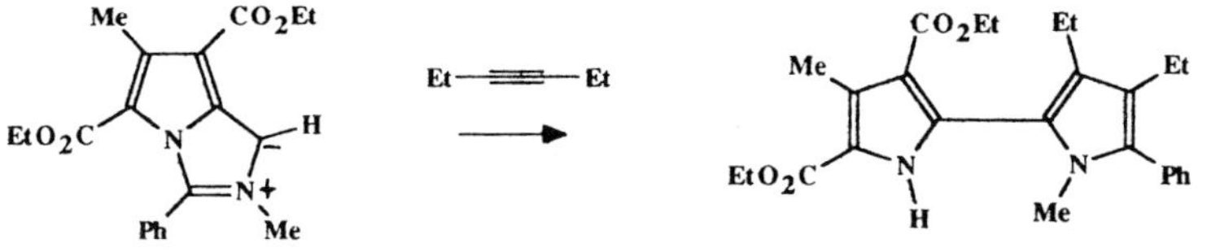

(186)

The reaction is believed to proceed *via* the expected cyclic adduct (187) from addition of the acetylene to the betaine, followed by rearrangement to the bipyrrole:

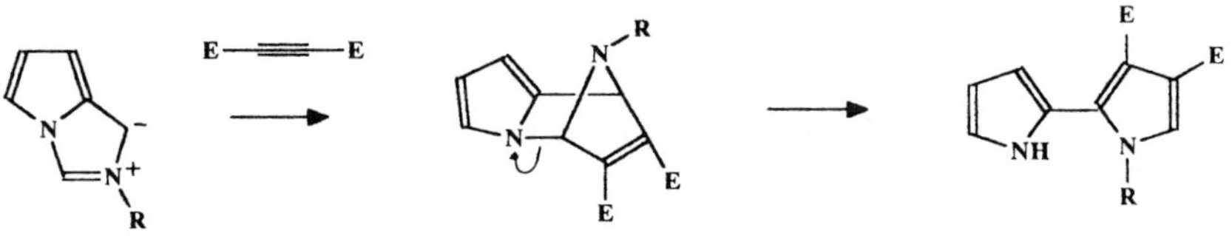

(187)

E = CO$_2$Me

Other fused systems which have been reported include benzimidazolo[2,1-e]imidazoles formed by the reaction of 2-aminomethylbenzimidazole dihydrochloride with aromatic aldehydes (Y.D. Reddy *et al.*, J.Indian chem.Soc., 1988, 65, 853).

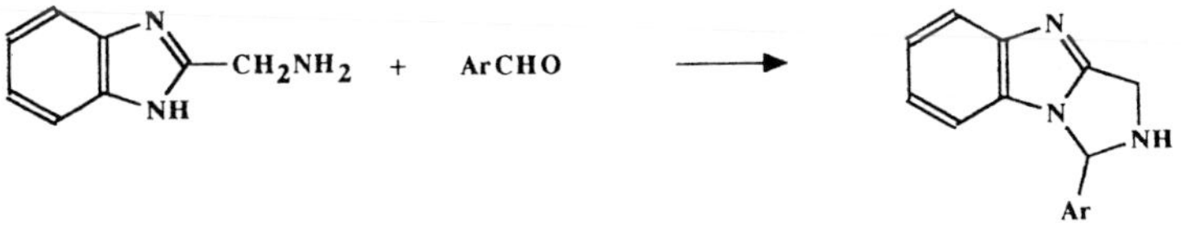

Imidazolo[1,2-a]pyridin-5-ones are of interest as anti-ulcer agents (J.J. Kiminski *et al.*, J.med.Chem., 1985, <u>28</u>, 876; 1987, <u>30</u>, 2047). They are readily prepared in a one-pot procedure by the reaction of 2-phenacylimidazole with an acetylenic ester (H.J. Knolker, R. Biese and R. Hitzemann, Heterocycles, 1989, <u>29</u>, 1551).

R = CO$_2$Me, 80%; R = H, 33%

Imidazolo[1,2-b][1,2,4]triazoles and triazines are of interest, as are many fused heterocycles containing bridgehead nitrogens, due to their activity as anti-viral agents, adenosine isosteres and antifungal agents. The imidazole-triazole fused system is readily prepared by the reaction of 1-amino-2-methylthio-4-phenylimidazole with an isothiocyanate (P. Molina, A. Lorenzo, and E. Aller, Synthesis, 1989, 843).

60-87%

Naphthoimidazoles have received attention due to their antihypertensive activity. 1-(1-Methylethyl)-2-(2-[4-(3-trifluoromethylphenyl)-1-piperazinyl]-ethyl)-1H-naphth[1,2-a]imidazole citrate (**188**) is a long-lasting antihypertensive agent (E. Tojo, D.Barone and E. Baldoli, U.S.P. 4,594,348, 1986; Chem.Abs. 1989, 104, 75025m).

A original synthetic route (E. Tojo *et al.*, Eur.J.med.Chem., 1987, 22, 221) is unsuitable for large scale production, however compound (**188**) has now been prepared on a kilogramme scale by the route shown below (E. Tojo and A. Trani, Org.prep.Procedure Intern., 1988, 20, 253).

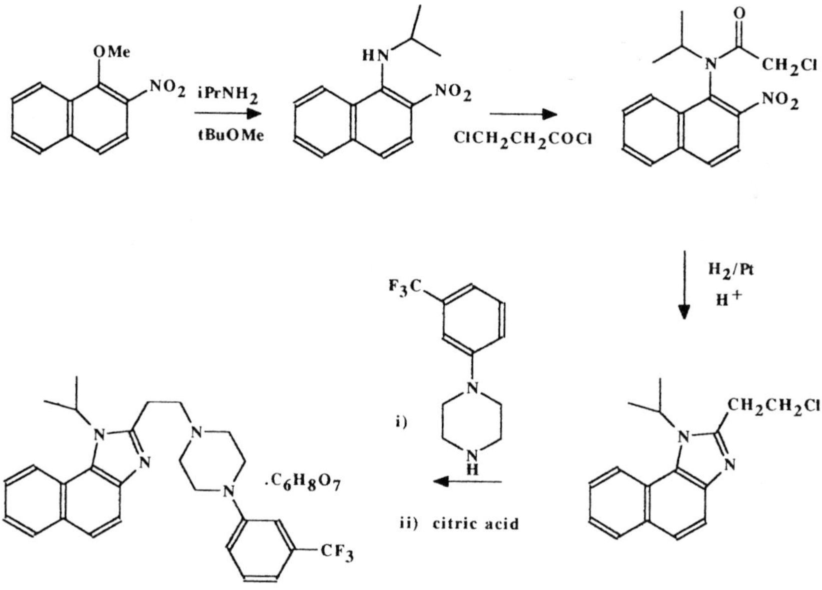

(188)

Other reports of fused imidazoles include a paper detailing the reactions of 2,3-dihydro-2-thioxo-1H-naphth[2,3-d]imidazole-2,9-diones(T.Nakamori, T. Saito, and T. Kasai, Bull.chem.Soc.Japan, 1988, 61, 2019). Tetracyclic compounds, for example (**189**), in which thiazolidinone or thiazinone rings are fused to 1H-naphth[2,3-d]imidazole-4,9-diones have been prepared (H.H. Zoorob *et al.*, Indian.J.Chem., 1990, 29, 1329).

(189)

The novel thiopyrano[2,3-d]imidazole derivatives are obtained by the reaction of 3-aryl-2-cyanothioacrylamides (190) with hydantoins (191) (M.A. Abdelaziz, Sulphur Letters, 1990, 11, 1).

(190)   (191)

(h) Reduced Imidazoles

2-Methylamino-1H-imidazole-4,5-dione (193) is the product from oxidation of creatinine (192), which is present in the muscular tissue of many vertebrates. This redox process is believed to be of importance in energy transfer and storage in organisms and so the oxidation reactions of (193) have been studied (H. Yamamoto, et al., Bull.Soc.Chem.Japan, 1987, 60, 4115). Creatinine is readily oxidised to (193) by mercury (II) acetate, creatone (193) is also formed, under mild conditions, from a number of other precursors.

Imidazolones are readily prepared by quenching N-protected-C-lithiated imidazoles with peroxides (B.H. Lipshutz et al., Tetrahedron Letters, 1988, 29, 3411). The N-protecting group used is 2-trimethylsilyl-ethoxymethyl (SEM) introduced by the reaction of N-H imidazoles with SEM-chloride, lithiation is then achieved using butyllithium. The

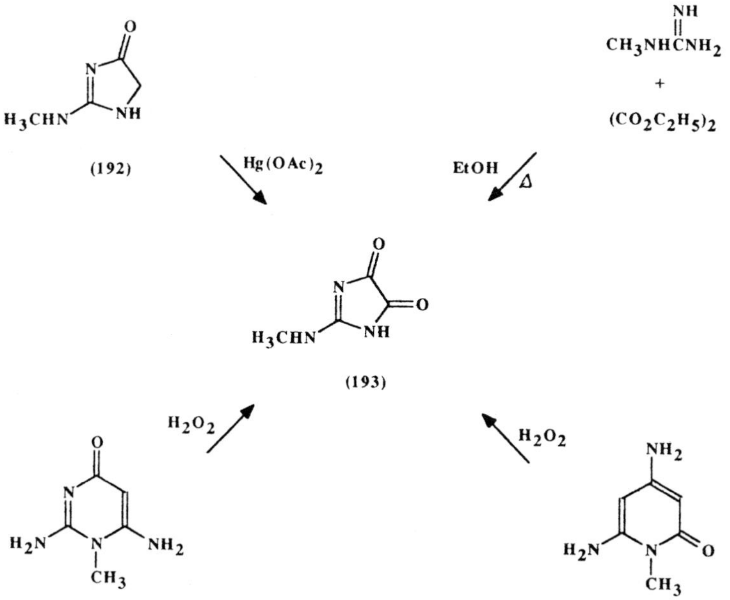

O
N
H3CHN
NH
(192)
Hg(OAc)2
NH
CH3NHCNH2
+
(CO2C2H5)2
EtOH
Δ
O
N
O
H3CHN
NH
(193)
H2O2
H2O2
O
N
H2N
N
NH2
CH3
NH2
H2N
N
O
CH3

oxidation to the imidazolone is effected by addition of a slight excess of benzoyl peroxide. Yields are in excess of 90%.

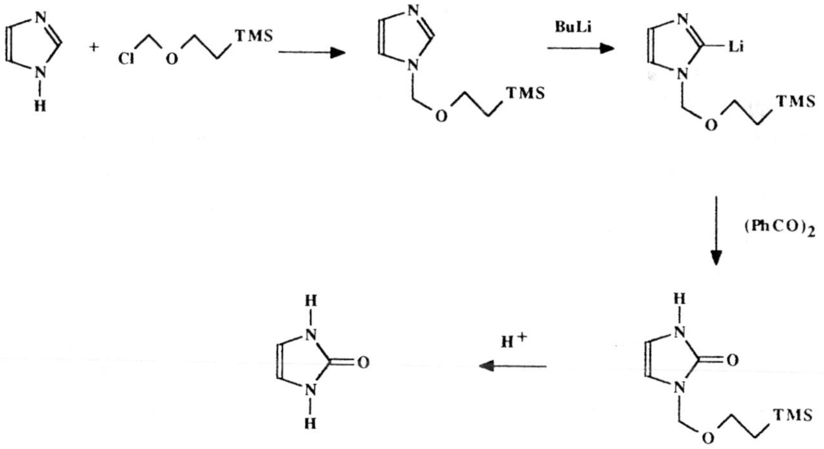

N
N
H
+
Cl
O
TMS
N
N
O
TMS
BuLi
N
Li
N
O
TMS
(PhCO)2
H
N
O
N
O
TMS
H+
H
N
O
N
H

Dihydro-2-thioxoimidazoles (**196**) are of interest as intermediates in the synthesis of the biologically important molecules thiolhistidines and thiolhistamines.   They are the products of the condensation reaction between *N*-isopropylisocyanoacetamide (**194**) and arysulphenyl thiocyanates (**195**) (R. Bossio *et al.*, Heterocycles, 1989, <u>29</u>, 1843).   The yields of dihydro-2-thioimidazoles (**196**) are high, ranging from 65 to 75%.

ArSSCN   +   iPrNHĊCH$_2$NC $\longrightarrow$ (with O above the C)

(**194**)   (**195**)   (**196**)

Ar = Ph, 4-MeC$_6$H$_4$, 4-ClC$_6$H$_4$, 2-NO$_2$C$_6$H$_4$, 4-Cl-2-NO$_2$C$_6$H$_3$

Thioxo-imidazoles bearing carboxaldehyde substituents are of interest as precursors to thiohistidines.   They have been prepared by the standard routes to heterocylic carboxaldehydes, that is the degradation of polyhydroxyalkylheterocycles with lead (IV) acetate or sodium periodate (G. Gonzalez *et al.*, ACS Symp.Ser., 1977, <u>39</u>, 207).   Such routes require several steps and result in very low overall yields.   A new route generates the 2-thioxo-imidazole-4-carboxaldehydes in two steps from the polyhydroxyalkylimidazoles in about 55% yield (I. Robina, J. Fuentes, and J.F. Bolanos, J.org.Chem., 1990, <u>55</u>, 750).   1-Alkyl(aryl)-1,3-dihydro-4-(*D*-arabino-tetritol-1-yl)-2*H*-imidazole-2-thiones (**197**) are oxidised by lead (IV) acetate to the corresponding 2,2'-dithiobis[1-alkyl(aryl)-1*H*-imidazole-4-carboxaldehydes] (**198**).   Reduction of the S-S bond in the latter by sulphur dioxide gives the 1-alkyl(aryl)-2,3-dihydro-2-thioxo-1*H*-imidazole-4-carboxaldehydes (**199**).

(**197**)   (**198**)   (**199**)

4. Benzimidazoles

Interest in the chemistry of benzimidazole and its derivatives stems from their wide range of biological activities including anti-tumour (C. Atassi *et al.*, Eur.J.Cancer, 1975, 11, 509; 609), anthelmintic (P. Actor, E.L. Anderson, and D.J. Dicuollo, Nature, 1967, 215, 321) and fungal (P.L. Langcake, P.J. Kohn, and P.J. Wade, in "Progress in Pesticides, Biochemistry and Toxicology", 1983, 3, 1).

(a) Synthesis

The two main routes to benzimidazoles are shown below:

(i) Cyclisation of an *o*-phenylenediamine derivative with a one carbon synthetic-equivalent.

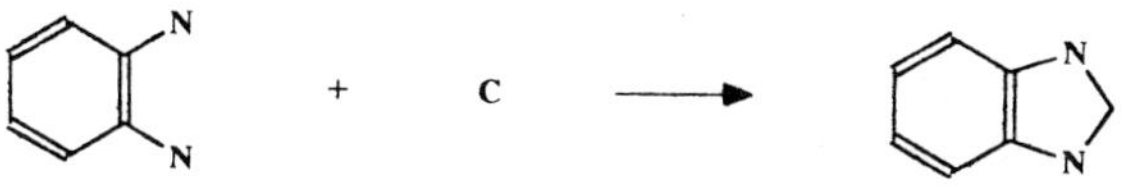

(ii) Formation of the five-membered ring by attack of nitrogen on the benzene ring.

Cyclisation of silyl derivatives of *o*-phenylenediamine is a useful, mild route to 2-substituted benzimidazoles. The mild reaction conditions and the absence of an aqueous workup give this new method advantages over the more robust traditional-methods of synthesis (B. Rigo, D. Valligny, and S. Taisne, Synth.Comm., 1988, 18, 167).

Condensation of an activated nitrile with o-phenylenediamine results in a 2-substituted benzimidazole. This method has also been used to prepare derivatives of thiazinone and quinazolinone (N.S.Ibrahim, Chem.Ind., 1988, 563).

X = CO₂Et or COPh   (75 - 80%)

Nitrogen ylides have been used in the synthesis of some unusual benzimidazoles. Thus an o-phenylenediamine derivative will react with the methyl ester of a dithiocarboxylic acid (A.M. Cusadro, J.A. Builla, and J.J. Vaquero, Heterocycles, 1988, 27, 1233).

1-[(methylthio)thiocarbonylmethyl] pyridinium iodide (**200**) (F. Krohnte and K. Gerlach, Ber., 1962, 1108) reacts with the o-phenylenediamine (**201**) to give the thioamide (**202**), which by loss of hydrogen sulphide yields the benzimidazole (**203**). The substituents R₁ and R₂ have considerable effect on the cyclisation. 1[(2'-Amino-5'-nitrophenylamino)-thiocarbonylmethyl]pyridinium iodide (**202**, R₁=H, R₂=NO₂) failed to

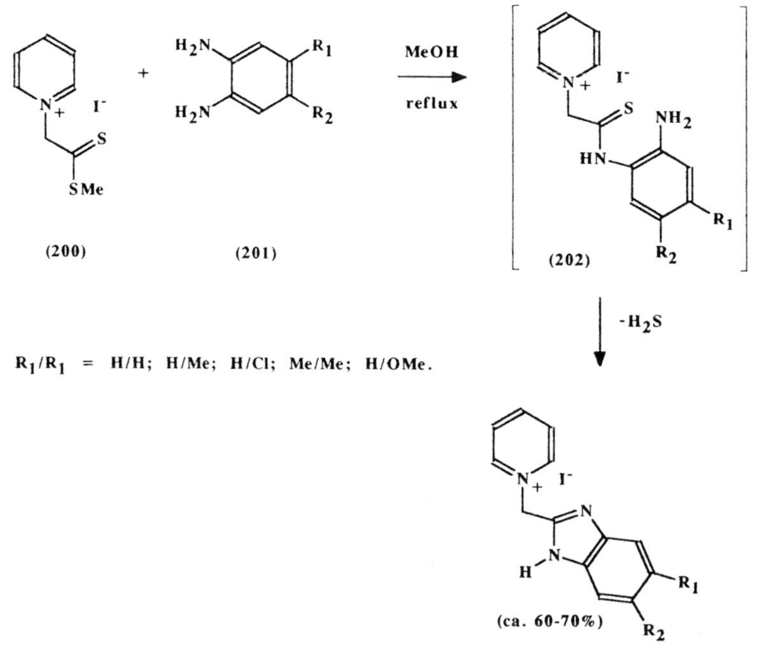

(200) (201) (202)

R_1/R_1 = H/H; H/Me; H/Cl; Me/Me; H/OMe.

$-H_2S$

(ca. 60-70%)

(203)

cyclise to the corresponding benzimidazole regardless of the conditions used.

The strong electron withdrawing group removes electron density from the amino function thus preventing it attacking the carbon of the thiocarbonyl group, similarly an electron withdrawing group at the 4-position of the pyridinium ring prevents the reaction (A.M. Cuadro et al., Heterocycles, 1989, 29, 52).

Functional groups may be easily introduced into benzimidazole (204) at the pyrrolic nitrogen and at the 2-position of the azole ring (A.M. Cuadro et al., Heterocycles, 1988, 27, 1233).

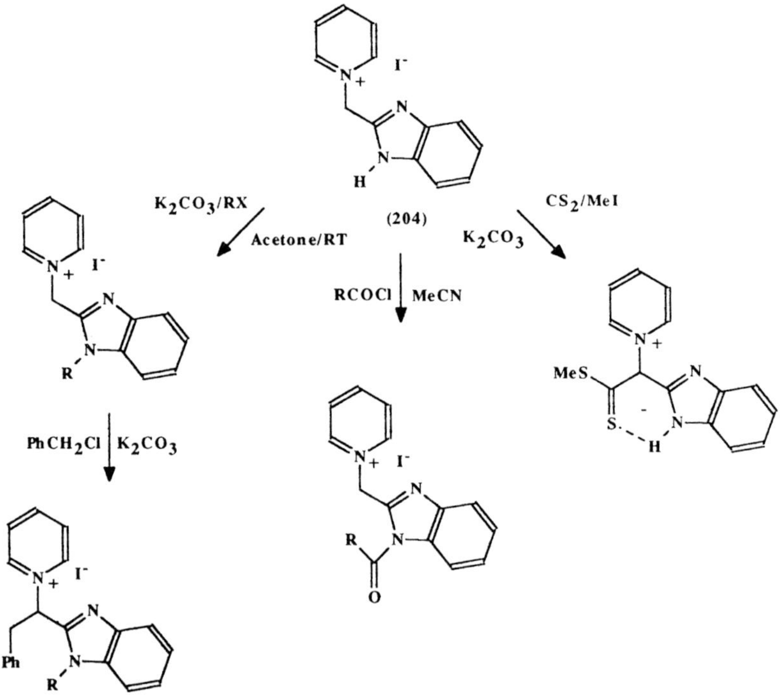

(204)

Oxidative ring closure, using barium manganate, of a Schiff's base prepared from *o*-phenylenediamine and an aromatic aldehyde, provides a benzimidazole in good yield (R.G. Srivastava and P.S. Venkataramani, Synth.Comm., 1988, 18, 1537).

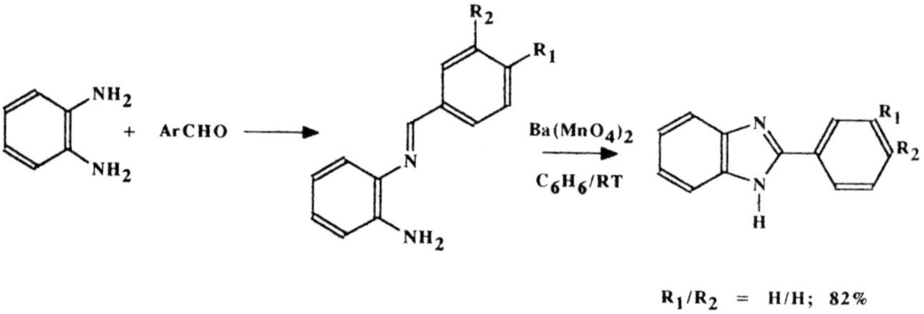

R_1/R_2 = H/H; 82%

R_1/R_2 = H/NO$_2$; 60%

There are few routes to benzimidazole which involve bond formation between nitrogen and a benzene ring carbon atom. Substituted benzimidazoles (206) may, however, be prepared by either thermal (DMSO/reflux) or acid-catalysed (acetic acid or silica) cyclisation of N-(N-arylbenzimidoyl)-1,4-benzoquinonimines (205) (T. Benicori and F. Sannicolo, J.heterocyclic Chem., 1988, 25, 1029).

(205)   (206)

Yield of (206)

R_2 = H, 46%; 4-NO$_2$, 31%; 4-MeO, 36%; 4-Cl, 57%

Yields are reduced by electron withdrawing groups on the phenyl ring, which reduces the effectiveness of the imidoyl nitrogen as a nucleophile.

(b) Benzimidazole Derivatives

(i) N-Oxides

The chemistry of benzimidazole-N-oxides has been extensively reviewed (D.M. Smith in "Benzimidazoles and Congeneric Tricyclic Compounds", ed. P.N. Preston, Wiley-Interscience, New York, 1981, chapter 2). Substituted benzimidazole-N-oxides (207) cannot be prepared by direct oxidation but are obtained by cyclisation of o-nitroaniline derivatives (208) (D.M. Smith et al., J.Chem.Soc.Perkin I, 1988, 681).

The same cyclisation method can be used to prepare 5- and 6-amino-1H-benzimidazole-3-oxide (D.M. Smith et al., J.chem.Soc.Perkin I, 1988, 691) and 4- and 7-amino benzimidazole-N-oxide (Idem., ibid., 1988, 1939).

(208)

X = 5-Me, 5-OMe, 5- or 6-F, 5-Cl, 4- or 5-NO$_2$

(ii) N-Alkyl Derivatives

N-Alkylbenzimidazoles are prepared by the reaction of benzonitrile oxide with 5(6)methyl-2-chloromethyl-1 H-benzimidazole (**209**) (P.S.N. Reddy, B. Ramamohan Rao, and G. Mohiuddin, Indian J.Chem., 1989, 28B, 24). The resulting benzimidazole oximes (**210**), undergo a facile dehydration in the presence of base, to give an unusual fused oxadiazino benzimidazole (**211**).

(209)   (210)   (211)

5(6)-Me	⟶	5 and 6-Me	⟶	7 and 8-Me
4(7)-Me	⟶	4 and 7-Me	⟶	9 and 6-Me

Biomimetic studies of dinuclear copper proteins (C.A. Reed *et al.*, J.Amer.chem.Soc., 1984, 106) require an efficient procedure for the *N*-alkylation of compounds (e.g. **212**) containing several benzimidazole groups. Methods used for the *N*-alkylation of simple benzimidazoles (Y. Kikugaw, Synthetic Comm., 1981, 124) are not suitable, because of the poor solubility of the substrate (**212**). Powdered potassium hydroxide in DMSO is a suitable medium for *N*-alkylation of these systems (C.M. Reed *et al.*, Tetrahedron Letters, 1988, 3033). Complete *N*-alkylation occuring usually within 1 hour at room temperature.

(212)

(iii) C-Alkyl Derivatives

5-Azido-2*H*-benzimidazole-2-spirocyclohexane (**213**) behaves as a stable benzoquinone diimine (H. Suschitzky *et al.*, J.chem.Soc.Perkin I, 1988, 983). It is best obtained (56%) from the 5-phenylsulphonyl derivative by nucleophilic displacement using sodium azide in DMSO.

(213)

U.V. irradiation of the azide in the presence of a secondary aliphatic amine leads to a 4-amino-5-dialkylamino-2*H*-benzimidazole. With morpholine 4-amino-5-morpholino-2*H*-benzimidazole (**214**) is obtained. The mechanism involves a nitrogen shift from position 5 to 4 and occurs *via* an azirine arising from a photo-induced nitrene. Ring opening of the azirine by nucleophilic attack leads to the product (**214**).

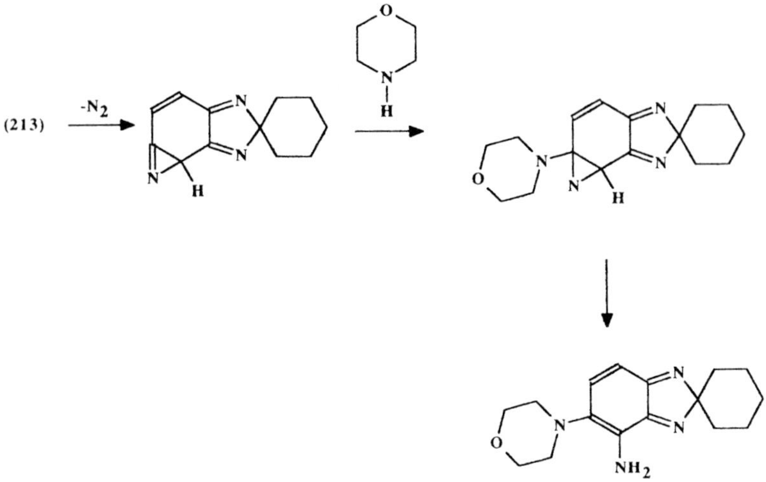

(214)

Piperidine gives the bis-2*H*-benzimidazole (**215**).

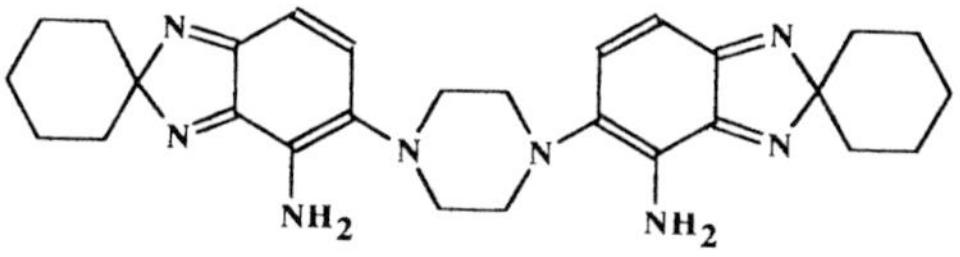

(215)

2*H*-Benzimidazoles undergo thermal isomerisation to give the 1*H*-isomers (H. Suschitzky *et al.*, J.chem.Soc. Perkin I, 1988, 991). A 1,5-sigmatropic shift takes place such that 2*H*-benzimidzole spirans thermally rearrange to 2,3-disubstituted-1*H*-benzimidazoles.

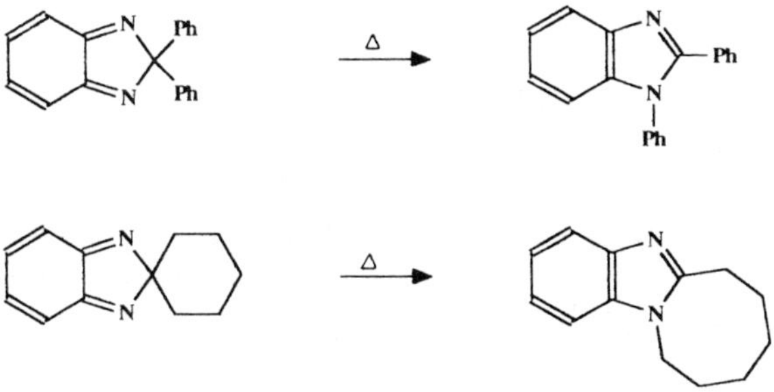

The synthetic versatility of 2*H*-benzimidazoles is further shown by ready nucleophilic displacement of chlorine in dichloro-2*H*-benzimidazole-2-spirocyclohexane (H. Suschitzky *et al.*, Synthesis, 1988, 871).

The preparation of 2-substituted benzimidazoles either involves cyclisation of a suitably substituted aldehyde with an *o*-phenylenediamine, or protection of the *N-H* group of the benzimidazole ring followed by substitution at the 2-position. A new method for 2-substitution of *N*-unsubstituted benzimidazoles involves transient protection by formaldehyde, lithiation of the 2-position and quenching of the carbanion by an electrophile. Deprotection by acidic workup then gives the 2-substituted-*NH*-benzimidazole (A.R. Katritzky and K.Akutagawa, J.org.Chem., 1989, 54, 2949).

(iv) Amino Derivatives

2-Aminobenzimidazoles are readily prepared from 2-isothiocyanato carboxylates by their reaction with *o*-phenylenediamine, followed by cyclisation of the resulting N,N'-substituted thiourea to a 2-substituted benzimidazole (L. Floch, M. Uher, and J. Lesko, Coll.Czech.chem.Comm., 1989, 54, 206).

Various *N*-substituted aminobenzimidazoles have been prepared in the hope of obtaining good antibiotics. Carbamates (216), in particular, have been prepared due to their suspected anthelmintic activity. The carbamate group is introduced by cyclising a substituted *o*-phenylenediamine with 1,3-

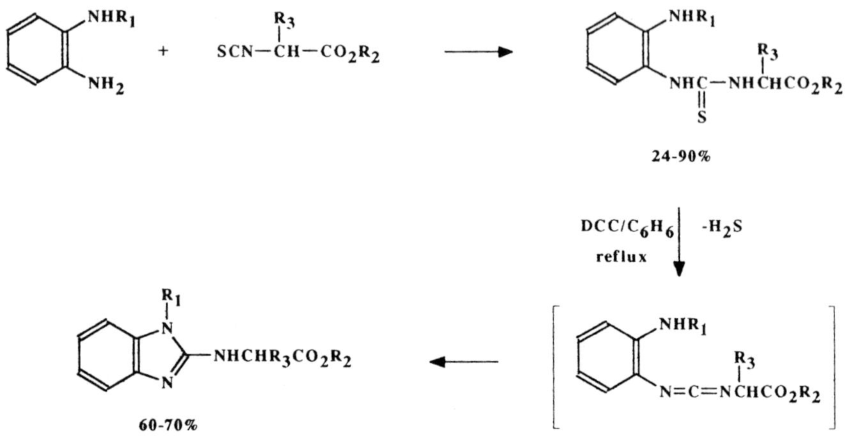

24-90%

DCC/C$_6$H$_6$ $\quad$ -H$_2$S
reflux

60-70%

(alkoxycarbonyl)-δ-methylisothioureas (C.V.R. Sastry *et al.*, Indian J.Chem., 1988, <u>27B</u>, 871).

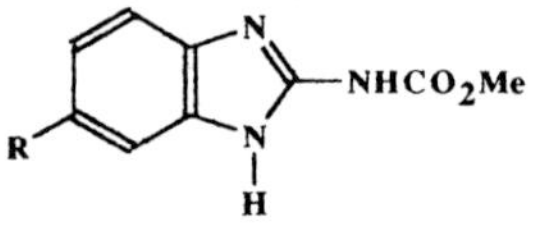

(216)

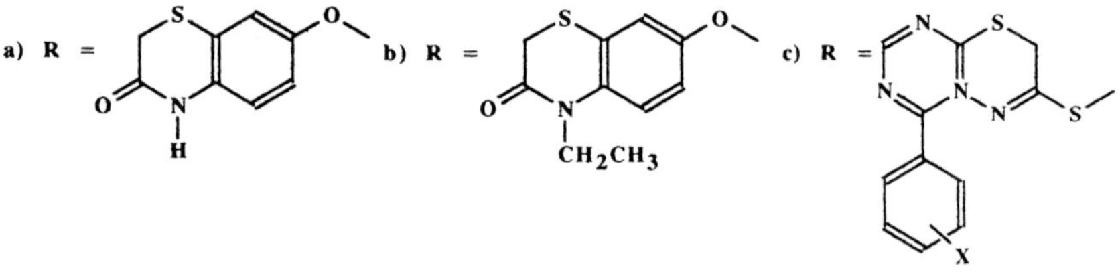

a) R =

b) R =

c) R =

Direct attachment of nitrogen heterocycles to the 2-position of benzimidazole is possible *via* the 2-amino derivative. 2-Aryl-1*H*-s-triazolo[1,5-a]benzimidazoles are prepared from 2-aroylaminobenzimidazoles by treating with phosphorus pentachloride, followed by sodium azide in aqueous acetone. Pyrolysis of the resulting tetrazole gives the triazolobenzimidazole (P.J. Rao and K.K. Reddy, Synth.Comm., 1988, <u>18</u>, 1995).

2-Hydrazino derivatives of benzimidazole have been reported and may be prepared by displacement of -SMe or -Cl by hydrazine (R.V. Venkataratnam and K. Sujatha, Synth.Comm., 1988, 18, 805). The *NH* of the benzimidazole ring must first be protected. If the protecting group terminates in a hydrazine function then intermolecular displacement may occur to give a tricyclic compound (217).

(217)

2-Hydrazinobenzimidazoles are precursors of other benzimidazole derivatives. They react with ß-diketones in neutral solution to give 2-(3,5-disubstituted-1H-pyrazol-1-yl)benzimidazoles and in acidic media to give hydrazones (K.C. Jochi *et al.*, J.heterocyclic Chem., 1988, <u>25</u>, 1641).

(v) Sulphur Containing Derivatives

There have been several reports relating to the synthesis of sulphur derivatives of benzimidazoles and as with many other derivatives the reason for these reports is the attempted preparation of compounds with improved biological activity. Sulphur derivatives can, however, be used as intermediates in the preparation of other benzimidazoles, particulary amine derivatives.

The benzimidazole-2-thioacetic acid ester (**218**) has been prepared from *o*-phenylenediamine and carbon disulphide, chloroacetic acid, and potassium hydroxide in ethanol solution, substitution of the *NH* group is then effected using standard procedures for Mannich reactions, formalin and a secondary amine, to give the Mannich base (**218**).

(218)

These derivatives show antiinflammatory and analgesic activity (P.B. Trivedi *et al.*, J.Indian chem.Soc., 1988, 65, 296).

An interesting development in benzimidazole chemistry has been the synthesis of transition-metal chelating-ligands such as (219) (C.G. Wahlgren and A.W. Addison, J.heterocyclic Chem., 1989, 26, 541).

(219)

2-Substituted amino-5(6)-arylsulphonobenzimidazoles show anthelmintic properties and are prepared by permanganate oxidation of 5-alkylthiobenzimidazoles.

83%

Thiobenzimidazoles are used as precursors in the preparation of other derivatives, for example, 2-mercaptobenzimidazole yields benzimidazole-2-hydrazone when heated with hydrazine hydrate at reflux (N. Srivastava and V.S. Misra, Indian J.Chem., 1988, 27B, 298).

(vi) Aldehydes and Ketones

Benzimidazole ketones are obtained by methods analogous to those for the preparation of the parent heterocycle. 6-Benzoyl-1-isopropylsulphonyl benzimidazole (220) is obtained by condensation of cyanogen bromide with the amino-4-nitrobenzophenone (221) in THF/H$_2$O at room temperature (S.J. Dominanni, Oppi Briefs, 1990, $\underline{22}$, 106).

(221)   **(220)**

5-Acyl-1H-benzimidazole-2-carbamates may be reduced to the corresponding alcohol by sodium borohydride in methanol. They possess anthelmintic activity (D.S. Bhakuni *et al.*, Indian J.Chem., 1989, $\underline{28B}$, 702).

(c) Fused Benzimidazole Derivatives

Condensation of either 1H-benzimidazole-2-acetonitrile or 1H-benzimidazole-2-acetates with bis-2,4,6-trichlorophenyl malonates gives pyrido[1,2-a]benzimidazoles (E.El-Ghazzawi, O. Kader, and T. Kappe, J.heterocyclic Chem., 1988, $\underline{25}$, 1087 and T. Kappe *et al.*, ibid., 1988, $\underline{25}$, 1725).

Likewise heating benzimidazole-2-acetonitrile first with an aromatic aldehyde and then with malononitrile or ethyl cyanoacetate leads to reasonable yields (60-80%) of a pyrido[1,2-a]benzimidazole (M. Hammad *et al.*, Egypt.J.Chem., 1986, $\underline{29}$, 549).

X = CN or CO_2Et

Tetracyclic systems have been prepared by reaction of 2-(cyanomethyl)benzimidazole with ethyl cyclopentanone-2-carboxylate (T. Kappe *et al.*, Monatsch., 1989, 120, 73).

Similar condensation reactions to those above are used to prepare pyrimidobenzimidazoles (A.M. El-Sayed *et al.*, J.heterocyclic Chem., 1989, 25, 405), including complex structures such as benzopyran[4',3':4,5]pyrimido[2,1-b]benzimidazole (**222**) (S. M. Ladenovic, *J.Serb.chem.Soc.*, 1989, 169). Condensed systems of this type exhibit antimicrobial activity.

(**222**)

Chapter 17

FIVE-MEMBERED HETEROCYCLIC COMPOUNDS WITH TWO DIFFERENT HETERO-ATOMS IN THE RING

JOHN D. HEPWORTH and MARK WAINWRIGHT

The material in this chapter covers the period 1981-1990 which saw the publication of a major work on heterocyclic chemistry (Comprehensive Heterocyclic Chemistry, ed. A.R. Katritzky and C.W. Rees, Pergamon Press, Oxford, 1984). In volume 6, individual chapters are devoted to isoxazoles and their benzologues (S.A. Lang, Jr. and Y.-i Lin, p. 1), the isothiazoles and benzologues (D.L. Pain, B.J. Peart and K.R.H. Wooldridge, p. 131), oxazoles and benzologues (G.C. Boyd, p. 177), thiazoles and benzologues (J. Metzger, p. 235) and five membered selenium-nitrogen heterocycles (I. Lalezari, p. 333). Studies of the ^{15}N nmr spectra of azoles incorporates many of the heterocycles discussed in this chapter (B.C. Chen, W. von Philipsborn and K. Nagarajan, Helv., 1983, 66, 1537; L. Stefaniak *et al.*, Org. mag. Reson., 1984, 22, 215), as does a review of the synthesis of functionalised azoles (B. S. Drach, Khim. geterosikl. Soedin., 1989, 723).

A comprehensive review of electrophilic substitution of heterocycles includes the reactions of 1,3-azoles with various electrophiles (A.R. Katritzky and R. Taylor, Advances in Heterocyclic Chemistry, Vol. 47, Academic Press, 1990).

1. Isoxazole and its Derivatives

The value of isoxazole and its reduced derivatives in synthesis is recognised in a number of reviews. The application of isoxazole to the synthesis of natural products has been reviewed (P. Baraldi *et al.*,

Synthesis, 1987, 857; F.A. Lackhvich, E.V. Koroleva and A.A. Akhrem, Khim. geterosikl. Soedin., 1989, 435), whilst other reviews concentrate on the value of isoxazolines as intermediates for the synthesis of naturally occurring compounds (V. Jäger *et al.,* Bull. Soc. chim. Belg., 1983, 92, 1039; A. P. Kozikowski, Acc. Chem. Res., 1984, 17, 410) and on aspects of stereoselectivity in reactions of isoxazolines (V. Jäger *et al.,* Lect. heterocyclic Chem., 1985, 8, 79).

The preparation of isoxazolines by cycloaddition reactions of nitrile oxides with alkenes and their use in the synthesis of unsaturated ketones has been discussed (S. Kwiatkowski, Pr. Nauk-Politech. Warsz. Chem., 1987, 38, 3). A review of 4-isoxazolines surveys their preparation and the extensive rearrangements which they undergo (J.P. Freeman, Chem. Rev., 1983, 83, 241) and two reviews examine the photochemistry of isoxazolines (L.Fisera, Stud. Org. Chem. (Amsterdam), 1988, 35, 12 and 441).

(a) Isoxazoles

(i) Synthesis

The major routes to isoxazoles involving ring closure either by formation of the O-C bond or by the concerted formation of the C4-C5 and the O-C bonds continue to attract attention.

The former approach is essentially concerned with the reaction of a 1,3-dicarbonyl compound with hydroxylamine, although considerable variation in reactants is possible.

Treatment of either triformylmethane (1) or the masked aldehyde salt (2) with hydroxylamine gives access to 4-substituted isoxazoles which are the synthetic equivalents of cyanoacetaldehydes. The initially formed 4-oxime can be sequentially dehydrated and hydrolysed to give high overall yields of 4-cyanoisoxazole, m.p. 63-65 ^{o}C, and 4-ethoxycarbonyl-isoxazole, b.p. 30 ^{o}C at 3 mmHg (R.O. Angus, Jr. *et al.,* Synthesis, 1988, 746).

Acylpyruvates afford 3,5-disubstituted isoxazoles on reaction with hydroxylamine. Reduction of the 3-glyoxylate function and oxidation of the resulting 3-hydroxymethyl group gives access to the synthetically useful 3-formylisoxazoles (P.G. Baraldi *et al.*, J. heterocyclic Chem., 1982, *19*, 557).

The 3-hydroxyisoxazole unit occurs in some physiologically active compounds. Muscimol (3), a potent agonist of the neurotransmitter γ-aminobutyric acid, can be prepared in 17% overall yield from 3-chloropropyne. Careful control of the pH is essential. Some secondary amine is produced in the reaction of the 5-chloromethylisoxazole with ammonia (B.E. McCarry and M. Savard, Tetrahedron Letters, 1981, 5153).

$$\text{Li}-\text{C}\equiv\text{C}-\text{CH}_2\text{Cl} \xrightarrow{\text{ClCO}_2\text{Et}} \text{EtO}_2\text{C}-\text{C}\equiv\text{C}-\text{CH}_2\text{Cl}$$

aq MeOH, NH_2OH
-35 °C, pH 8 - 9

(3)

Although β-keto esters react with hydroxylamine to give mainly isoxazolin-5-ones, buffering the reaction mixture at pH 10-11 allows the predominance of 3-hydroxyisoxazole. Quenching the reaction mixture with concentrated hydrochloric acid is essential (N. Jacobsen, H. Kolind-Andersen and J. Christensen, Can. J. Chem., 1984, 62, 1940).

The complex reaction of β-keto esters with hydroxylamine has been studied using ^{13}C nmr spectroscopy. At pH 10-12, a mixture of isoxazolin-5-ones and 5-hydroxyisoxazolidin-3-ones is initially formed in which the latter is predominant and is the source of the isolated products. Rapid acidification results in its dehydration to 3-hydroxyisoxazoles. However, gradual reduction of the pH causes it to ring open and subsequent recyclisation leads to the isoxazolin-5-one (A.R. Katritzky et al., J. org. Chem., 1986, 51, 4037).

Oxidation of 1,3-dioximes with phenyliodine (III) bistrifluoracetate PhI(O$_2$CCF$_3$)$_2$ yields a mixture of pyrazole dioxides and isoxazoles, whilst 1,4-dioximes yield pyridazine dioxide and the isoxazoloisoxazole (4) (S. Spyroudis and A. Varvoglis, Chem. Chron., 1982, 11, 173).

(4)   (5)

There are several examples of the formation of isoxazoles from oxygen heterocyclic precursors. The reactions are particularly susceptible to changes in reaction conditions and structural features. Thus, chromone-3-carboxylic acids undergo conjugate addition with hydroxylamine to give 5-(2-hydroxyphenyl)isoxazoles (6), whereas the corresponding products from the 3-carboxylic esters undergo an acid catalysed lactonisation to the benzopyrano[3,4-d]isoxazoles (5).

(6)

The reaction of 3-acetyl-4-hydroxycoumarin with hydroxylamine is more complex. The initially formed oxime cyclises with concomitant opening of the lactone ring to give a 4-aroylisoxazolin-5-one (7) in the presence of potassium acetate when a slight excess of hydroxylamine is used. Further reaction occurs when a large excess of hydroxylamine is present, the initially formed isoxazole ring opening as a new 3-phenylisoxazolin-5-one (8) is formed. Reformation of the pyran ring completes the reaction, leading to the benzopyrano[4,3-c]isoxazole (9). The course of the reaction is markedly different under acid catalysis, the products comprising of a mixture of fused isoxazole and oxazole derivatives (B. Chantegrel, A.I. Nadi and S. Gelin, J. org. Chem., 1984, 49, 4419).

The reaction of chromone with hydroxylamine is also dependent on pH and temperature and yields a variety of products. Initial ring opening by cleavage of the O-C2 bond leads to either the aldoxime (10) or the dioxime (11) and the reactivity of these two compounds influences the nature of the final product which may be the 5-(2-hydroxyphenyl)isoxazole (12) derived from (10) or the 3-substituted isomer formed from (11) via the isoxazoline (13) (V. Szabo et al., Tetrahedron, 1984, 40, 413).

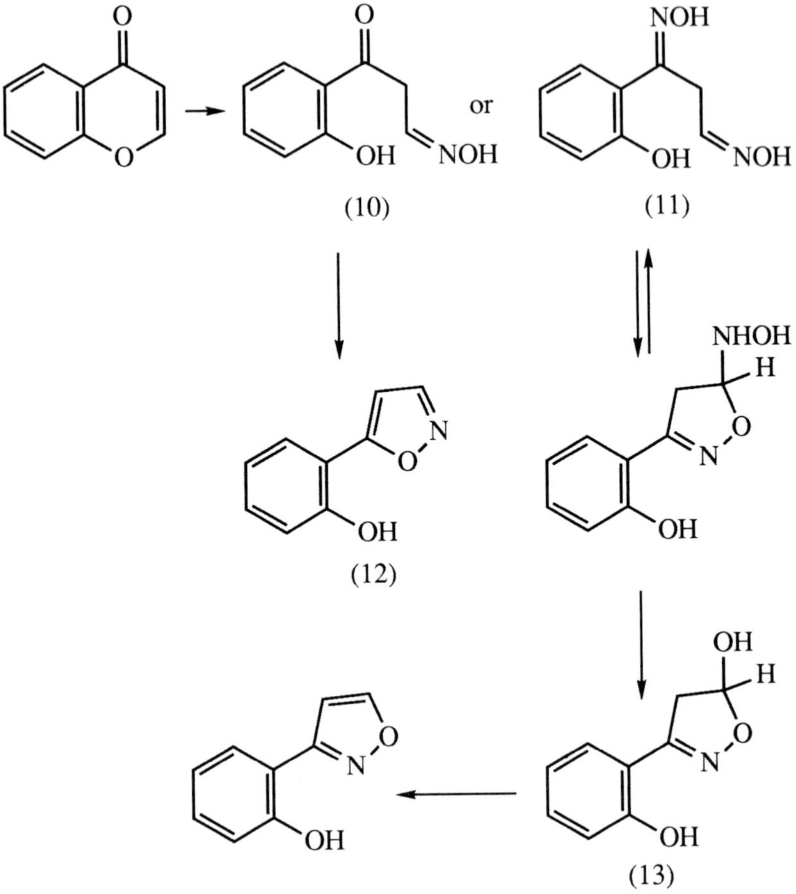

(10)   (11)

(12)

(13)

Further work on the base-induced cyclisation of 1,3-diketone dioximes has confirmed the complexity of the reaction. When a 1-(2-hydroxyphenyl) substituent is present, 3-aryl-1-hydroxypyrazoles are formed together with some 5-aryl-4-aminoisoxazoles. The key intermediate is possibly (14) which is prone to base catalysed ring opening. Although, the 1-(2-thienyl) derivative yields the 5-thienylisoxazole, 1-phenylbutan-1,3-dione dioxime gives the 3-phenylisoxazole. Reaction of the diketones with NH_2OH gives the 5-arylisoxazole (A.O. Fitton, R.N. Patel and R.W. Millar, J. chem. Res. (S), 1986, 124).

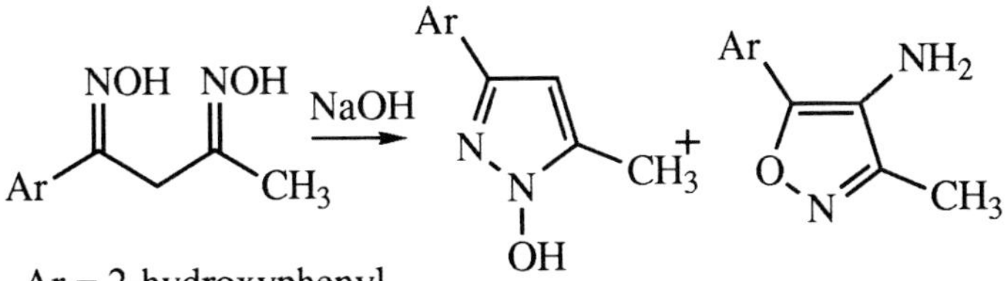

Ar = 2-hydroxyphenyl

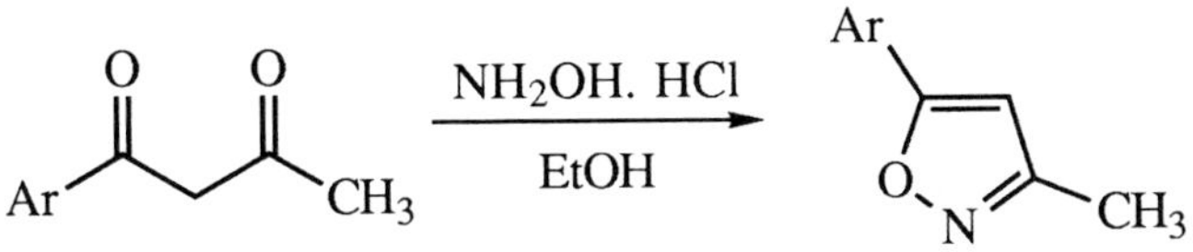

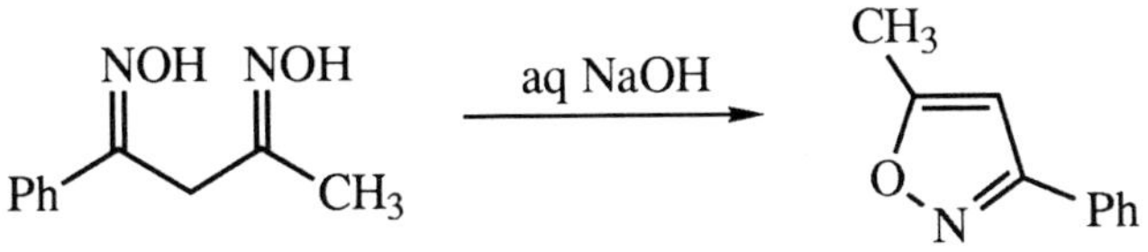

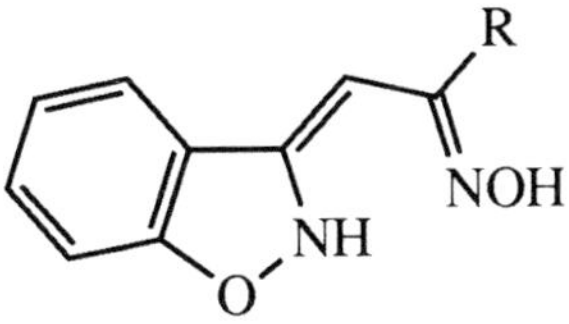

(14)

Pyranooxazoles (16), which result from the reaction of α-acetylhomotetronic acids (15) with hydroxylamine, undergo hydrazinolysis of the lactone ring to give the highly substituted isoxazole (17). The derived azide undergoes a Curtius rearrangement to give the [1,2]oxazolo[4,5-d][1,3]oxazepine (18) (B. Chantegrel and S. Gelin, Synthesis, 1981, 315).

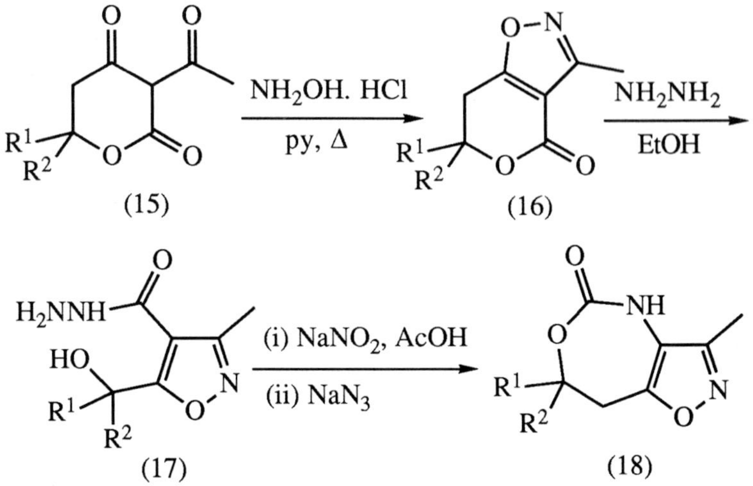

(15) → (16)

(17) → (18)

A reliable, moderate yielding one-pot synthesis of 4-nitroisoxazoles is based on the reaction of acetylenic secondary bromides with sodium nitrite in aqueous ethanol. The proposed mechanism also accounts for the formation of 3-nitroisoxazoles from the corresponding primary bromides (E. Duranti *et al.*, J. org. Chem., 1988, <u>53</u>, 2870). 3-Alkylisoxazoles can be obtained from 1,1,3-tribromoalkanes by treatment with sodium nitrite in DMF (L. Dasaradhi *et al.*, Synth. Commun., 1990, <u>20</u>, 2799).

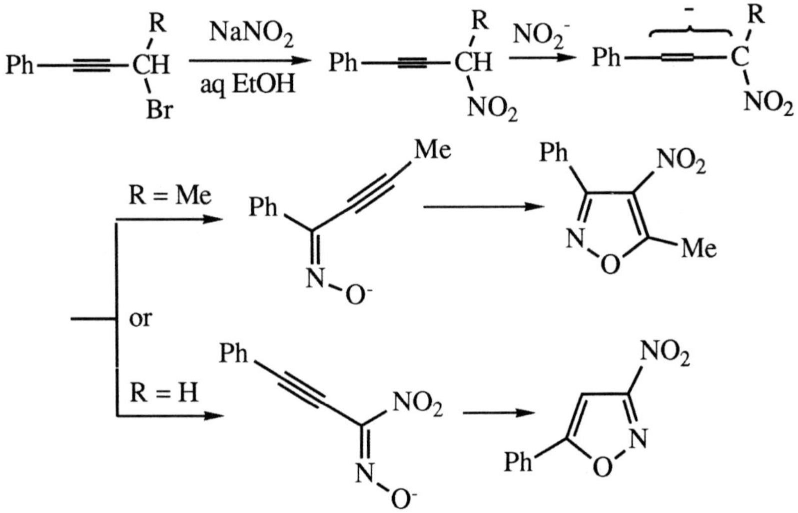

The synthesis of isoxazoles and their reduced derivatives by 1,3-dipolar cycloaddition is still frequently used. Some of the published work refers mainly to the generation of the dipolar intermediate, with the heterocyclic cycloadduct being of minor concern. For example, a short review which examines the generation of nitrile oxides from nitroparaffins includes their reactions with unsaturated compounds to yield isoxazolines (S.B. Markofsky, Spec. Chem., 1989, 9, 172). The generation of (phenylthio)acetonitrile oxide enables 3-phenylthio substituted isoxazoles and isoxazolines to be prepared (S. Kanemasa et al., Bull. chem. Soc. Japan, 1988, 61, 3973).

A detailed account of the reactivity of furoxans, 1,2,5-oxadiazole 2-oxides, towards dipolarophiles concludes that both nitrone-type 1,3-dipolar adducts and nitrile oxide like derivatives are formed, with the latter route favoured when dimethylformamide is used as co-solvent (T. Shimizu et al., Tetrahedron, 1985, 41, 727).

Fulminic acid, HCNO, conveniently generated by the hydrolysis of trimethylsilylcarbonitrile oxide, is trapped by alkynes and by alkenes to afford good yields of isoxazoles and isoxazolines, respectively, which are unsubstituted at the 3-position (F. De Sarlo et al., Tetrahedron, 1985, 41, 5181).

Trimethylsilated isoxazoles result from the reaction of nitrile oxides with vinyl silanes or silylacetylenes (A. Padwa and J.G. MacDonald, Tetrahedron Letters, 1982, 3219).

Reagents: (i) PhCHCl=NOH, NEt$_3$, CCl$_4$; (ii) PhC≡N$^+$-O$^-$

5-Ethoxyisoxazoles can be obtained directly by the use of 1,1-diethoxypropene as the dipolarophile or indirectly from the reaction of hydroxylamine with diethyl 2-oxobutandioates and alkylation of the resulting isoxazolinone (S. Auricchio, A. Ricca and O. Vajna de Pava, J. org. Chem., 1983, <u>48</u>, 602). The particular interest is in the ring opening and subsequent cyclisation to 5,7-dihydroxy-4-methylphthalide.

3-Arylprop-2-ynoates and 1-aryl-3-phenylprop-2-yn-1-ones react with arylnitrile oxides to afford good yields of 3,4,5-trisubstituted isoxazoles (F.A. Fouli *et al.*, J. prakt. Chem., 1984, <u>329</u>, 1116), whilst 5-chloroisoxazoles are formed when 1,1-dichloroethene is used as the dipolarophile, the reaction proceeding through the 5,5-dichloroisoxazoline. The 5-halogen is susceptible to nucleophilic displacement (R.V. Stevens and K.F. Albizati, Tetrahedron Letters, 1984, 4587).

(ii)   Reactions

A major feature of the chemistry of isoxazoles is their susceptibility to undergo ring opening by cleavage of the N-O bond. The resulting acyclic species, the nature of which depends on the structure of the isoxazole derivative and the method of ring cleavage, is then used *in situ* to synthesise a variety of compounds. The isoxazole ring thus behaves as a masked equivalent for a variety of functional groups. The major transformations of isoxazoles into acyclic materials are summarised in Scheme 1.

Scheme 1

Hydrogenolysis of isoxazoles gives β-aminoenones which react regiospecifically with hydrazine derivatives to give pyrazoles (A. Alberola *et al.*, An. Quim., 1987, 83C, 55) and 5-acylimidazoles result from the base catalysed cyclisation of α-(acylamino)enaminones which are formed by the hydrogenation of 4-amidoisoxazoles (L.A. Reiter, J. org. Chem., 1987, 52, 2714). Samarium diiodide also effects cleavage of the N-O bond, a reaction that can give an aminoenone in high yield (N.R. Natale, Tetrahedron Letters, 1982, 5009). Similarly, metal carbonyls cleave the ring following complexation of the isoxazole unit to the metal (M. Nitta and T. Kobayashi, J. chem. Soc., Perkin I, 1985, 1401). When this reaction is carried out in the presence of dimethyl acetylenedicarboxylate, pyridine derivatives are formed. It is suggested that an initial cycloaddition to the C4-C5 bond is followed by cleavage of the N-O and C1-C5 bonds (T. Kobayashi and M. Nitta, Bull. chem. Soc. Japan, 1985, <u>58</u>, 152).

The catalytic reduction of isoxazoles to enaminoketones proceeds *via* a rearrangement to a 2*H*-azirine. Using a poisoned catalyst in dioxan, it is possible to stop the reaction at the intermediate stage and obtain a high yield of the azirine (S. Auricchio and O. Vajna de Pava, J. chem. Res. (S), 1983, 132).

Alkylation of 5-substituted isoxazoles and basic cleavage of the isoxazolium salt yield β-ketoamides which undergo a base catalysed cyclisation to 3-acyltetramic acids (P. DeShong *et al.*, J. org. Chem., 1983, <u>48</u>, 1149; 1988, <u>53</u>, 1356).

Isoxazolium salts react with hydroxylamine and with hydrazine derivatives in a ring opening-ring closure sequence initiated by abstraction of the acidic C-3 proton to yield isoxazoles and pyrazoles, respectively. Hydride ion reduction of the salts yields 4-isoxazolines (A. Alberola *et al.*, Synthesis, 1982, 1067; 1988, 203).

Protected forms of the precursor hydroxy acid sub-units of the macrocyclic antibiotics pyrenophorin and vermiculine can be derived from the aminoenone (19), which is available by the reaction sequence given below, in which preferential ring opening of the isoxazole moiety is a key feature (P.G. Baraldi *et al.*, J. org. Chem., 1983, 48, 1297).

(19)

Base catalysed ring opening of isoxazoles in the presence of aldehydes leads to the oxonitrile (20). Michael addition of malononitrile then gives aminopyrans, whilst the reaction with aromatic aldehydes gives substituted furans (J.A. Ciller *et al.*, J. chem. Soc., Perkin I, 1985, 2581).

An electron withdrawing substituent in the isoxazole 4-position not only allows further functionalisation to be achieved but also influences the nature of the acyclic product resulting from the basic cleavage of the isoxazole ring. Enolates, β-diketones and esters may be formed (A. Alberola *et al.*, J. heterocyclic Chem., 1982, 19, 1073; An. Quim., 1987, 83C, 182). 3,5-Dimethylisoxazole is lithiated and alkylated regiospecifically at the 5-methyl group. Further treatment results in attack at the 3-substituent. Combined with a ring opening procedure, access to a variety of β-diketones is available (D.J. Brunelle, Tetrahedron Letters, 1981, 3699).

Other instances where the initial ring opened product is of value in its own right include the formation of alkynes from 5-aminoisoxazoles *via* diazotisation (E.M. Becalli, A. Manfredi and A. Marchesini, J. org. Chem., 1985, 50, 2372) and the Z-β-siloxyacrylonitriles produced by treatment of isoxazoles with lithium diisopropylamide and silylation of the ring opened enolate (A. Alberola *et al.*, Tetrahedron Letters, 1986, 2027).

A royal-blue solution containing PhC≡C-O⁻ is formed when 3,4-diphenylisoxazole is treated with *n*-butyl lithium at -78 °C. Cycloaddition of an electron deficient imine discharges the colour and yields a β-lactam (A.G.M. Barrett and P. Quayle, J. chem. Soc., Perkin I, 1982, 2193).

Reagents: (i) *n*-BuLi, THF, -78 °C; (ii) $Ar^1N=CHAr^2$

Substituted isoxazoles are directly useful in the synthesis of a variety of fused heterocyclic systems. Thus, 3,5-diaminoisoxazole reacts with 1,3-diketones $RCOCH_2COMe$ to give 2-aminoisoxazolo[2,3-a]-pyrimidinium salts (21), the free bases of which are unstable and rearrange to 2-cyanomethylenepyrimidine 1-oxides (G. Zvilichovsky and M. David, J. org. Chem., 1983, 48, 575).

(21)   (22)   (23)

Dihydroisoxazolo[5,4-*b*]pyridines (22) are formed from 5-aminoisoxazoles and 3-arylidene derivatives of pentan-2,4-dione or other dicarbonyl compounds (T. Yamamori *et al.*,Tetrahedron, 1985, 41, 913) and the pyrano[3,2-*d*]isoxazole system (23) is formed on reaction with ethyl propenoate (E. Zayed *et al.*, Arch. Pharm., 1983, 316, 105).

The highly substituted isoxazole (24) reacts with dimethyl acetylenedicarboxylate to afford the new isoxazolodiazepine system (25) (R. Nesi *et al.*, Heterocycles, 1987, 26, 2419).

(24)   (25)

The acid catalysed rearrangement of the aryloxymethylisoxazoles (26, X = O) leads to the isomeric benzoxocinoisoxazolones (27, X = O) (C. Deshayes, M. Chabannet and S. Gelin, J. heterocyclic Chem., 1986, 23, 1595). The corresponding benzocyclooctaisoxazolones (27, X = CH_2) result from the analogous (2-arylethyl)isoxazoles (26, X = CH_2).

(26)   X = O or CH$_2$ (27)

The electron deficient isoxazole (28) reacts with a diazoalkane to give the cyclopropane derivative (29) and the pyrazoloisoxazole (30) which is too unstable to isolate. With an excess of diazomethane, the polycyclic systems (31) and (32) are formed (R. Nesi et al., J. Org. chem., 1989, 54, 706).

(28)   (29)   (30)

(31)   (32)

Cyclophanes containing an isoxazole unit have been prepared from 3,5-dichloromethylisoxazole and 1,4- and 1,3-dithiomethylbenzene. Extrusion of sulfur from the paracyclophane (33) occurs on photolysis to give the cyclophane (34), m.p. 162 °C (S. H. Mashraqui and P.M. Keehn, J. org. Chem., 1983, 48, 1341).

(33)

(34)

Developments in the synthesis of functionalised isoxazoles include the preparation of the 4-aldehydes by a neutral dichromate oxidation of 4-hydroxymethylisoxazoles (N.R. Natale and D.A. Quincy, Synth. Comm., 1983, 13, 817). Various functionalities can be introduced at C-4 via the lithio derivative obtained from the 4-bromoisoxazole (A. Alberola et al., Heterocycles, 1989, 29, 667).

The reduction of 4-cyanoisoxazoles leads to 5-aminoisoxazoles through a ring-opening ring-closure sequence (A. Alberola et al., J. org. Chem., 1984, 49, 3423), whilst the 3-amino derivatives can be obtained by nucleophilic displacement of a 3-chlorine atom from isoxazolium salts (S. Sugai et al., Chem. pharm. Bull., 1984, 32, 530). The reductive dechlorination of 5-chloroisoxazoles can be achieved using tetrabutylammonium borohydride in dichloromethane or sodium borohydride in dimethylsulfoxide (F. Ponticelli and P. Tedeschi, Synthesis, 1985, 792).

Introduction of halogens into isoxazoles has been achieved using N-halosuccinimides or N-haloacetamides (T. Sakakibara, T. Kume and T. Hase, Chem. Express, 1989, 4, 85) and 4-isoxazolin-3-ones are converted into 3-chloroisoxazoles on treatment with phosphorus oxychloride (G. Schlewer and P. Krogsgaard-Larsen, Acta Chem. Scand., 1984, B38, 815). Chlorine is also introduced during the Vilsmeier reaction on 3-phenyl-5-isoxazolinone which yields 5-chloro-3-

phenylisoxazole-4-carboxyaldehyde (D.J. Anderson, J. org. Chem., 1986, 51, 945).

A palladium-catalysed coupling of isoxazoles with benzene or iodobenzene results in phenylation at C-4 (N. Nakamura, Y. Tajima and K. Sakai, Heterocycles, 1982, 17, 235) and palladium catalysis is also utilised in the reaction of 4-iodoisoxazoles with alkenes and alkynes which leads to 4-styryl and 4-phenylethynyl derivatives of isoxazole (H. Yamanaka *et al.*, Chem. pharm. Bull., 1981, 29, 3543).

(b) Reduced Systems

(i) Synthesis of Isoxazolines

The well-established route to 2-isoxazolines, the cycloaddition of nitrile oxides to alkenes, has been extended and developed to encompass newer routes to the 1,3-dipolar species and to synthesise new derivatives of isoxazolines.

In the former category, mention can be made of the electro-generation of nitrile oxides from oximes (T. Shono *et al.*, J. org. Chem., 1989, 54, 2249). The generation of nitrile oxides from nitro-compounds is attractive since useful functions can be incorporated into the precursors of the 1,3-dipole. Thus 3-carbethoxy- and 3-acyl- isoxazolines can be converted into 3-hydroxyalkyl- and hence 3-alkenyl- isoxazolines. The latter take part in a [4+2]-cycloaddition with tetracyanoethylene (P.A. Wade *et al.*, J. org. Chem., 1984, 49, 4595).

When cyanogen *N*-oxide (35) is used as the 1,3-dipolar species, 3-cyanoisoxazolines result, from which isoxazoline-3-carboxylic acids may be obtained. The ethyl esters are formed in high yield from carbethoxyformonitrile oxide (37) which is readily available from the stable oxime (36).

The protected 3-hydroxyisoxazoline (39) is formed from the nitrile oxide (38). Deprotection and cleavage of the isoxazoline ring yields the β-hydroxycarboxylic acid in what is effectively a *cis*-carboxyhydroxylation of the alkene (A.P. Kozikowski and M. Adamczyk, J. org. Chem., 1983, 48, 366). Further functional group interconversion provides a route to α-methylene lactones (A.P. Kozikowski and A.K. Ghosh, Tetrahedron Letters, 1983, 2623), providing a further example of the equivalence of 2-isoxazolines to β-hydroxy-ketones and hence aldol adducts (D.P. Curran, J. Amer. chem. Soc., 1982, 104, 4024).

The thermolysis of ethyl 2-nitroalkanoates at 230 °C in the presence of dipolarophiles provides a route to 3-alkylisoxazolines and the corresponding isoxazoles (T. Shimizu *et al.*, Bull. chem. Soc. Japan, 1987, 60, 1948), whilst 3-acylisoxazolines arise from the reaction of α-nitroketones and alkenes under acidic conditions in the presence of ceric ammonium nitrate (T. Sugiyama *et al.*, Chem. Abs., 1988, 109, 230867).

3-Vinylisoxazolines result when the nitroalkane precursor of the nitrile oxide contains an unsaturated function. A second cycloaddition is now possible and the bisisoxazoline (40) results. Similarly, the 5-(nitromethyl)isoxazoline (41) yields (40) on treatment with styrene in the presence of triethylamine (A. Baranski, Polish J. Chem., 1984, <u>58</u>, 425).

$$H_2C = CCH_2NO_2$$

R

$$\xrightarrow[\substack{NEt_3, C_6H_6 \\ PhCH=CH_2}]{PhNCO}}$$

(41)

$$\xrightarrow[\substack{PhNCO, NEt_3 \\ C_6H_6}]{PhCH=CH_2}}$$

(40)

Benzenesulphonylcarbonitrile oxide can be generated from the substituted nitromethane (42) or the oxime (43) and readily undergoes cycloaddition reactions with alkenes to give 3-phenylsulfonylisoxazolines (44). These compounds undergo a variety of nucleophilic substitution reactions leading to variously 3-substituted isoxazolines (P.A. Wade et al., J. org. Chem., 1983, <u>48</u>, 1796). Refluxing a chloroform solution of $PhSO_2CH=N(O)OMe$ with electron-rich dipolarophiles also yields (44), but electron-poor dipolarophiles afford N-methoxyisoxazolidines (C. Bellandi et al., Heterocycles, 1984, <u>22</u>, 2187).

3-Trifluoromethyl -isoxazoles, -isoxazolines and -isoxazolidines have been obtained from trifluoroacetonitrile oxide and N-methyl-C-trifluoromethylnitrone (K. Mitsuhashi and K. Tanaka, Chem. Abs., 1989, 110, 154225).

Use of 1,2-disubstituted ethylenes in the cycloaddition reaction with nitrile oxides can lead to a mixture of regioisomers. Cinnamaldehyde yields predominantly the 4-formylisoxazoline (45) with either acetonitrile oxide or benzonitrile oxide. An excess of the 1,3-dipolar species leads to the bis-adduct (46) by addition to the formyl group. A facile dehydrogenation of the initial mono-adduct gives the 4-formylisoxazole (47), which reacts only slowly at the carbonyl function (F. de Sarlo, A. Guarna and A. Brandi, J. heterocyclic Chem., 1983, 20, 1505).

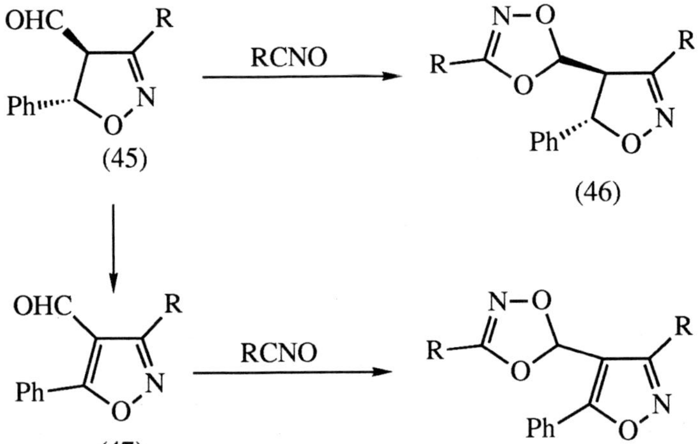

(45) → (46)

(47) →

The reaction with cinnamonitrile also yields a predominance of the 4-regioisomer (48), but in this case the bis-adducts which are formed arise by cycloaddition to the C=C bond of the styryloxadiazole side product rather than to the nitrile group (A. Corsaro *et al.*, J. chem. Res. (S), 1984, 402).

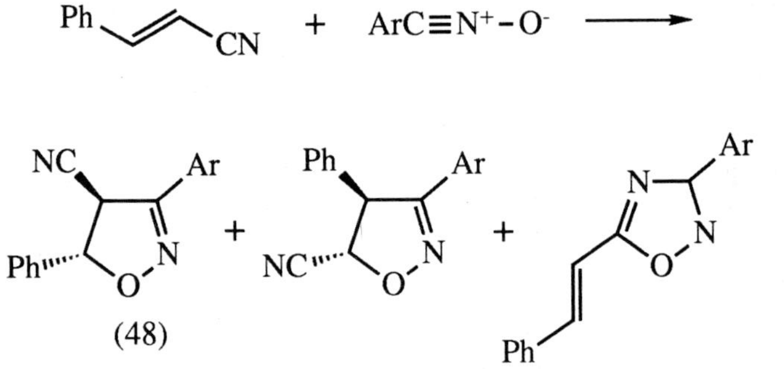

$$Ph\diagdown\diagup_{CN} + ArC\equiv N^+-O^- \longrightarrow$$

(48)

Cycloaddition of benzonitrile oxide to β-substituted nitroethylenes also favours formation of the *trans*-4-nitroisoxazoline (49). However, in addition to the 5-nitroisoxazoline (50), small amounts of the *cis*-4-nitroisoxazoline (51) and the 4-nitroisoxazole (52) are also formed (A. Baranski, Polish J. Chem., 1982, <u>56</u>, 257; 1986, <u>60</u>, 107).

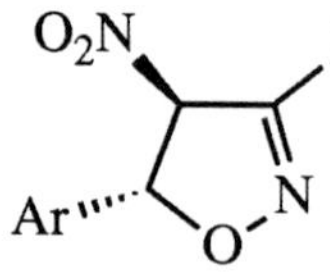

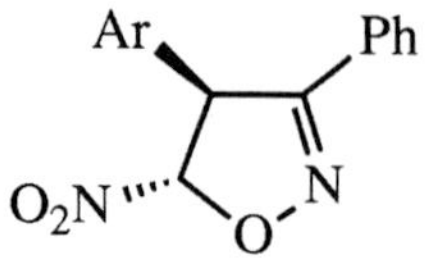

(49) (50)

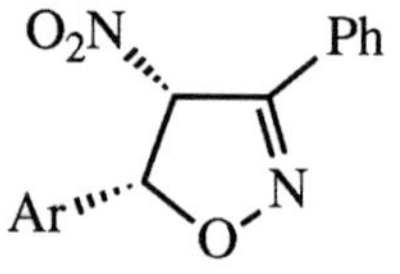

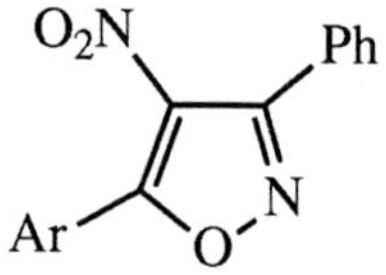

(51) (52)

Thiophene 1,1-dioxide undergoes a highly regioselective cycloaddition with benzonitrile oxide to give the fused isoxazoline (53) in high yield. The product reacts a further time to give the bis-adduct (54) with only a minor amount of the isomer (55) (A. Marinone *et al.*, Tetrahedron, 1982, <u>38</u>, 3629).

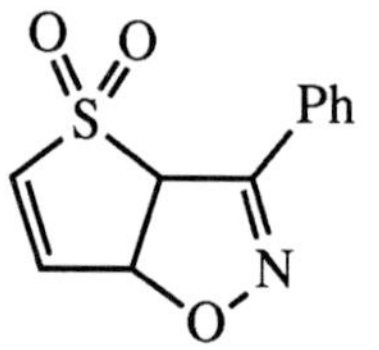

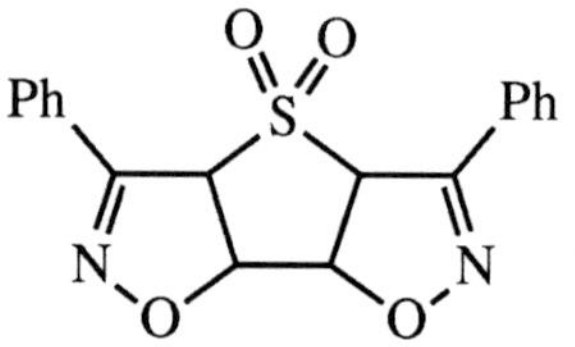

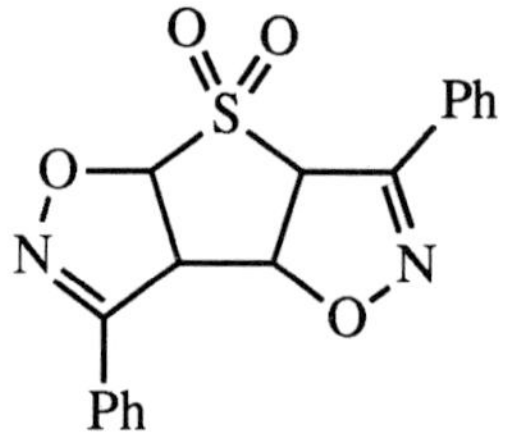

(53) (54) (55)

A fused isoxazoline (56) results from the use of 1,5-cyclooctadiene as the dipolarophile in cycloaddition reactions. Dehydrogenation and oxidative cleavage affords the isoxazole (57) which was utilised in the synthesis of 3,4-symmetrically substituted 2-carbethoxy-5-methylpyrroles (P.D. Williams and E. LeGoff, Heterocycles, 1984, <u>22</u>, 269).

(56)　　　　　　　　　　　　　(57)

Incorporation of an alkene function into a nitrile oxide precursor can result in an intramolecular cycloaddition. Such a reaction has been achieved as part of a synthesis of the anti-tumour agent sarkomycin, in which the isoxazoline acts as a masked α,β-unsaturated ketone (A.P. Kozikowski and P.D. Stein, J. Amer. chem. Soc., 1982, 104, 4023).

An initial intermolecular 1,3-dipolar reaction between dimethyl acetylenedicarboxylate and the nitrone (58) is followed by an intramolecular cycloaddition to give the benzopyranopyrrole (59). The 4-isoxazoline thermally rearranges to the acylaziridine which itself undergoes ring cleavage to the more stable azomethine ylide (O. Tsuge, K. Ueno and S. Kanemasa, Chem. Letters, 1984, 797).

4-Isoxazolines are also produced when DMAD reacts with α,β-unsaturated nitrones (60) derived from chalcones and these thermally rearrange to the ylides (61) from which pyrroles may be obtained by hydrogenation (N. Khan and D.A. Wilson, J. chem. Res., (S), 1984, 150).

A conversion of 2-isoxazolines into 3-isoxazolines can be accomplished by lithiation, alkylation and deprotonation of the resulting isoxazolium salts (S. Shatzmiller *et al.*, Ann. 1983, 906).

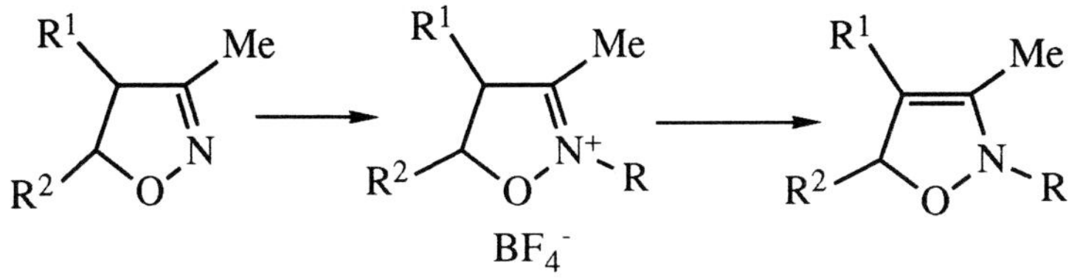

(ii) Synthesis of Isoxazolidines

The regio- and stereo- selectivity of the reaction of *N*-methyl-*C*-trifluoromethylnitrone with alkenes is influenced by the nature of the alkene. Monosubstituted alkenes yield the 5-substituted isoxazolidines exclusively, with an 85:15 preference for the *trans*-disposition of the 3- and 5-substituents in the reaction with styrene. Dimethyl maleate affords the 3,4-*trans*-stereoisomer almost exclusively, whereas the fumarate yields equal amounts of the 3,4-*trans*- and the 3,4-*cis*- compound (Scheme 2). Maleimides also yield a stereoisomeric mixture (K. Tanaka *et al.*, J. heterocyclic. Chem., 1989, 26, 381).

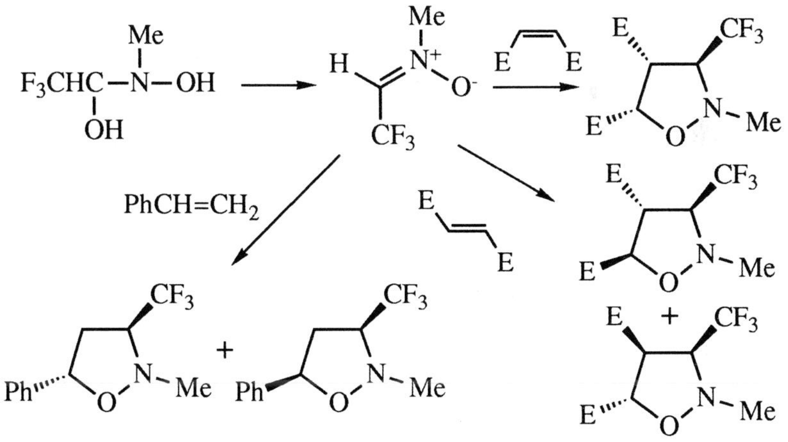

Scheme 2

The reaction of phenylnitrones with *trans*-1-cyano-2-nitroethene shows some regioselectivity, but its importance lies in the ability to convert the resulting isoxazolidines into β-lactams with some stereochemical control (A. Padwa, K. Koehler and A. Rodriguez, J. Amer. chem. Soc., 1981, 103, 4974).

Nitroxides (62, R = Ph) dimerise to the bisisoxazolidine (63) by a 1,3-dipolar cycloaddition involving the nitrone and alkene moieties. A competing formation of the tricyclic dimer (64) is favoured for the *t*-butyl nitroxide (62, R = CMe₃) (H.G. Aurich *et al.*, J. org. Chem., 1988, 53, 4997).

$$RN(O \cdot)CH=C-COR^2$$

(62)   (63)   (64)

The cycloaddition of nitrones to allenes has been investigated. Whereas both methoxy- and phenoxy- allene yield predominantly the *anti*-product, fluoroallene gives predominantly the *syn*-adduct (65) (W.R. Dolbier *et al.*, J. Amer. chem. Soc., 1985, 107, 7183).

(65)

However, N-phenylnitrones react with allenes containing electron withdrawing groups to yield substituted benzazepin-4-ones (67) by way of cycloaddition to the more activated π-bond and formation of the 5-methylideneisoxazolidine (66). Products derived from addition to the less active double bond can be obtained by the reaction of nitrones with 2,3-bis(diphenylsulfonyl)-1-propene or ethyl 3-phenylsulfonylbut-3-enoate, both of which behave as allene equivalents (A. Padwa, D.N. Kline and B.H. Norman, J. org. Chem., 1989, 54, 810).

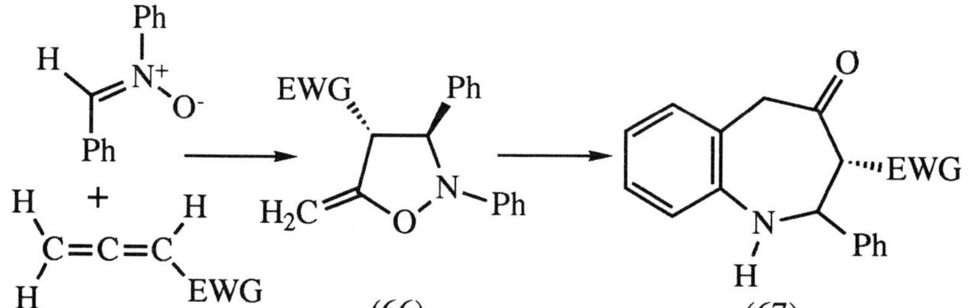

(66)   (67)

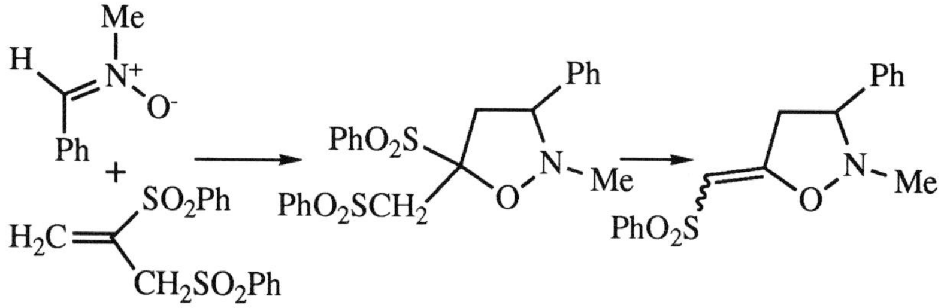

The reaction of diphenylnitrone with 1,3-dienes is complex, giving oxazepines (69) from a [4+3]-cycloaddition, as well as the 1,3-dipolar product, the isoxazolidine (68). It appears that the reaction is at the borderline between a concerted and a diradical process (J. Baran and H. Mayr, J. org. Chem., 1989, 54, 5774).

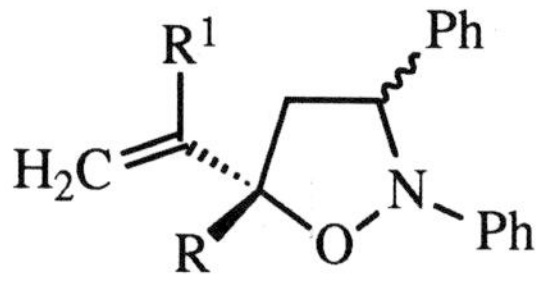

(68) 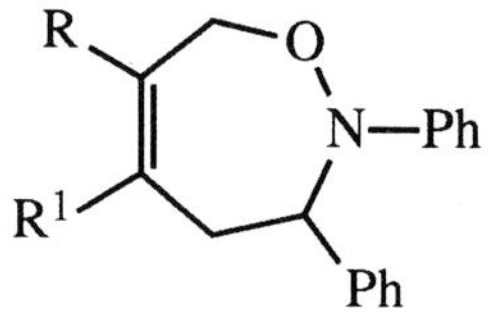   (69)

The 3,3-disubstituted isoxazolidines (70) obtained from the reaction of silyl nitronates with alkenes can be rearranged to the blue tertiary nitroso compounds on treatment with fluoride ion, whilst those derived from acrylonitrile afford α,β-unsaturated aldehydes (S.K. Mukerji and K.B.G. Torsell, Acta Chem. Scand., 1981, B35, 643).

Aryl silyl-enol ethers react with nitrosobenzene to yield silylated α-hydroxyaminoketones, which on heating with oxalyl chloride give the new isoxazolidin-4,5-dione system (71). The 3-benzoyl (m.p. 183-185 °C) and 3-*p*-anisyl (m.p. 220-222 °C) derivative each exhibit carbonyl stretching frequencies at 1770, 1710 and 1690 cm^{-1} and the C-3 proton resonates at 5.5 ppm (M. Ohno *et al., Synthesis*, 1986, 666).

(71)

(iii) Reactions of isoxazolines and isoxazolidines

Isoxazolines can be cleaved under similar conditions to those described above for isoxazoles. Thus, hydrogenolysis of the bicyclic isoxazoline (72) leads stereospecifically to the *cis*-β-hydroxyketone in excellent yield (A.P. Kozikowski and M. Adamczyk, Tetrahedron Letters, 1982, 3123).

(72)

The diastereoisomeric 3-sulfinylmethylisoxazolines (75), formed by sequential metallation of racemic isoxazolines (73) and treatment with (-)-(*S*)-menthyl toluene-*p*-sulfinate (74), can be separated and individually converted into the stereochemically pure isoxazoline. The latter on stereoselective reduction yield optically active γ-aminoalcohols and β-ketols (R. Annunziata *et al.*, J. chem. Soc., Perkin I, 1985, 2289).

The 5-nitroisoxazolidine (76) undergoes a base catalysed ring contraction on heating to give the thermodynamically favoured *cis*-azetidinone as a result of deprotonation, cleavage of the N-O bond and cyclisation of the resulting acyl nitro intermediate (A. Padwa, K.F. Koehler and A. Rodriguez, J. org. Chem., 1984, 49, 282).

(76)

The reaction of isoxazolidinium salts with lithium aluminium hydride proceeds by nucleophilic attack at C-3 and the resulting C-N bond cleavage affords a substituted hydroxylamine (A. Liguori, G. Sindona and N. Uccella, Tetrahedron, 1983, 39, 683).

The reaction of benzal halides with isoxazolones in the presence of base leads to 1,3-oxazinones (77) through an initial alkylation at nitrogen, followed by cleavage of the N-O bond and ring closure (E. Beccalli, T. Benincori and A. Marchesini, Synthesis, 1988, 630).

(77)

The photolysis of isoxazolines in the presence of iron carbonyls results in fragmentation to an aldehyde and a ketone by cleavage of both the N-O and C4-C5 bonds. In certain cases, elimination of the 5-substituent occurs in preference to C4-C5 bond cleavage, in which case an aminoenone results (M. Nitta and T. Kobayashi, J. chem. Soc., Perkin I, 1984, 2103).

The acid catalysed thermal rearrangement of 4-isoxazolines affords substituted indoles (A. Liguori *et al.*, Heterocycles, 1988, 27, 1365).

A Lewis acid mediated cyclisation of the isoxazoline 2-oxide (78) leads to the benzofuroisoxazole (79) in good yield (S. Zen *et al.*, Chem. pharm. Bull., 1983, 31, 1814).

The product from the alkylation of 5-methylideneisoxazolidines depends on the alkylating agent. Thus (80) yields the α-methylated derivative (81) but allylation occurs at the γ-methylene site to give (82) (A. Padwa *et al.*, J. chem. Soc., Perkin I, 1988, 2639).

2. Benzisoxazoles

The chemistry of both 1,2- and 2,1-benzisoxazoles has been reviewed (R.K. Smalley, Advances in Heterocyclic Chemistry, Vol. 29, Academic Press, 1981) and the synthesis and rearrangements of

1,2-benzisoxazoles have been discussed (A.J. Boulton, Bull. Soc. chim. Belg., 1981, 90, 645).

(a)   1,2 - Benzisoxazoles (Indoxazenes)

(i) Synthesis

Further examples of the synthesis of 1,2-benzisoxazoles from o-hydroxybenzoyl oximes include the oxidative cyclisation using lead (IV) acetate or sodium hypochlorite which leads to 1,2-benzisoxazole 2-oxides (83) (A.J. Boulton, P.G. Tsoungas and C. Tsiamis, J. chem Soc., Perkin I, 1986, 1665) and the formation of the fused analogues, naphtho[2,1-d] isoxazoles (84) (M.A. Elkasaby and M.A.I. Salem, Indian J. Chem., 1980, 19B, 571). Cyclisation of salicylaldoximes can be achieved using Cl₃CCONCO under basic conditions (G. Stokker, J. org. Chem., 1983, 48, 2613), and fluorosalicylhydroxamic acids yield fluorobenzisoxazoles on treatment with thionyl chloride and triethylamine (T. Slawik and E. Domagalina, Acta Pol. Pharm., 1984, 41, 625). 3-Arylbenzisoxazoles result from the oxidation of halogenobenzophenone oximes by potassium persulphate (T.P.Tolstaya, L.D. Egorova and I. N. Lisichkina, Khim. geterosikl. Soedin, 1985, 474).

(83)   (84)

The reaction of nitrile oxides with substituted 1,4-benzoquinones is dependent on both the site and nature of the substituent. 2-Chloro-5(6)-methylbenzoquinones give two products by reaction at both of the C=C double bonds, although attack at the 2,3-site is followed by dehydrochlorination and 1,3-dipolar cycloaddition to the carbonyl function (S. Shiraishi, S.B. Holla and K. Imamura, Bull. chem. Soc. Japan, 1983, 56, 3457).

(ii) Properties and reactions

Under electron impact, 1,2-benzisoxazoles rearrange to benzoxazoles prior to fragmentation, since intense $MeCO^+$ ions (m/z 43) mass are observed (R.K. Kallury *et al.*, Org. Mass Spectrom., 1981, 16, 552). The 2-oxides behave similarly, but deoxygenation is an alternative initial process (C. Tsiamis and P.G. Tsoungas, J. heterocyclic Chem., 1985, 22, 687). The Raman and infrared spectra of both 1,2- and 2,1-benzisoxazoles have been recorded (G. Mille *et al.*, J. Raman Spectros., 1980, 9, 339). ^{13}C and ^{15}N chemical shift data have been reported for some 1,2-benzisoxazoles and their 2-oxides (P. G. Tsoungas and B.F. De Costa, Magn. Reson. Chem., 1988, 26, 8).

Although nitration of 1,2-benzisoxazole 2-oxides proceeds by simple substitution, both bromination and acylation cleave the heterocyclic ring to yield derivatives of phenol (A.J. Boulton, P.G. Tsoungas and C. Tsiamis, J. chem. Soc., Perkin I, 1987, 695).

In the absence of electrophiles, the anion derived from 3-methyl-1,2-benzisoxazoles (85, R = H or Me₃CMe₂Si) rearranges and then undergoes self-condensation to give the reduced pyridazine (86) (J.E. Oliver, R.M. Waters and W.R. Lusby, J. org. Chem., 1989, 54, 4970).

(85)   (86)

Cleavage of the isoxazole ring of methyl benzisoxazolylacetates (87) by various bases leads to azirines (88) or their ring opened derivatives, the iminobenzofurans (89) (S. Ueda *et al.*, J. chem. Soc., Perkin I, 1988, 1013).

On the other hand, benzisoxazolium salts (90) react with organometallic reagents to give the aziridine (91) or the simple addition product (92) (A. Albertola *et al.*, J. chem. Soc., Perkin I, 1988, 767).

(90) → RLi → (91)

and / or

(92)

On treatment with acyl chlorides, 3-benzisoxazolylacetic acid derivatives afford benzisoxazolo[2,3-a]pyridines (93) (S. Naruto *et al.*, Chem. pharm. Bull., 1982, 30, 3418).

1,2-Benzisoxazol-3-ol is converted into *N,N'*-carbonyldi[1,2-benzisoxazol-3(2H)-one] (94) on reaction with trichloromethyl chloroformate. The product is a useful condensing agent for peptide synthesis (M. Ueda *et al.*, Bull. chem. Soc. Japan, 1983, 56, 2485).

(93)   (94)

(b) 2,1- Benzisoxazoles (Anthranils)

(i) Synthesis

The formation of 1-hydroxymethyl-2,1-benzisoxazol-3(1H)-one (96, R = CH$_2$OH) by treatment of 2-nitrostyrene oxide (95) with acid probably proceeds through the intermediacy of 2-nitrosobenzaldehyde. This product is also formed by reduction of methyl 2-nitrobenzoate in the presence of formaldehyde (W. Wierenga *et al.*, J. org. Chem., 1984, <u>49</u>, 438).

(95)   (96)   R

2-Nitrosobenzoic acids, obtained by the electrolysis of the nitrobenzoic acids, react with sulfinic acids to give the *N*-sulfonyl-benzisoxazolones (96, R = SO$_2$R') (C. Gault and C. Moinet, Tetrahedron, 1989, <u>45</u>, 3429).

Both 4- and 7-methyl-2,1-benzisoxazoles, obtained from the corresponding 2-nitrobenzaldehyde by reduction with tin (II) chloride, have been converted into the formylbenzisoxazoles and thence into the dihydropyridine derivatives by Hantzsch methodology (B.T. Phillips and G. D. Hartman, J. heterocyclic Chem., 1986, <u>23</u>, 897).

Under basic conditions, 2-(2'-nitrophenyl)-2-phenylacetonitriles (97) are converted into 2,1-benzisoxazoles (Z. Vejdelek *et al.*, Coll. Czech. chem. Comm., 1986, <u>52</u>, 2545; 1988, <u>53</u>, 361). Some 2-nitrophenyl derivatives of acetophenone (98) behave similarly (M. Jawdosiuk *et al.*, Polish J. Chem., 1981, <u>55</u>, 379) and the 3-(2'-thienyl)-2,1-benzisoxazole (99) is formed from 4-chloronitrobenzene and 2-cyanomethylthiophene (L.H. Clemm, C.E. Klopfenstein and S.K. Nelson, J. heterocyclic Chem., 1982, <u>19</u>, 675).

(97)   (98)   (99)

3-Aminobenzisoxazoles are formed when activated 2-chlorobenzonitriles are treated with hydroxylamine in the presence of base (J. Wrubel and R. Mayer, Z. Chem., 1984, 24, 254).

EWG = NO$_2$ or CN

Photolysis of 2,2'-dinitrodiphenylmethanes in acidified ethanol yields 3-arylbenzisoxazoles and the same compounds can be obtained by the reduction of 2,2'-dinitrobenzophenones to the benzhydrols and cyclisation in sulphuric acid (M. Christudhas and C. P. Joshua, Austral. J. Chem., 1982, 35, 2377).

The formation of partially reduced benzisoxazoles by treatment of 2-aroylcyclohex-2-enones with hydroxylamine is also acid catalysed. The reaction is very sensitive to experimental conditions (Y. Tamura et al.,

Chem. pharm. Bull., 1981, 29, 3226). The related xanthenedione (100) leads to the reduced benzisoxazole (101) through anionic opening of the pyran ring (C.D. Gabbutt, J.D. Hepworth and M.W.J. Urquhart, unpublished results).

(100)   (101)

The synthesis of 2,1-benzisoxazoles by heating 2-azidobenzophenones is well documented and occurs by a pericyclic process rather than *via* a nitrene. Further examples include the preparation of benzisoxazole-4,7-quinones as depicted (R. Cassis *et al.,* Synth. Comm., 1987, 17, 1077) and of cycloalkan[*c*]isoxazoles (B. Tabyaouti *et al.,* Synth. Comm., 1988, 18,1475).

However, in basic solution loss of nitrogen from 2-azidophenyl *s*-alkyl ketones (102) proceeds via the enolate and an alternative mode of cyclisation leads to 1,2-dihydro-3*H*-indol-3-ones (103) (M. Azadi-Ardakani *et al.,* J. chem. Soc., Perkin I, 1986, 1107).

(103)   (102)

When benzyl 2-azidobenzoates (104) are pyrolysed, insertion of the nitrene into the C-O bond is observed and an N-benzylbenzisoxazolone (105) is produced. Other esters give products such as carbazoles and acridones derived from a spiro-intermediate (M.G. Clancy, M.M. Hesabi and O. Meth-Cohn, J. chem. Soc., Perkin I, 1984, 429).

(104)   (105)

A nitrene is also invoked in the formation of 3-phenylbenzisoxazole by the sodium hypochlorite oxidation of 2-aminobenzophenone (L.K. Dyall, Austral. J. Chem., 1984, 37, 2013).

(ii) Reactions

The major interest in 2,1-benzisoxazole chemistry continues to be associated with ring-opening reactions. The advantages of cleaving the heterocyclic ring with iodotrimethylsilane are the high selectivity of the reagent and the excellent yields of 2-aminobenzophenones achieved under mild conditions (D. Konwai *et al.*, Synth. Comm., 1984, 14, 1053). 2-Alkylaminobenzophenones are accessible from the benzisoxazolium salts using this reagent (D. Konwai, R. Boruah and J.S. Sandhu, Indian J. Chem., 1984, 23B, 975). When Vilsmeier reagents are used to open the isoxazole ring, N-formanilides (106) are produced (D. Konwai *et al.*, Tetrahedron Letters, 1987, 955). Organozinc reagents in the presence of nickel acetylacetonate convert 2,1-benzisoxazoles into 2-(substituted-amino)benzaldehydes (J.S. Baum, M.E. Condon and D.A. Shook, J. org. Chem., 1987, 52, 2983).

(106)

3-Methoxy-[2,1]-benzisoxazole-4,7-quinones (107) and the corresponding naphtho[2,3-c]isoxazoles are methylated by dimethyl sulfoxide and dimethyl formamide, leading to 1-methylbenzisoxazol-3(1H)-onequinones (108). Additionally, sulfoximidoquinones (109) are formed, presumably by nitrene addition to the DMSO (T. Torres, S.V. Eswaran and W. Schäfer, J. heterocyclic Chem., 1985, 22, 701 and 705).

(107)   (108)   (109)

Anthra[1,9-cd]isoxazol-6-ones (110) are cleaved by DMSO and by triphenylphosphine to give sulfoximido and triphenylphosphazo substituted anthraquinones (111), respectively (P. Sutter and C.D. Weis, J. heterocyclic Chem., 1982, 19, 997).

(110)   (111)

On occasions, the ring-opening process is followed by further reaction. The transformation of 2,1-benzisoxazoles into acridones has been studied in some detail and found to be critically dependent on reaction conditions and substrate structure (D.G. Hawkins and O. Meth-Cohn, J. chem. Soc., Perkin I, 1983, 2077). On thermolysis, 3-*p*-tolyl-2,1-benzisoxazole yields a mixture of 2- and 3- methylacridone and 2-aminobenzophenone, the proportions of which vary with the temperature and duration of heating, and particularly significantly with solvent. Various metals, metal salts and complexes catalyse the formation of the acridones. The rearranged product (113) is considered to arise from a nitrene-derived spiro-intermediate, whilst the expected product (112) is derived from attack of the nitrene at the *ortho*-position.

(112)   (113)

In the case of 3-(2,6-disubstitutedphenyl)anthranils, the spiro-intermediate is considered to be the source of the products, 4-methylacridone, 4,10-dimethylacridone and 4,5-dimethylacridone. The last compound is presumably derived by consecutive 1,5-methyl shifts and prototropy.

Thermolysis of 3-thienyl-2,1-benzisoxazoles (114) gives the corresponding thienoquinolones (115), which in some instances are aromatised.

(114)   (115)

2,1-Benzisoxazolium salts unsubstituted at the 3-position undergo a ring-opening ring-closure sequence on treatment with triethylamine to give N-alkylbenzoazetinones (116), which are themselves readily cleaved by nucleophiles (R.A. Olofson *et al.*, J. org. Chem. 1984, 49, 3367 and 3373).

(116)

Enamines convert 2,1-benzisoxazoles into quinoline derivatives probably as a result of a [4+2]-cycloaddition which leads to the quinoline 1-oxide (K.Ohta, H. Shimizu and Y. Nomura, Nippon Kagaku Kaishi, 1989, 846). Quinoline-1-oxides result from the base catalysed reaction of arylidenemalononitriles with benzisoxazoles (D. Konwar, R.C. Boruah and J.S Sandhu, Heterocycles, 1985, 23, 2557).

Although 2,1-benzisoxazole itself does not undergo a cycloaddition reaction with dimethyl acetylenedicarboxylate, both the 5-chloro- and 6-nitro- derivatives yield the adduct (117) (R.C. Boruah, J.S. Sandhu and G. Thyagarajan, J. heterocyclic Chem., 1981, 18, 1081).

(117)

3. Oxazole and its Derivatives

A major treatise on oxazoles is available covering their synthesis, reactions, spectroscopic properties and synthetic uses (I.J. Turchi, 'Oxazoles', Chemistry of Heterocyclic Compounds, Vol. 45, Wiley, 1986). Other reviews include recent advances in oxazole chemistry (I.J. Turchi, Ind. Eng. Chem. Prod. Res. Dev., 1981, 20, 32) and the chemistry of 2-aminoxazoles and their derivatives (L. Peshakova, V. Kalcheva and D. Simov, Khim. geterosikl. Soedin, 1981, 1011). The reactions of C-silylated and C-stannylated oxazoles with carbon electrophiles have been reviewed. The regio- and chemo-selectivity of the reactions of these stable species give access to new oxazole derivatives (A. Dondoni et al., Gazz. chim. Ital., 1988, 118, 211). Electrophilic substitution of protonated oxazoles and of the free bases has been discussed (L. I. Belen'kii et al., Chem. Scripta, 1985, 25, 266 and 295).

4-Alkoxycarbonyloxazoles behave as latent β-hydroxy-α-aminoacids and their value in natural product synthesis has been reviewed (T. Shioiri and Y. Hamada, Heterocycles, 1986, 27, 1035). The synthesis of vitamin B_6 via oxazoles has been discussed (H. Zhou, Yiyao Gongye, 1985, 16, 265).

The value of oxazolines in synthesis is recognised by reviews covering their use in aromatic substitution (M. Reuman and A.I. Meyers, Tetrahedron, 1985, <u>41</u>, 837) and the application of chiral oxazolines to asymmetric carbon-carbon bond formation (A.I. Meyers, ACS Symp. Ser., 1982, <u>185</u>, 83). The synthesis of isoquinolines based on protonated derivatives of 2-oxazoline has been discussed (T. Kopczynski, Chem. Inz. Chem., 1986, <u>16</u>, 143). Aspects of the chemistry of 2-oxazolin-5-ones have been reviewed (A.K. Mukerjee and P. Kumar, Heterocycles, 1981, <u>16</u>, 1995; F. Adler, I. Thondorf and M. Strube, Wiss. Z-Martin Luther Univ. Halle-Wittenberg, Math.-Naturwiss Reihe, 1986, <u>35</u>, 19), as has the chemistry of 2-oxazolidones (V.A. Pankratov, Ts. M. Frenkel and A.M. Fainleib, Usp. Khim., 1983, <u>52</u>, 1018).

(a) Oxazoles

(i) Synthesis

Many syntheses of the oxazole ring proceed through formation of the O-C2 or O-C5 bonds, typified by the cyclisation of α-acylaminoketones.

There are various routes to these precursors and so the method is attractive, allowing the production of a wide range of substituted oxazoles. Acylation of α-aminoketones by heterocyclic acid chlorides followed by cyclodehydration with phosphorus oxychloride leads to the mixed heterocycles (1) (A.P. Shkumat *et al.*, Ukr. Khim. Zh. 1987, <u>53</u>, 529) and 2,5-diaryloxazoles are obtained similarly (O.P. Shvaika *et al.*, Khim. geterosikl. Soedin, 1985, 193).

(1) X = O, S

The reaction of aromatic aldehydes with hydroxyimino-ketones leads to salts of oxazole N-oxides through cyclodehydration of the N-substituted α-acylaminoketone. The 2,5-disubstituted oxazoles which result from reduction of the free N-oxides are of interest in connection with the enzymatic hydroxylation of 2,5-diphenyloxazole (N.W. Jacobsen and A. Philippides, Austral. J. Chem., 1985, 38, 1335).

Cyclisation of the methyl esters of N-acyl-α-benzoylglycines (2) derived by acylation of the aminoacid ester leads to trisubstituted oxazoles. Further structural elaboration is possible through the ester function (Y. Ozaki et al., Chem. pharm. Bull., 1983, 31, 4417).

(2)

Aminoacid esters also feature in a synthesis of oxazole-4-carboxylic acid derivatives. Ethyl glycinate and the imidate hydrochloride of acetonitrile react in the presence of base to yield the N-substituted imidate ester (3). Acylation of the enolate and cyclisation in boiling acetic acid gives the oxazole (5) (A.I. Meyers et al., J. org. Chem., 1986, 51, 5111). A feature of this route is that it is possible to metallate and thus further substitute the acyclic oxazole precursor (4). Cyclisation then occurs under milder conditions and the overall process is equivalent to substitution at C-2 of the oxazole-4-carboxylic acid, a normally impossible task (Scheme 1).

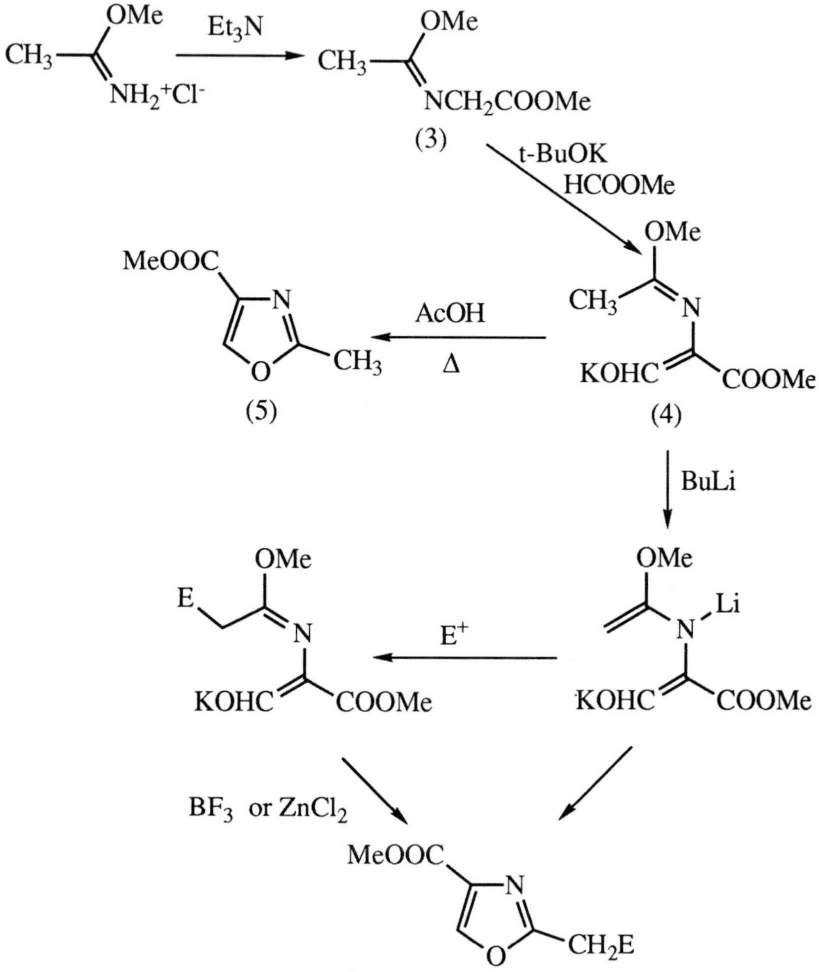

Scheme 1

This approach has been incorporated in the synthesis of the oxazole dienylamine moiety of griseoviridin utilising the chiral aldehyde (6) as the electrophilic species and yielding (7) as a mixture of diastereoisomers.

(4) + (6) → (7)

α-Acylaminoketones probably feature in the synthesis of 4-aminooxazoles from α-oxonitriles by the reaction of the latter with anhydrous ammonium acetate (a source of ammonia) and hexamethylenetetramine (a formaldehyde precursor) (R. Lakhan and R.L. Singh, Org. prep. Proced. Int., 1989, 21, 141).

Reagents: $(CH_2)_6N_4$, $AcONH_4$, AcOH, 120 °C, 3-5 h

Aromatic hydrocarbons are acylaminoacylated on treatment with 2-(fluoroaryl)azlactones in the presence of aluminium chloride and the resulting acylamino-ketones are readily dehydrated to 2,5-diaryloxazoles (I. Schiketanz et al., Rev. Roum. Chim., 1985, 30, 969).

Rather different mechanistically is the Lewis acid catalysed cyclisation of the N-acylamines of hexafluoroacetone (8), which leads to fluorinated derivatives of oxazole (K. Burger et al., Ber., 1982, 115, 2494). Other heterocyclic units can be introduced through the N-acyl function and fused heterocycles such as (9) are accessible from the corresponding N-heterocyclic imine (K. Burger, K. Geith and D. Hübl, Synthesis, 1988, 199).

(8)   (9)

Several oxazole syntheses involve the use of isocyanides or related compounds.

Acylation of the anion derived from methyl isocyanoacetate and subsequent cyclisation leads to a methyl 5-substituted oxazole-4-carboxylate (Y. Ozaki *et al.,* Chem. pharm. Bull., 1983, $\underline{31}$, 4417).

The formation of oxazoles from tosylmethylimine derivatives (10) follows a similar pattern, involving the acylation of the derived 2-azaallyl anion with an aromatic aldehyde. Yields are high, indeed quantitative when X = Cl, R = OMe and Ar = p-NO_2phenyl, and the method is a valuable route to 2-methoxyoxazoles. The synthesis of analogues of the mould metabolite pimprinine (11) by this method is of interest, though it is preferable to use α-tosylethylisocyanide, TosCH(Me)NC, instead of the imine in the reaction with the substituted indole-3-aldehyde (H.A. Howing, J. Wildeman and A.M. van Leusen, J. heterocyclic Chem., 1981, $\underline{18}$, 1133).

(11)

The trioxazolometacyclophane, m.p. 249-250 °C, has been obtained from the reaction of 3-(isocyanotosylmethyl)benzaldehyde with base (H. Sasaki *et al.*, Chem. Letters, 1988, 1531).

N-(Tosylmethyl)carbodiimides (12), prepared from the corresponding thioureas by a Mannich reaction with toluene-p-sulfinic acid and formaldehyde, can also be acylated by an aromatic aldehyde under either phase transfer or simple basic conditions, providing a direct entry to

2-alkylamino- and hence 2-amino- oxazoles (A.M. van Leusen *et al.*, J. org. Chem., 1981, **46**, 2069).

$$TosCH_2N=C=NCPh_3 \xrightarrow[ArCHO]{NaH, DME}$$

(12)

4-Aminoisoxazoles may be converted into 2-aminooxazole-4-carbonitriles in excellent yield *via* their thiourea and carbodiimide derivatives. The rearrangement occurs under basic conditions which promote opening of the isoxazole ring and subsequent cyclisation to the oxazole (A. Pascual, Helv., 1989, **72**, 556).

Reagents: (i) $CSCl_2$, $CaCO_3$, CH_2Cl_2, H_2O, 0 °C 2h; (ii) R^2NH_2, toluene, RT, 1-2h; (iii) 2-chloro-1-methylpyridinium iodide, Et_3N, CH_3CN, RT, 1-5h; (iv) Et_3N, CH_3CN, 80 °C, 1-4h.

The isothiocyanate (13) prepared from chloro(phenylthio)methyl-trimethylsilane is similar to an isocyanide in that aromatic aldehydes can acylate the derived anion. Cyclisation of the product then provides a new route to 5-aryloxazoles with the sulfur moiety retained as a 2-thiol substituent (I. Yamamoto *et al.*, J. chem. Soc., Perkin I, 1984, 435).

β-(Acyloxy)vinyl azides, formed by the selective enol acylation of α-azidoketones, undergo a Staudinger reaction with triphenylphosphine or triethyl phosphite to give an azaphosphorane. An intramolecular Wittig reaction then yields an oxazole. This new route to oxazoles is compatible with acid-labile substituents (H. Takeuchi *et al.*, J. org. Chem., 1989, 54, 431).

Preparative flash vacuum pyrolysis of 1-acyl derivatives of 1,2,4-triazoles gives access to 5-substituted oxazoles by elimination of nitrogen (A. Maquestiau *et al.*, Heterocycles, 1989, 29, 1).

Loss of nitrogen also occurs in the formation of highly functionalised oxazoles from dimethyl diazomalonate and a nitrile, catalysed by rhodium (II) acetate. The intermediate carbenoid species generally undergoes a 1,3-dipolar addition in preference to cyclopropane formation (R. Connell *et al.*, Tetrahedron Letters, 1986, 5559).

A similar reaction occurs when an α-diazocarbonyl compound reacts with a nitrile in the presence of boron trifluoride. 2-Chloroacetonitrile yields a chloromethyloxazole which undergoes facile nucleophilic substitution with a variety of amines leading to *N*-substituted 2-(aminomethyl)oxazoles (T. Ibata *et al.*, Bull. chem. Soc. Japan., 1984, 57, 2450; 1989, 62, 618).

Oxidation of a ketone, dissolved in a nitrile, by an iron (III) nitrile complex yields a 2-substituted oxazole. One electron oxidation of the enolic form of the ketone is followed by reaction with the solvent nitrile. (E. Kotani *et al.*, Chem. pharm. Bull., 1989, 37, 606). The oxazole (14) has been derived from cholestanone in this manner using acetonitrile. The iron (III) nitrile complex may be replaced by copper (II) trifluoromethanesulfonate in such syntheses (K. Nagayoshi and T. Sato, Chem. Letters, 1983, 1355).

(14)

(ii)   Reactions

The reaction of oxazoles with dienophiles provides access to other heterocyclic systems. Bipyridyls are the final products from the reaction of 2,4-dimethyloxazole with vinylpyridines (P.B. Terent'ev *et al.*, Khim. geterosikl. Soedin, 1980, 1255). The regioselectivity of the reaction is influenced by the substitutents in the vinyl group. Thus 2-vinylpyridine affords the 2,3-bipyridyl derivative, whereas 3-(2-pyridyl)propenoic acid yields the 2,4-isomer.

A hydroxypyridine results directly from the reaction of an oxazole with either propenonitrile or propenoic acid. However, with ethyl propenoate or an acyl ethylene, the initial adducts (15) can be isolated. The adduct may be converted into either the expected pyridine or the unexpected pyrrole (B.A. Johnsen and K. Undheim, Acta Chem Scand., 1983, B37, 127).

The pyrrole system (16) rather than the expected Diels-Alder adduct is formed from reaction of oxazoles with tetracyanoethylene. Its formation is thought to involve a zwitterionic intermediate which ring opens to a second zwitterion (T. Ibata *et al.*, Chem. Comm., 1986, 1692).

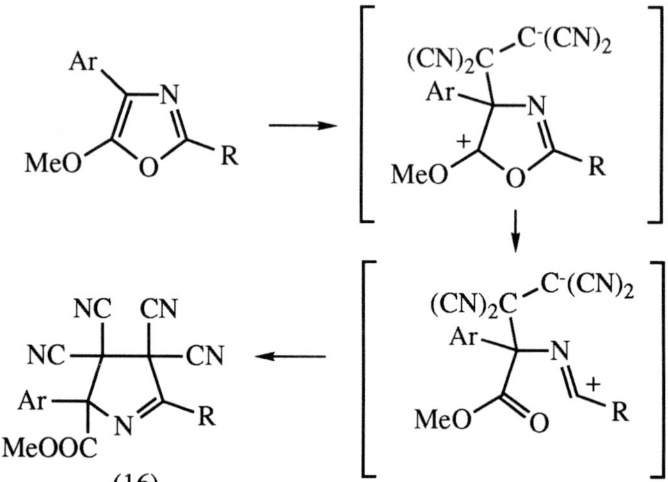

(16)

A similar reaction occurs following the cycloaddition reaction of oxazoles with 4-phenyl-4H-1,2,4-triazole-3,5-dione (PTAD), when the product is a triazolotriazole (17) (A. Huth *et al.*, Ann., 1984, 641; T. Ibata *et al.*, Chem. Letters, 1988, 1551).

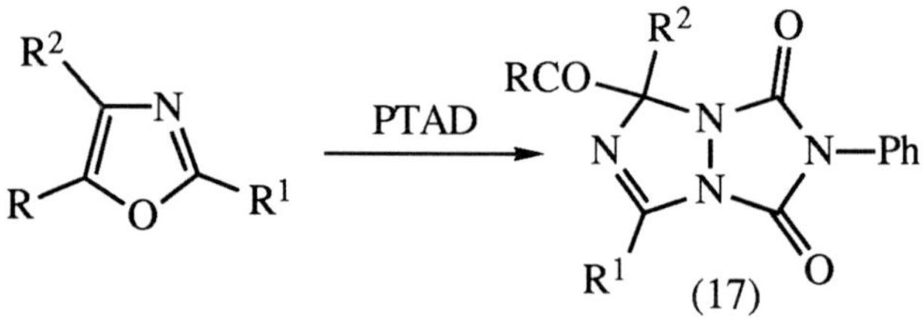

(17)

The silyloxazolium salts (18) derived from 4- and 5- phenyloxazole by treatment with trimethylsilylmethyl trifluoromethane sulfonate are desilylated by caesium fluoride. The resulting oxazolium methylide (19) is only trapped by acetylenic esters and a facile 1,4-elimination from the adduct (20) leads to a pyrrole (A. Padwa, U. Chiacchio and M.K. Venkatraman, Chem. Comm., 1985, 1108).

Mesoionic oxazolones react with electron deficient alkenes to give a pyrroline (21) by loss of carbon dioxide from the first formed adduct. When a dicyanoethylene is used (X = CN), subsequent loss of hydrogen cyanide gives the aromatic heterocycle (D. Yebdri, O. Henri-Rousseau and F. Texier, Tetrahedron Letters, 1983, 369).

A useful route to pyrroles is based on the treatment of an oxazolium salt (22) with trimethylsilylcyanide to yield a cyanooxazoline.

154

The latter spontaneously ring opens to the azomethine ylide (23). In the presence of a dipolarophile, a [3+2]- cycloadduct is formed from which hydrogen cyanide may be lost (E. Vedejs and J.W. Grisson, J. org. Chem., 1988, 53, 1876).

However, the reaction is not always so straightforward. In the absence of cyanide, some oxazolines resist ring opening and are trapped as the [2+2] adduct (24) (E. Vedejs and J.W. Grisson, J. Amer. chem. Soc., 1988, 110, 3238).

Me R² COOMe

N

R¹ O R³ COOMe

(24)

Further examples of the reaction of oxazoles with alkynes confirm that the initial adduct loses a nitrile to give a furan derivative. Thus 4-methyloxazole with 3,3,3-trifluoropropyne leads to 3-trifluoromethylfuran (K. Kawada, O. Kitagawa and Y. Kobayashi, Chem. pharm. Bull., 1985, 33, 3670), whilst trisubstituted oxazoles and ethyl propiolate give a mixture of the isomeric furans (25) and (26) in which the former predominates (T. Jaworski and T. Mizerski, Polish J. Chem., 1981, 55, 47).

Me EtOOC COOEt

N →

EWG O Me EWG O Me EWG O Me

(25) (26)

The adducts (27) from the reaction of 2-aminooxazoles with dimethyl acetylenedicarboxylate can be isolated, but the oxazolo-[3,2-a]pyrimidines (28) are also formed (G. Crank and H.R. Khan, J. heterocyclic Chem., 1985, 22, 1281).

H₂N O MeOOC

N COOMe R N

R R¹ COOMe R¹ O N O

(27) (28)

Under carefully controlled experimental conditions involving the simultaneous addition of 1-aminobenzotriazole and lead (IV) acetate to 4-phenyloxazole, the benzyne-oxazole adduct (29) can be isolated. The tan solid, m.p. 105-110 °C, is stable, but in solution thermolysis readily leads to the isobenzofuran through a retro-Diels-Alder elimination of benzonitrile. The adduct rearranges to 3-phenyl-4-isoquinolinol under mildly acidic conditions (S.E. Whitney and B. Rickborn, J. org. Chem., 1988, 53, 5595).

(29)   (30)

Endoperoxides (30) result from the photochemical [4+2]-cycloaddition of singlet oxygen to oxazole and its alkyl and aryl derivatives. The structural assignment is based on an analysis of the ^{13}C nmr spectra (K. Gollnick and S. Koegler, Tetrahedron Letters, 1988, 1003).

The reaction of oxazoles with ozone is complex, since not only do the products obtained depend on the isolation technique but also the reaction exhibits varying stoichiometry. Addition of ozone to both double bonds of 2,5-diphenyloxazole leads to benzoic anhydride and isocyanic acid, or products derived from them, presumably by the mechanism shown. Mixed aromatic anhydrides can be obtained in this way (C. Kashima et al., J. chem. Soc., Perkin I, 1988, 529).

The reaction of oxazoles with ketenes depends on the substitution pattern of the heterocycle and on the nature of the ketene. Thus, with dichloroketene, oxazole and 4-methyloxazole yield the derived 2-chloroacetyloxazoles, presumably *via* the *N*-acyloxazolium ylids. When the 2-position is blocked, acylation at C-5 is accompanied by cycloaddition.

When 2-(*N*-benzyl-*N*-methylamino)-4,5-dimethyloxazole (31) is the substrate, the 2,5-dihydrofuran (32) is the product, arising from ring opening of the initial addition product (A. Dondoni *et al.*, J. org. Chem., 1984, 49, 3478).

(31)　　(32)

2-Allyloxy-4,5-diphenyloxazole (34) undergoes a concerted [3,3]-sigmatropic shift on thermolysis to give the 3-allyl-4-oxazolin-2-one (35). Inversion of the allyl unit is observed with appropriately substituted allyloxy functions. However, no such inversion is seen during photolysis and it seems possible that this rearrangement to the 3-oxazolin-2-one (33) involves radicals derived from C-O bond scission (A. Padwa and L.A. Cohen, Tetrahedron Letters, 1982, 915; J. org. Chem., 1984, 49, 399).

(33)　　(34)　　(35)

2-Aminooxazoles react with isothiocyanates to give thioamides and thioureas in a ratio which is influenced by the 4-substituent in the oxazole. Thus, 2-aminooxazole with methyl isothiocyanate yields only the thiourea, whereas 2-amino-4-phenyloxazole gives the thioamide exclusively (G. Crank and H.R. Khan, Austral. J. Chem., 1985, 38, 447).

Nucleophilic displacement of the 5-fluorine atom from the fluorinated oxazole (36) occurs readily and various substituted 4-trifluoromethyloxazoles have been synthesised in this way (K. Burger, D. Hübl and K. Geith, Synthesis, 1988, 194).

(36)

5-Nitrooxazoles are accessible by displacement of halogen by nitrogen dioxide. Replacement of the 2- and the 4-halogen atom is also possible, but the yield of the corresponding nitro-compound is low (W.J. Hammar and M.A. Rustad, J. heterocyclic Chem., 1981, 18, 885).

The halogen in 2-bromomethyloxazoles is labile, a property which has potential in the synthesis of the virginiamycins (Y. Nagao, S. Yamada and E. Fujita, Tetrahedron Letters, 1983, 2291).

The fluorine in 2-(2-fluorophenyl)oxazoles is activated by the oxazole ring towards nucleophilic displacement. Treatment with an aryl Grignard reagent leads to a biphenyl derivative of oxazole and indeed to a terphenyl from 2-(2,6-difluorophenyl)oxazole. Subsequent conversion of the oxazole moiety into a carboxyl group, or its derivative, provides a new approach to unsymmetrical aryl-aryl couplings. The oxazole ring thus plays a dual role, acting as an activating group and as a masked carboxyl function (D.J. Cram, J.A. Bryant and K.M. Doxsee, Chem. Letters, 1987, 19).

The presence of an ester function normally interferes with the Sharpless asymmetric epoxidation of an allylic alcohol. However, a 4,5-diphenyloxazole unit is compatible with this process, provided it is sufficiently remote from the reaction site. The oxazole unit can subsequently be converted into an ester group and the formation of the epoxide (37) with better than 99% enantiomeric excess further illustrates the value of the oxazole system as a masked ester function (L.N. Pridgen, S.C. Shilcrat and I. Lantos, Tetrahedron Letters, 1984, 2835).

(37)

There are further examples of the lithiation of oxazoles at C-2 and subsequent reaction of the organolithium species with electrophiles. The aroylation of 2-lithio-5-phenyloxazole with N-methyl-N-(2-pyridinyl)-4-cyanobenzamide is illustrative (L.N. Pridgen and S.C. Shilcrat, Synthesis, 1984, 1048).

The transition-metal phosphine complex catalysed coupling of Grignard reagents with 2-methylthiooxazoles provides a route to 2-alkyl and 2-aryl oxazoles (L.N. Pridgen, Synthesis, 1984, 1047). The air and moisture sensitive trimethylstannane derivatives of both oxazoles and oxazolines undergo a palladium catalysed cross coupling with aryl and heteroaryl halides in high yield (A. Dondoni et al., Synthesis, 1987, 693).

Oxazoles may be converted into their 2-trimethylsilyl derivatives through isomerisation of the α-isocyanosilyl ethers (38) which involves a 1,5-shift of the silyl group from oxygen to carbon.

The stable trimethylsilyloxazoles function as 2-oxazolyl anions and react with electrophiles such as aldehydes, acid chlorides, activated ketenes and heteroaryl cations (Scheme 2) (A. Dondoni et al., J. org. Chem., 1987, 52, 3413).

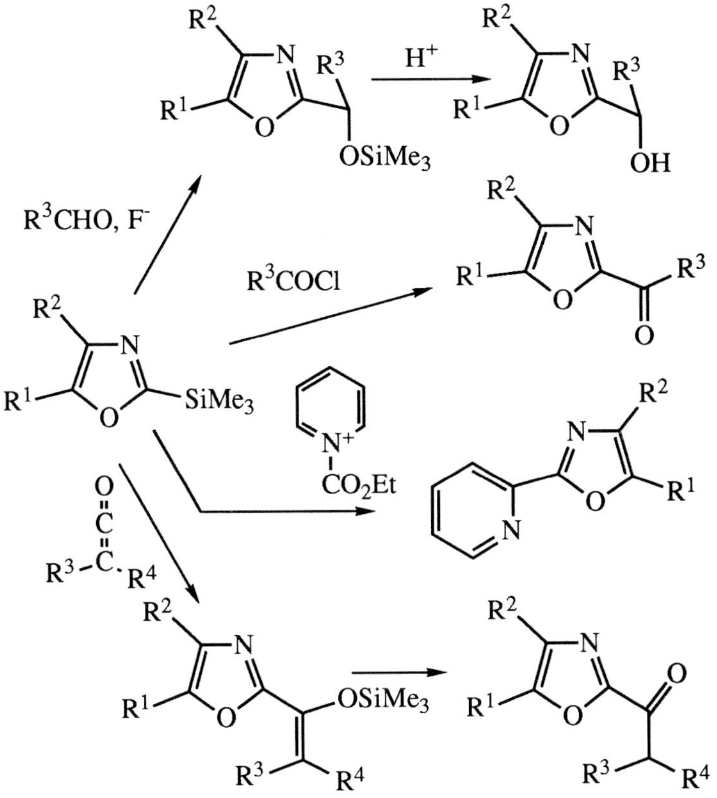

Scheme 2

Attempts to metallate the methyl group in 2-methyloxazole-4-carboxylic acid results in metallation of the oxazole ring at the 5-position. The directing influence of the carboxyl function has been confirmed by the deuteriation of the metallated oxazole-4-carboxylic acid (A. I. Meyers and J.P. Lawson, Tetrahedron Letters, 1981, 3163).

The directing influence of the carboxyl group does not overcome the effects of the heteroatoms during the metallation of 2,4-dimethyloxazole-5-carboxylic acid by either butyl lithium or lithium diisopropylamide and only attack at the 2-methyl group is observed. A similar reaction with 2,5-dimethyloxazole-4-carboxylic acid is not regiospecific and metallation occurs at both methyl groups. However, the corresponding *N,N*-diethylamide is metallated exclusively at the 5-methyl group (P. Cornwall, C.P. Dell and D.W. Knight, Tetrahedron Letters, 1987, 3585).

(b) Reduced systems

New reagents have been reported for the conversion of *N*-acyl derivatives of α-aminoalcohols into 2-oxazolines. Phosphorus reagents are effective under mild conditions (D.M. Roush and M.M. Patel, Synth. Commun., 1985, 15, 675; C. Sund, J. Ylikoski and M. Kwiatkowski, Synthesis, 1987, 853). Polyphosphoric ester has been used in the preparation of 2-styryloxazoline-4-carboxylic acid derivatives, which exhibit potent hypoglycaemic activity (K. Meguro *et al.*, Chem. pharm. Bull., 1986, 34, 2840). Tin (II) octanoate and toluene-*p*-sulfonic acid have been used to prepare some 2-(α-hydroxyalkyl)oxazolines (Scheme 3) (L.N. Pridgen and G. Miller, J. heterocyclic Chem., 1983, 20, 1223). Amide acetals convert α-hydroxyaminoalcohols into oxazoline *N*-oxides, which are nitrones trapped as the cycloadduct and as such react with DMAD to give a reduced oxazoloisoxazole. Acid treatment affords the trisubstituted isoxazole (S. P. Ashburn and R. M. Coates, J. org. Chem., 1985, 50, 3076).

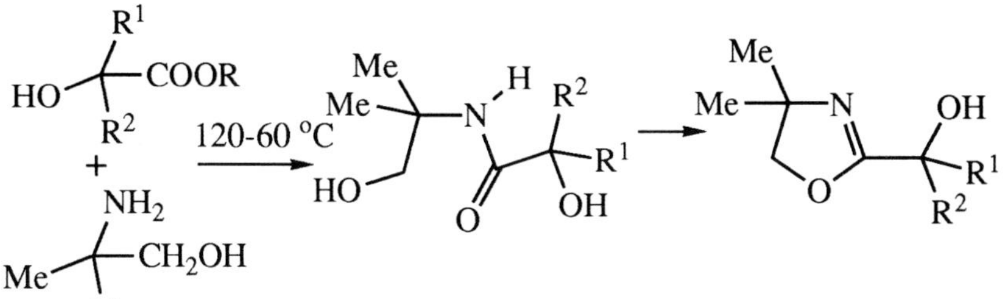

Scheme 3

N-Acyl derivatives of aminoacids undergo a similar cyclisation to yield the oxazolin-5-ones, a source of diacylamines through a base catalysed oxygenative decarboxylation (K. Suda, F. Hino and C. Yijima, Chem. pharm. Bull., 1985, 33, 882).

The reaction of Grignard reagents and organoaluminium compounds with iminoethers provides a new route to oxazolines (M. Bellassoued *et al.,* Bull. Soc. chim. Belg., 1986, 95, 65).

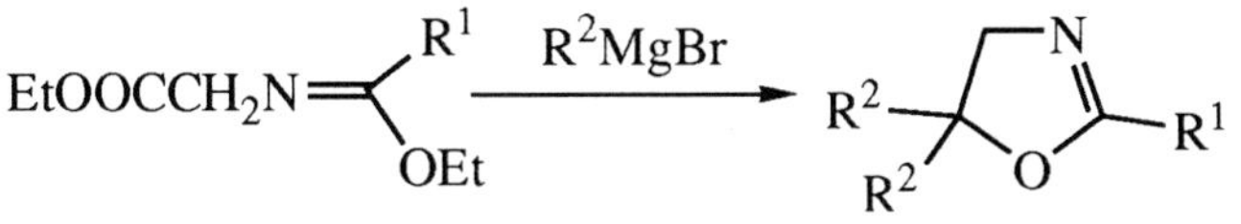

The fused oxazoline (39) can be obtained indirectly from cyclohexene. The addition of a nitrogen nucleophile to the alkene is assisted by initial reaction with the electrophilic dimethyl(methylthio)sulfonium fluoroborate. Subsequent alkylation at sulfur creates a good leaving group to facilitate cyclisation to the oxazoline. The overall process is the equivalent of *cis*-hydroxyamination of the alkene (B.M. Trost and T. Shibata, J. Amer. chem. Soc., 1982, 104, 3225).

(39)

The synthesis of oxazolines from *vic*-iodoisocyanates is quite similar in approach (R.C. Cambie *et al.*, Austral. J. Chem., 1983, <u>36</u>, 2569).

Formation of the oxazolidinone from a substituted cyclohex-2-en-1-ol by iodocyclisation enables hydroxyamination of the alkene to be achieved in either orientation (S. Knapp and D.V. Patel, J. org. Chem., 1984, <u>49</u>, 5072).

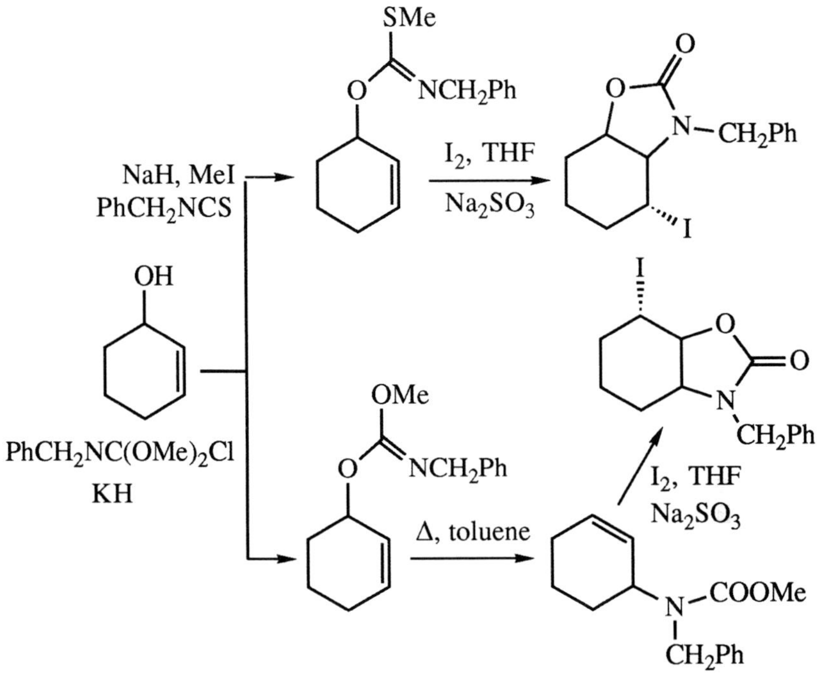

Activation of alkenes towards nucleophilic attack by a nitrile can be achieved using the electrophilic palladium complex $Pd(CH_3CN)_4(BF_4)$. Intramolecular trapping of the generated nitrilium ion by an unsaturated alcohol leads, as shown below, to a 2,4-disubstituted oxazoline (L.S. Hegedus, T.A. Mulhern and H. Asada, J. Amer. chem. Soc., 1986, 108, 6224).

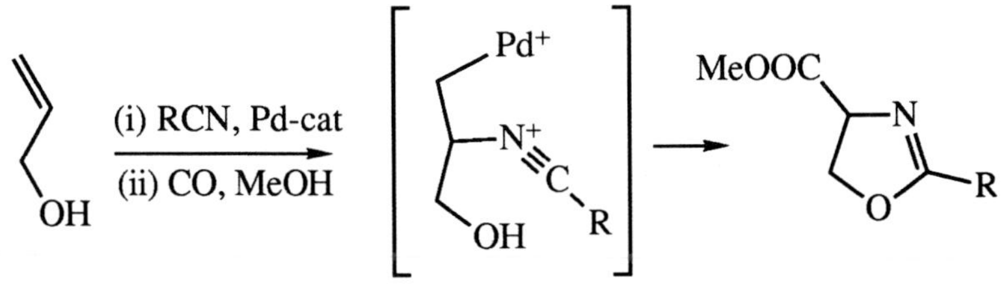

2-Oxazolines and 2-oxazolidinones can be obtained from alkenes by an organotellurium-mediated reaction with acetonitrile and ethyl carbamate, respectively. The BF_3-catalysed amidotelluration is a simple direct route to *cis*-2-aminoalcohols (N.X. Hu *et al.*, Chem. Comm., 1987, 1447; Tetrahedron Letters, 1988, 1049).

The reaction of a 1-alkoxy-2,3-epoxypropane with sodium cyanamide is a neat route to a 2-aminoxazoline (C. Jarry and R. Golse, Ann. Pharm. Fr., 1985, 43, 183).

The use of isocyanides in oxazoline synthesis has increased appreciably since the initial use of tosylmethyl isocyanide (tosmic) for this purpose, much of the interest stemming from the facile oxazole ring opening to amino acids. Isocyanoacetates undergo aldol reactions with aldehydes and ketones to form the acyclic oxazoline precursor. Zinc chloride or CuCl/Et₃N promote this reaction with α,β-unsaturated carbonyl compounds (Y. Ito, T. Matsuura and T. Saegusa, Tetrahedron Letters, 1985, 5781), whilst chiral ferrocenylphosphine-gold (I) complexes catalyse the reaction with aldehydes to give optically active oxazolines with high enantio- and diastereo-selectivity (Y. Ito *et al.*, J. Amer. chem. Soc., 1986, 108, 6405; Tetrahedron, 1988, 44, 5253). α-Chloroketones react with ethyl isocyanoacetate in the presence of calcium, giving eventually β,γ-unsaturated α-aminocarboxylic acids (F. Heinzer and D. Bellus, Helv., 1981, 64, 2279.

Chiral derivatives of tosmic, such as (40), may be converted into an oxazoline and thence into an optically active aldehyde (F.J.A. Hundscheid et al., Tetrahedron, 1987, 43, 5073).

(40)

N-Silylmethylthioimidates on reaction with fluoride ion form the 2-azaallyl anion which can be trapped with carbonyl compounds to yield 2,5-disubstituted oxazolines (O. Tsuge et al., J. org. Chem., 1987, 52, 2523; A. Padwa et al., ibid., 1027).

The dihydrooxazolium salts resulting from the reaction of 3-amido-propynes with methyl(bismethylthio)sulfonium hexachloroantimonate are readily converted into the 5-alkenyloxazolines (G. Capozzi, G. Caristi and M. Gattuso, J. chem. Soc., Perkin I, 1984, 255).

The key step in the synthesis of the 4-oxazoline (42) from serine and pivaldehyde is the electrochemical conversion of the carboxyl group in (41) into an acetoxy function. Loss of acetic acid from the intermediate oxazolidine completes the sequence (D. Seebach, G Stucky and P. Renaud, Chimia, 1988, 42, 176).

(41)   (42)

An improved synthesis of 4-oxazolin-2-one is based on the electrochemical hydroxylation of oxazolidin-2-one and subsequent elimination of methanol from the derived 4-methoxyoxazolidin-2-one. The N-formyl derivative results when this methoxy compound is treated with acetic formic anhydride (P.C. Wang, Heterocycles, 1985, 23, 2237; D. Tavernier et al., Bull. chim. Soc. Belg., 1988, 97, 859). A detailed preparation of the N-acetyl derivative from ethanolamine has been described and some cycloaddition reactions of the product have been reported (K.H. Scholz, H.G. Heine and W. Hartmann, Org. Synth., 1984, 62, 149).

Ring opening of aziridines with carbonyl compounds gives rise to oxazolidinones. Organoantimony and organotin compounds in a polar aprotic solvent catalyse the specific reaction between an aziridine and carbon dioxide (H. Matsuda, A. Ninagawa and H. Hasegawa, Bull. chem. Soc. Japan, 1985, 58, 2717).

The nitrile-ylide generated by photolysis of the 3-aminoazirine (43) undergoes a [3+2]-cycloaddition with methyl trifluoroacetate to yield the oxazoline (44) (K. Dietliker and H. Heimgartner, Helv., 1983, 66, 262).

(43)

CF₃COOCH₃

(44)

Cycloaddition of a 1,1-difluoroalkene to the fluorinated oxaziridine (45) leads to the oxazolidine (B.A. O'Brien, W.Y. Lam and D.D. DesMarteau, J. org. Chem., 1986, 51, 4466).

(45)

Aryloxiranes are ring opened by some nitrogen nucleophiles to give a mixture of the cis-and the trans-oxazolidin-2-one (A. Huth and E. Frost, Ann., 1983, 261), but the reaction of vinyl epoxides with 2-methoxy-1-naphthyl isocyanate in the presence of palladium (0) as a catalyst exhibits good diastereoselectivity to give the cis-disubstituted oxazolidinones (46). The N-aryl function is readily removed with ceric ammonium nitrate (B.M. Trost and A.R. Sudhakar, J. Amer. chem. Soc., 1988, 110, 7933).

(46)

The cycloaddition of heterocumulenes to oxiranes is catalysed by organotin halides complexed with Lewis bases such as $Bu_3SnI-Ph_3PO$. Thus phenyl isocyanate and carbodiimides yield 2-oxazolidinones and 2-oxazolidinimines, respectively. The adduct generated from the reaction of $(Bu_3Sn)_2O$ and an ω-haloalkyl isocyanate is readily cyclised to the N-tributylstannyl-2-oxazolidinone. Insertion of the isocyanate into the Sn-O bond is followed by the thermal cyclisation of the carbamate. Displacement of the tributylstannyl unit occurs readily on treatment with an acyl or other activated halide (I. Shibata *et al.*, J. org. Chem., 1986, <u>51</u>, 2177).

An extension of this approach to oxazolidinones involves treatment of an alkene with iodine isocyanate followed by bis(tributyltin)oxide to afford the adduct (47). Cyclisation of the latter in HMPA involves an S_N2-like displacement of iodine by the oxygen function leading stereospecifically to the *cis*-4,5-disubstituted 2-oxazolidinone (I. Shibata *et al.*, Bull. chem. Soc. Japan, 1989, <u>62</u>, 853).

(47)

Diphenylsulfilimine is acylated by diphenylketene, but the product undergoes a second reaction with the ketene to yield the oxazolin-4-one, from which the corresponding oxazolidinone can be obtained by hydride reduction (D.M. Ketcha *et al.,* Tetrahedron Letters, 1983, 2811; J. org. Chem., 1985, 50, 2224).

Reagent: (i) Ph$_2$C=C=O

Various tosmic derived reagents yield oxazolidinones when they react with aldehydes or ketones through a fluoride-induced desilyation which generates an azaallyl anion (O. Tsuge, S. Kanemasa and K. Matsuda, Chem. Letters, 1984, 1827). Similarly, isothiocyanates afford oxazolidine-2-thiones (T. Hirao *et al.,* Bull. chem. Soc. Japan, 1982, 55, 1163). Azomethine ylides are formed when α-amino-acids undergo a decarboxylative condensation with carbonyl compounds. When the ylides are trapped by carbonyl dipolarophiles, oxazolidines and sometimes oxazolidinones are obtained (O. Tsuge *et al.,* Bull. chem. Soc. Japan, 1987, 60, 4079).

Ring contraction of a dioxazine to an oxazolidinone occurs on heating the former in the presence of amines (G. Schwarz and D. Geffken, Ann., 1988, 35) and thermolysis of a 7-aminoperhydro-1,4-oxazepine (48) gives an oxazolidine (H. Griengl, G. Prischl and A. Bleikolm, Ann., 1980, 1573).

(48)

The intramolecular insertion reaction of singlet nitrenoformates yields an oxazolidinone. For example, the nitrene generated from the azide (49) gives a mixture of the *cis-*(50) and the *trans-*(51) oxazolidinone, the composition of which depended on the reaction temperature and hence the conformational composition of the precursor (P.C. Marais and O. Meth-Cohn, J. chem. Soc., Perkin 1, 1987, 1553).

(49)   (50)

(51)

However, the intramolecular conjugate addition of a carbamoyloxy nitrogen to a chiral vinyl sulfoxide or sulfone proceeds with high diastereoselectivity. Treatment of the vinylsulfoxide (52) with base gives almost exclusively the *trans*-oxazolidinone (M. Hirama *et al.,* Tetrahedron Letters, 1988, 3121 and 3125).

(52)

Cyclisation of protected *t h r e o* - a m i n o a l c o h o l s $RCH(OH)CH(NHR)CH_2CH=CH_2$ to the *cis*-4,5-disubstituted oxazolidinone is effected by thionyl chloride. Thus the diastereo-conversion of a *threo*-2-aminoalcohol to the *erythro*-isomer can be effected *via* cyclocarbamation (S. Kano *et al.,* Heterocycles, 1988, 27, 1241).

Treatment of an *N*-protected 2-aminosulfide with *N*-chloro-succinimide provides a highly diastereoselective synthesis of a *trans*-4,5-disubstituted oxazolidinone. Subsequent photolytic allylation using allyl tributylstannane gives a single diastereoisomer (S. Kano, T. Yokomatsu and S. Shibuya, J. org. Chem., 1989, 54, 513).

Reagents: (i) NCS, CCl$_4$; (ii) SnCl$_4$

Vinyl ethers of alkanolamines are cyclised to 2-methyloxazolidines by mercury (II) acetate (B.F. Kukharev *et al.*, Khim. geterosikl. Soedin, 1986, 536) and *N*-protected norephedrines on treatment with the dimethylacetal of an α,β-unsaturated aldehyde cyclise to the isomeric 2-alkenyloxazolidines with *cis*-isomer predominating (A. Bernardi *et al.*, J. org. Chem., 1988, 53, 1600).

(-)-Phenylglycinol reacts with formaldehyde in the presence of cyanide ion to form (-)-*N*-cyanomethyl-4-phenyloxazolidine, mono-alkylation of which affords a mixture of the separable diastereoisomers (J.L. Marco, J. Royer and H.P. Husson, Tetrahedron Letters, 1985, 3567).

The nitrosation of substituted (2-hydroxyethyl)ureas leads to 3-nitroso-2-oxazolidinones, which can be denitrosated by treatment with methanolic hydrogen chloride. The reaction probably involves an isocyanate formed following the initial nitrosation of the primary amino group (M. Miyahara, M. Nakadate and S. Kamiya, Chem. pharm. Bull., 1985, $\underline{33}$, 497).

4. Benzoxazoles

(a) Synthesis

The synthesis of benzoxazoles by heating 2-aminophenol with carboxylic acids or their derivatives continues in use. The reaction of trifluoropyruvic acid hydrate with a 4-substituted 2-aminophenol gives a mixture of two 1,4-benzoxazin-2-ones and a 2,3-dihydro-2-trifluoromethylbenzoxazole. The nature of the substituent in the 4-position of the phenol influences the ratio in which the products are formed. When it is electron withdrawing, a larger proportion of the benzoxazole results, suggesting that it is predominantly the phenolic group which attacks the α-carbon of the pyruvic acid. The second benzoxazinone arises from initial attack by the amino group (M.E. Mustafa, A. Takaoka and N. Ishikawa, Heterocycles, 1986, $\underline{24}$, 593.

Boric acid catalyses the direct route to benzoxazoles from 2-aminophenols, possibly *via* a 1,3,2-benzoxazaborole, since it is known that 2-phenylbenzoxazole is formed when 2-phenyl-1,3,2-benzoxazaborole is heated in acetic acid (M. Terashima, M. Ishii and Y. Kanaoka, Synthesis, 1982, 484). 2-Benzoxazolinones result when 2-aminophenols react with 1,1-carbonyldiimidazole. The 6-methoxy derivative is apparently a reproductive stimulant for voles! (R.J. Nachman, J. heterocyclic Chem., 1982, 19, 1545).

The reaction of 2-aminophenol with a dithiocarbonimidate (53) leads to a 2-(2-benzothiazolylamino)benzoxazole (54) (J. Garin *et al.*, Synthesis, 1987, 368).

(53)　　　　　　　　　　(54)

Both *N*-(2-hydroxyphenyl) -thioamides and -thioureas undergo cyclodesulfurisation on treatment with potassium superoxide to form a benzoxazole. The key intermediate appears to be the peroxysulfur species (55) (Y.H. Kim *et al.*, Chem. Letters, 1986, 1291; Heterocycles, 1989, 29, 213).

(55)

The formation of the 2-carbomethoxybenzoxazole from the reaction of 2-amino-5-methylphenol with dimethyl acetylenedicarboxylate may be rationalised as shown below (N. Kawahara and T. Shimamori, Heterocycles, 1986, 24, 2803). The same compound has been obtained by the reaction of the copper (II) complex of *o*-benzoquinone monoxime with DMAD (C.B. Castellani and R. Millini, J. chem. Soc., Dalton, 1984, 1461).

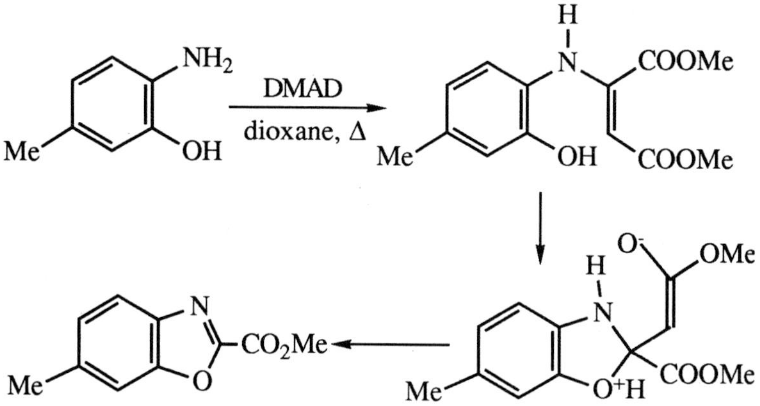

An improved synthesis of benzoxazoles utilises a mixture of phosphorus oxychloride and dimethylacetamide to effect the Beckmann rearrangement of oximes at room temperature. The method is particularly suitable when hydroxy substituents are present in the benzene ring (S. Fujita, K. Koyama and Y. Inagaki, Synthesis, 1982, 68).

The formation of benzoxazoles by photolysis of aromatic Schiff bases proceeds through the benzoxazoline. Aldehyde derived Schiff bases then absorb a second photon to give the benzoxazole. However, those derived from ketones are converted into the oxazole by the absorption of only one photon and a radical intermediate is proposed (E. Taner and K.H. Grellman, J. org. Chem., 1981, 46, 4252).

The oxazaphosphole (56) reacts with 4-nitrobenzaldehyde to give 2-(4-nitrophenyl)benzoxazole *via* a Schiff's base (J.I.G. Cadogan, J.B. Husband and H. McNab, Chem. Comm., 1981, 1054).

α-*N*-Diarylnitrones yield 2-arylbenzoxazoles on heating with *O*-methyl diphenylphosphinothioate, a radical derived from the Schiff's base being a likely intermediate. Substituted benzoxazoles are formed by the silver oxide oxidation of the related Schiff's base (M. Yoshifuji, R. Nagase and N. Inamoto, Bull. chem. Soc. Japan, 1982, 55, 873).

Oxidation of an appropriate Schiff's base to a 2-substituted benzoxazole can be effected by oxygen in the presence of copper (I) chloride in pyridine solution (G. Speier, J. mol. Catal., 1987, 41, 253).

Although the decomposition of azides in tetrachlorothiophene results in formation of an S,N-ylide, tetrabromothiophene is attacked at the α-position and rearrangement leads to benzoxazolone (O. Meth-Cohn and G. van Vauren, Chem. Comm., 1984, 190).

The carbonylation of 2-azidophenol catalysed by rhodium (I) complexes yields benzoxazolone quantitatively through the intermediacy of the isocyanate. A nitrene is not implicated (G. La Monica *et al.*, J. mol. Catal., 1986, <u>38</u>, 327).

Benzonitrile oxides afford 2-aryl-4,6-dihalobenzoxazoles on reaction with *S,S*-dimethyl-*N*-(2,4,6-trihalophenyl)sulfimides, possibly through the involvement of a nitroso intermediate (S. Shiraishi, T. Hayakawa and T. Shigemoto, Bull. chem. Soc. Japan, 1983, <u>56</u>, 1514).

Photolysis of 3-phenyl-2*H*-1,4-benzoxazin-2-one yields 2-phenylbenzoxazole (E. Tauer and K.H. Grellmann, Ber., 1986, <u>119</u>, 215) and 2-nitrophenyl alkyl ethers are photocyclised to 2-substituted benzoxazoles (S. Oguchi and H. Torizuka, Bull. chem. Soc. Japan, 1980, <u>53</u>, 2425).

The thiolactim (57) derived from a benzoxazepinedione yields a 2-substituted benzoxazole on reaction with morpholine. The reaction of the benzoxazepinone (58), or the analogous thione, with phosphorus pentachloride and subsequent treatment with a nucleophile results in ring contraction to 2-vinylbenzoxazole. The latter then undergoes a Michael reaction with the nucleophile (H. Bartsch and T. Erker, Ann., 1988, 795; 1989, 177).

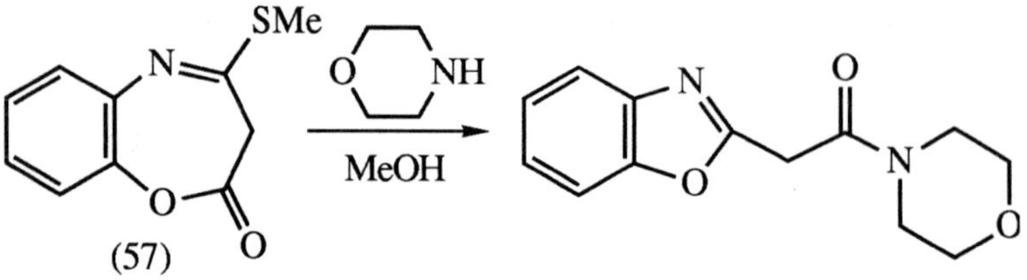

(57)

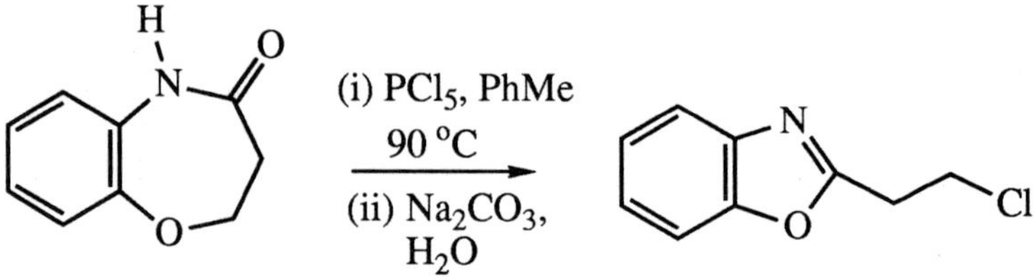

(58)

The anodic oxidation of 2,6-di-t-butylphenols in acetonitrile leads to generally good yields of 7-t-butyl-2-methylbenzoxazoles (E.-L. Dreher *et al.*, Ber., 1982, 115, 288) and a similar oxidation of N-methylcarbanilides in methanol gives N-methylbenzoxazolium salts (H. Ohmori *et al.*, J. chem. Soc., Perkin II, 1981, 1599).

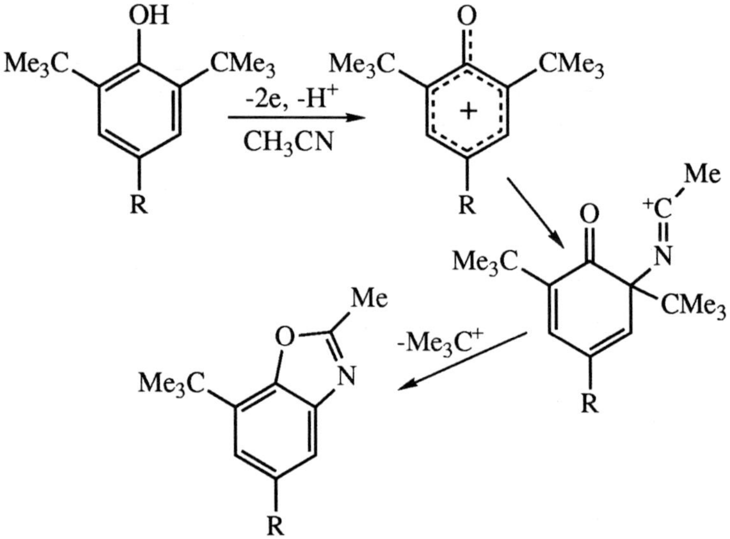

3-Substituted benzoxazolin-2-thiones result from the reaction of 2-alkylaminophenols with potassium *O*-methyldithiocarbonate. Reduction of Schiff's bases with sodium borohydride is a convenient route to the alkylaminophenol. The thiones may also be prepared from the corresponding benzoxazolinones by alkylation and subsequent treatment with phosphorus pentasulfide (M. Yamato *et al.*, Chem. pharm. Bull., 1983, 31, 1733).

Benzoxazolin-2-selenone is obtained from 2-aminophenol and carbon diselenide. The reaction proceeds through the isoselenocyanate (F. Cristiani *et al.*, Phosphorus and Sulfur, 1984, 20, 231).

(b) Reactions

The reactions of benzoxazolin-2-ones with electrophiles and nucleophiles have been reviewed (S.N. Ivanova *et al.*, in Reakts. Metody Issled. Org. Soedin, ed. I.L. Knunyants, N.N. Mel'nikov and V.D. Simonov, Khimiya, Moscow, 1983, p. 72) and a review of 2-aminobenzoxazoles covers their synthesis, reactions and pharmacology (A. Hetzheim and B. Hehmke, Wiss. Z. Ernst-Moritz-Arndt-Univ. Greifsw., 1985, 34, 81).

Metallation of benzoxazole with butyl lithium occurs at C-2 but is accompanied by the ring opened product, the isocyanide. The reaction of this mixture with electrophilic species is illustrated in Scheme 4 (P. Jutzi and U. Gilge, J. organomet. Chem., 1983, 246, 159).

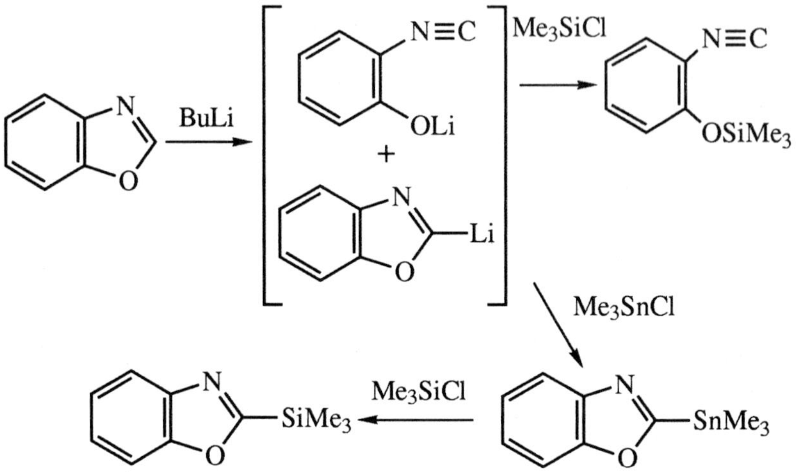

Scheme 4

2-Methylbenzoxazole is ring opened by allylic Grignard reagents (S. Florio *et al.*,Tetrahedron, 1984, 40, 5089), but undergoes a Claisen like self condensation with butylmagnesium bromide to give the enamine or ketone on work-up (S. Florio, G. Ingrosso and R. Sgarra, Tetrahedron, 1985, 41, 3091).

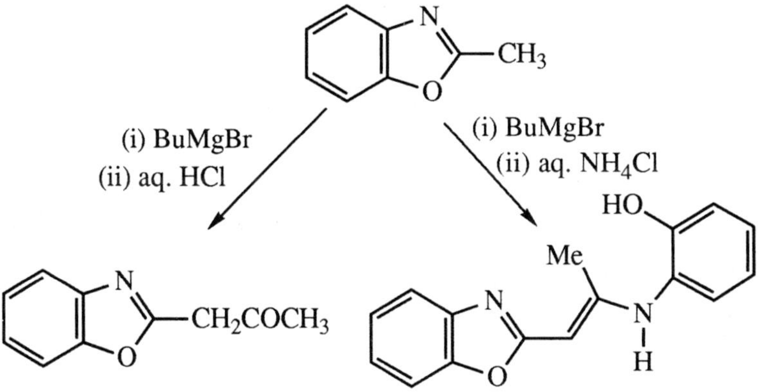

2-Allyl-, 2-allenyl- and 2-propargyl benzoxazoles can be formed by the reaction of the appropriate unsaturated Grignard reagent with

2-chlorobenzoxazole (F. Babudri, S. Florio and L. Ronzini, Tetrahedron, 1986, 42, 3905). The cross-coupling of 2-iodobenzoxazole with per-fluorovinylzinc species results in the formation of 1-(2-benzoxazolyl)-1,2,2-trifluoroethylene (J.P. Gillet, R. Sauvetre and J.F. Normant, Tetrahedron Letters, 1985, 3999).

Of the three 1,3-azoles, benzoxazole is the most reactive towards oxidation. Lead (IV) acetate attacks at C-2 and C-6 competitively and the benzoxazole is rapidly destroyed. Seven products can be isolated which are considered to arise from the 2- and 6-acetoxy derivatives (E.R. Cole, G. Crank and D. Sumantri, Austral. J. Chem., 1986, 39, 295).

When a solution of a benzoxazole (59; R = H, Cl, F, OMe) in hexane is poured into distilled water, the photoproduct (60) forms at the interface of the deoxygenated system. In the head-to-tail dimer arising from a [2+2]- cycloaddition of two C=N bonds, the azetidine ring is planar (N. Paillous et al., Chem. Comm., 1987, 578; J. org. Chem., 1986, 51, 672).

(59)　　　　　　　　　(60)

Benzoxazole reacts with trimethylsilylcyanide and acid chlorides to provide the first Reissert compounds in this ring system (B.C. Uff et al., Chem. Comm., 1984, 1245).

5.   Isothiazole and its derivatives

Isothiazoles are important in pharmaceutical research, where many biological activities have been ascribed to them, notably the anti-cancer properties of the fused pyrimidine systems.

(a) Isothiazoles

(i) Synthesis

A "one pot" method for the synthesis of isothiazoles involves the reaction of an acetylenic aldehyde or ketone with hydroxylamine-O-sulphonic acid and sodium hydrosulphide (F. Lucchesini et al., Heterocycles, 1989, 29, 97).

A similar synthesis uses a reactive methylene compound, an isothiocyanate and chloramine to afford a monosubstituted-diaminoisothiazole (1) (C.J. Shishoo et al., J. heterocyclic Chem., 1988, 25, 759).

$$EtO_2CCH(Na)CN \ + \ PhNCS \ + \ ClNH_2 \longrightarrow$$

(1)

Thiocarbamates react with bromine at room temperature to yield an isothiazole and also a benzothiazoline derivative (2) (H. Schaefer and K. Gewald, J. prakt. Chem., 1987, 329, 355).

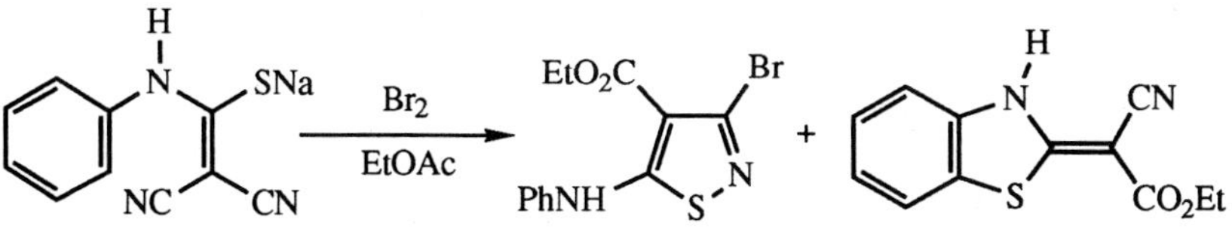

(2)

2-(Hydroxyimino)alkyl-*N,N*-dialkylthiocarbamates may be cyclised with tosyl isocyanate thus yielding a disulphide-bridged diisothiazole (M. Ishida *et al.*, Synthesis, 1987, 849).

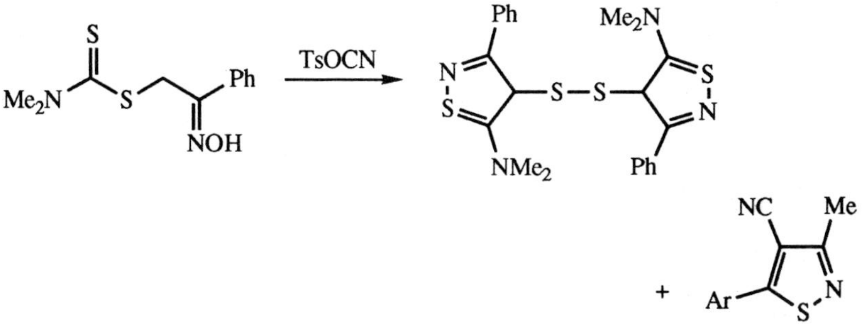

(3)

Isothiazoles (3) are produced by the electrochemical oxidation of alkylarene dithioates in acetonitrile using benzyltriethylammonium chloride as the supporting electrolyte (J. Voss *et al.*, Phosphorus and Sulfur, 1986, 27, 261).

Ring transformations may be utilised to furnish the isothiazole nucleus. An example of this is the extrusion of acetylene from an azidothiophene, the thermally-formed nitrene coordinating with sulphur and causing ring cleavage (C.J. Moody *et al.*, J. chem. Soc., Perkin I, 1984, 915).

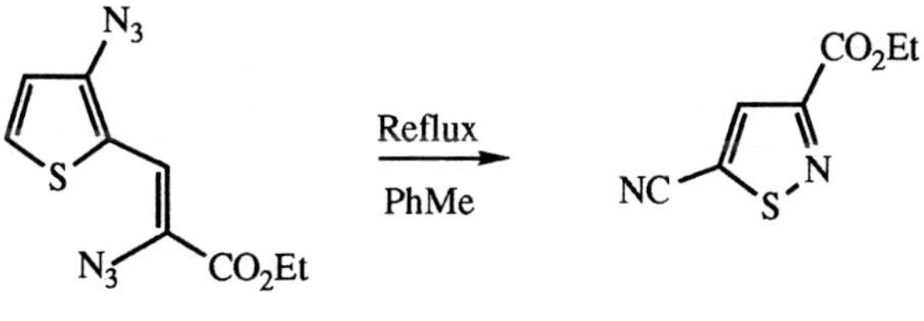

3-Hydroxy-4-phenyl-1,2,5-thiadiazole also undergoes ring transformation to an isothiazole *via* a ring-opened intermediate (S. Makata and K. Takahashi, J. heterocyclic Chem., 1985, 22, 1499).

Ring cleavage of a pyrazolo[4,3-c]isothiazole by an amine yields a 3,4-disubstituted isothiazole (T. Ueda, Y. Shibata and J. Sakikabara, J. heterocyclic Chem., 1986, 23, 1773).

Dihydrothiazepine *S*-oxides undergo a Pummerer reaction with sodium acetate and acetic anhydride to give a 3(2*H*)-isothiazolone (K. Yamamoto *et al.*, J. org. Chem., 1987, 52, 5239).

Mono-substituted 3(2*H*)-isothiazolones may be synthesised by a dearoylation reaction of a 5-aroyl-3(2*H*)-isothiazolone derived by cyclisation of an *N*-substituted propanamide (A. Tsolomitis and C. Sandris, Heterocycles, 1987, 25, 569).

$$PhCOCH_2CH_2CONHMe \xrightarrow{SOCl_2}$$

Fused isothiazolopyridines are prepared by cyclisation of dicyanopyridines (T. Zawisza and W. Malinka, Acta Pol. Pharma, 1987, 44, 32).

Isothiazole-1,1-dioxides may be obtained from chlorosulphonyl compounds. Thus, ethyl 2-(chlorosulphonyl)acetate is amidated and then condensed with diethyl oxalate. Subsequent chlorination and ethanolysis yields 3,4-diethoxyisothiazole-1,1-dioxide (Chem. Abs., 1983, 98, 160577).

The simpler isothiazolidine-1,1-dioxides are produced by the reaction between propane sultone and a primary aromatic amine, e.g. 4-chloroaniline (I.I. Ismail, E. El-Shereafy and A.H. Abd-Elalleim, Comm. Fac. Sci. Univ. Ankara, 1986, 32, 55).

Such compounds are saccharin analogues. The hexahydro derivative of saccharin (4) is synthesised by peroxide oxidation of 4,5,6,7-tetrahydrobenzisothiazole (N.B. Jones and K.W. Lumbaid, Tetrahedron Letters, 1983, 1647).

(4)

(ii) Reactions

Isothiazolidinones on chlorination followed by ring contraction give
β-lactams. The process may be analogous to that occurring in the *in vivo*
transformation of L-aminoadipoyl-L-cysteinyl-D-valine to isopenicillin
N (C.J. Easton, Chem. Comm., 1983, 1349; J. chem Soc., Perkin I, 1985,
153).

R= Ph, Me

Ring transformations not involving ring contraction include the
conversion of 5-aminoisothiazoles into 1,2,4-thiadiazoles (6). The amino
group allows coordination of the imino group to the ring sulphur after
initial reaction with an aromatic nitrile. Ring opening and cyclisation
ensue (K.Y. Akiba *et al.*, Phosphorus and Sulfur, 1983, 16, 111).

(5)

(6)

1,2,5-Thiadiazoles are produced in the reaction of 5-aroyl-2-aryl-3(2*H*)-
isothiazolones with phenylhydrazine. With hydroxylamine, 1,2,5-
thiaoxazoles result (A. Tsolomitis and C.J. Sandris, J. heterocyclic
Chem., 1984, 21, 1679).

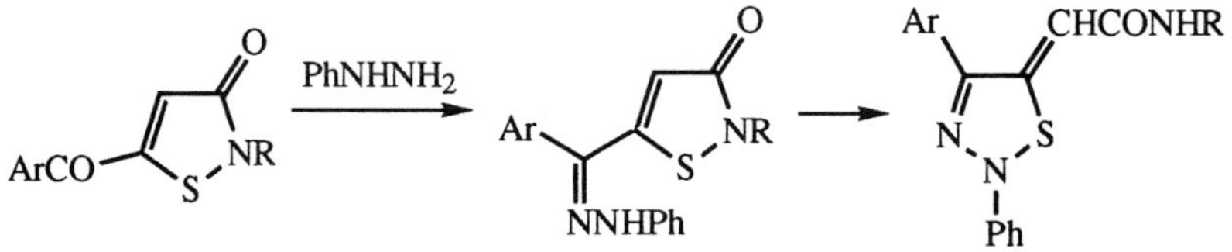

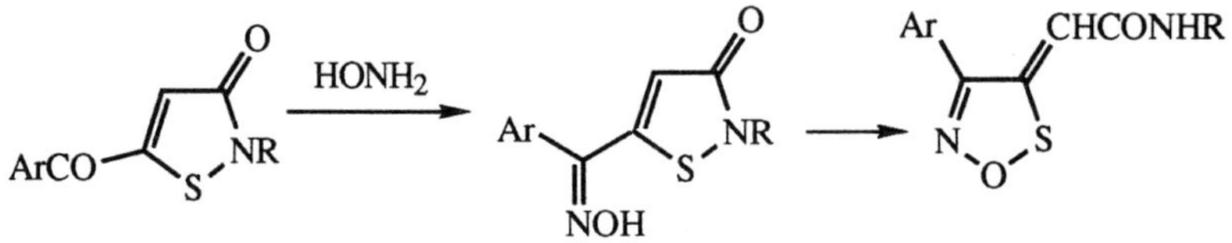

When the reaction with phenylhydrazine is carried out in refluxing ethanol, ring opening to the hydrazone (7) occurs. Cyclisation of the latter in aqueous acid results in the formation of the imidazolinone (8) (A. Tsolomitis and C. Sandris, J. heterocyclic Chem., 1984, 21, 1679; 1985, 22, 1465).

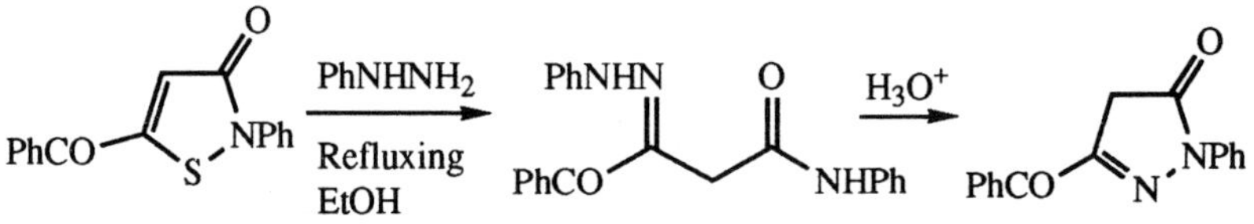

(7) (8)

Treatment of (9) with sodium cyanide in ethanol at 5 °C gives the tetra-substituted pyrrole (10) (L. Rajappa, Chem. Abs., 1983, 99, 212375).

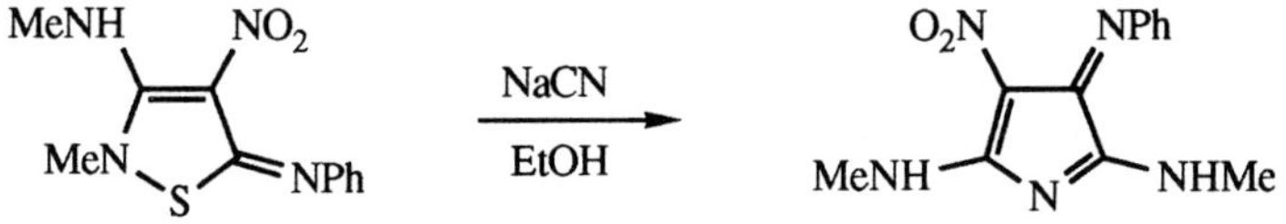

(9) (10)

Similarly, 2-phenyl-5-benzoylisothiazol-3(2H)-one reacts with dimethyl malonate and sodium methoxide at room temperature to form the hydroxypyrrolidone (R.J.W. Beer and D. Wright, Tetrahedron, 1981, 37, 3867).

The action of sodium ethoxide on 2-benzyl-5-mesitoyl-3(2*H*)-isothiazolone causes a rearrangement by the attack of the intermediate benzylic carbanion on the ring sulphur to give 2,3-dihydro-1,3-thiazin-4-one (A. Tsolomitis and C.J. Sandris, J. heterocyclic Chem., 1985, 22, 1635).

A benzylic function is also involved in the sigmatropic rearrangement of the isothiazolidinium species (11) to the benzothiazepine derivative (12) (S. Sato *et al.*, Heterocycles, 1984, 22, 1045).

(11)   (12)

The isothiazole system undergoes a [2+2]-cycloaddition with diphenylketene to yield a β-lactam (M.E. Hassan, Bull. Soc. Chim. Belg., 1985, 94, 149).

An amino group facilitates the introduction of halogens into the ring system. Non-aqueous diazotisation of an ester of 4-amino-3-

arylisothiazole-5-carboxylic acid is the first step in the synthesis of the 4-deamino-, 4-chloro-, 4-bromo-and 4-iodo-compounds (J.R. Beck and R.P. Gajewski, J. heterocyclic Chem., 1987, 24, 243).

X = H, Cl, Br, I

Diazotisation of 4-amino-3-phenylisothiazole-5-carboxaldehyde is the initial step in the synthesis of the triazino[5,6-d]isothiazole derivative (13). The latter has found use as a xanthine oxidase inhibitor (M.R. Attwood Chem. Abs., 1988, 109, 170465).

(13)

3-Aminoisothiazolo[5,4-d]pyrimidines (14) are converted into the novel thienopyrimidines (15) by ring opening with chloroacetone and subsequent cyclisation. Under similar conditions, the benzisothiazole analogue of (14) gives a tricyclic heteroaromatic compound (16) containing a bridgehead nitrogen, rather than the ring-opened product (J.M.C. Gore and R.M. Scrowston, J. chem. Res. (S), 1988, 46).

(14) → ClCH₂COMe → (15)

BrCH₂COPh → (16)

The Grignard reagent derived from 4-bromo-3,5-dimethylisothiazole reacts with allylic halogen derivatives to introduce an unsaturated group at C-4
(A. Alberola *et al.*, Synth. Comm., 1987, 17, 1207).

4-Isothiazolyl ketones may be prepared by treatment of the 4-acetonitrile derivatives with alkyl or aryl magnesium halides (A. Alberola *et al.*, Org. Prep. Proced. Int., 1988, 20, 377).

6. Benzisothiazoles

The benzisothiazoles, important in many commercial spheres such as pharmaceuticals and dyestuffs, are perhaps best known in the form of the S-dioxides, the saccharins. A new family of anti-rheumatic drugs, based on the 1,2-benzothiazole Piroxicam, have been synthesised from saccharins.

(a) Synthesis

5-Nitro-1,2-benzisothiazole (17) is formed by treatment of 2-chloro-5-nitrobenzaldehyde with sulphur and 25% aqueous ammonia in DMF (H. Hagen, Chem. Abs., 1982, 96, 68980).

(17)

The 7-chloro- and 7-nitro- derivatives are prepared in similar fashion (L.K.A. Rahman and R.M. Scrowston, J. chem. Soc., Perkin I, 1984, 385).

Condensation of a primary amine, e.g. 2-aminopentane, with $3H$-1,2-benzodithiole-3-thione yields a 2-substituted benzisothiazole-3(2H)-thione (P. Borgna *et al.*, Farmaco Ed. Sci., 1983, 38, 801).

The cycloaddition - elimination reaction between 1,2,5-thiadiazole and benzyne yields benzisothiazole (M.R. Bryce *et al.*, J. chem. Soc., Perkin I, 1988, 2141).

6,7-Diethoxy-4-phenyl-2H-1,3-benzothiazine on periodate oxidation undergoes ring contraction to give 5,6-diethoxy-3-phenyl-1,2-benzisothiazole (J. Szabo *et al.*, Tetrahedron, 1988, 44, 2985).

Benzothietane with benzylamine in boiling toluene, undergoes a ring expansion to yield 2-benzyl-2,3-dihydro-1,2-benzisothiazole (K. Kanakarajan and H. Meier, Angew. Chem., 1984, 96, 220).

The reaction of 2,2'-diaminodiphenylmethane with N-sulfinyl methanesulphonamide yields the 2,1-benzisothiazole (18). Methylation of the latter yields the 2,1-benzisothiazolo[2,3-b]-2,1-benzisothiazole (D.M. McKinnon and K.A. Duncan, J. heterocyclic Chem., 1988, <u>25</u>, 1095).

(18)

(b) Synthesis of benzisothiazole-1,1-dioxides

New methods for the industrial production of saccharin (19), probably the best known benzisothiazole sulphoxide, have been reported. The original preparation, the oxidative cyclisation of 2-carboxybenzenesulphonamide may be replaced by the oxidation of toluene-2-sulphonamide (Chem. Abs., 1983, <u>99</u>, 53744; 1984, <u>100</u>, 121054).

(19)

Treatment of 2-chlorothiobenzenecarbonyl chloride with 2,4-dinitroaniline gives a 1,2-benzisothiazol-3(2H)-one derivative (20) which on peroxide oxidation yields the saccharin derivative (21), a selective protease inhibitor (H. Jones, Chem. Abs., 1981, <u>95</u>, 203929).

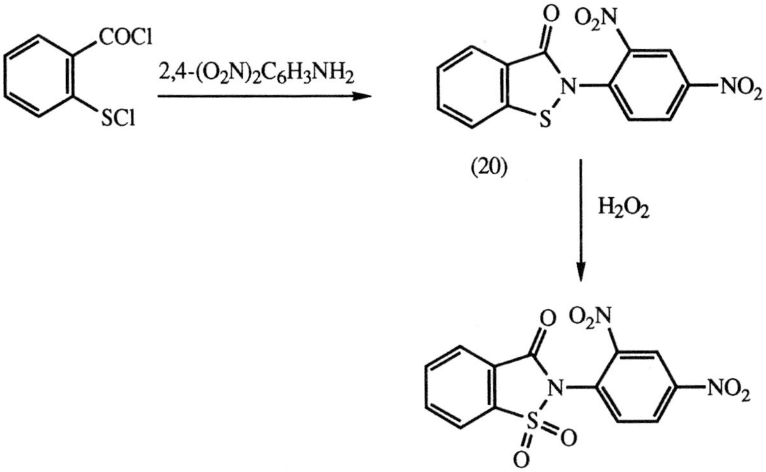

(20)

H₂O₂

(21)

The 3-deoxo-saccharin (22) is formed on oxidation of 6-nitrotoluene-2-sulphonamide with chromium (VI) oxide in acetic acid, followed by sulphuric acid cyclisation. The use of the selective oxidant avoids the formation of the 3-carbonyl function (R.A. Abramovitch, B. Mavunkel and J.R. Stowers, Chem. Comm., 1983, 520).

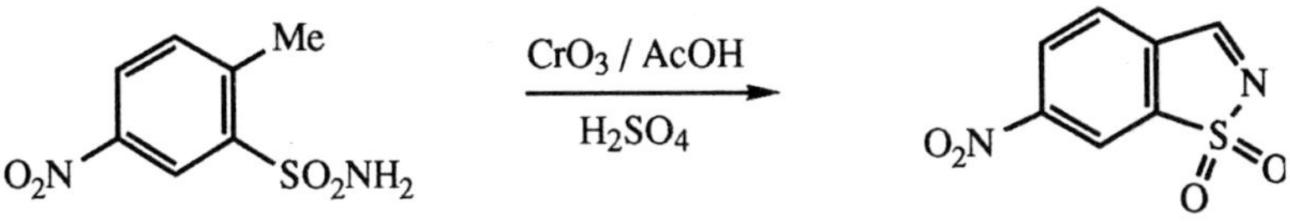

(22)

2,1-Benzisothiazoline-2,2-dioxides (23) may be prepared by cyclisation of either (2-chlorophenyl)methanesulphonamide or sodium 2-anilino methanesulphonate (D. Chiarino and A.M. Contri, J. heterocyclic Chem., 1986, 23, 1645).

(23)

(c) Reactions

3-Chlorobenzisothiazoles are important starting materials in several syntheses. For example, ammonolysis of (24) with dry gaseous ammonia in formamide at elevated temperatures gives the 3-amino analogue (25) (A.B. Korzhenevskii, Chem. Abs., 1988, 109, 110418).

(24)   (25)

Treatment of (24) with a substituted phenylacetonitrile followed by oxidative decarboxylation gives a 3-(substituted benzyl)-1,2-benzisothiazole (26) (F. Bordi et al., Acta Nat. Ateneo Parmence, 1983, 19, 35; G. Morini et al., Farmaco Ed. Sci., 1983, 38, 794).

3-Amino-1,2-benzisothiazole (25) reacts with bromoacetone in refluxing alcohol to yield imidazo[1,2-b]-1,2-benzisothiazole (27). The pyrimidino analogue (28) is formed from (25) and pentane-2,4-dione in trifluoroacetic acid (V.A. Chuigule and E.L. Komar, Khim. geterosikl. Soedin, 1983, 1134).

(26, R, R^1 = H, Cl, OMe)   (27)   (28)

3-Cyano-1,2-benzisothiazole is converted into the antipyretic 5-(1,2-benzisothiazolyl)tetrazole (29) on treatment with sodium azide and ammonium chloride in DMF (P. Vicini *et al.*, Farmaco Ed. Sci., 1986, <u>41</u>, 111).

(29)

2,1-Benzisothiazole salts (30) are susceptible to nucleophilic attack. Stabilised carbanions condense at the 3-position which is activated by the positively charged sulphur. Thus (30) on treatment with diethyl sodiomalonate yields (31) (D.M. McKinnon, K.A. Duncan and L.M. Millar, Canad. J. Chem., 1982, <u>60</u>, 440).

(30)   (31)

However, treating 2,1-benzisothiazolium salts such as (32) with ethyl cyanoacetate in pyridine gives 3-cyano-2-quinoxalones (33) (M. Davis and M.J. Hudson, J. heterocyclic Chem., 1983, <u>20</u>, 1707).

(32)   (33)

The substituted 1,2-benzisothiazol-3(2*H*)-one (34) is ring-opened with sodium methoxide in dimethyl sulphoxide and then cyclised to yield 1,2-benzothiazine-1,1-dioxide (35) (Patent, Chem. Abs., 1985, 103, 123493).

(34)   (35)

As with other members of the thiazole group, benzisothiazoles react with acetylenes to produce a wide variety of products. For example, 7-cyano-1,2-benzisothiazole reacts with dimethyl acetylenedicarboxylate (DMAD) to give the benzothiophene (36) (M. Sindler-Kulyk and D.C. Neckers, Tetrahedron Letters, 1981, 525), whereas 2,1-benzisothiazole with DMAD gives the quinolinecarboxylate ester (37) (M.R. Bryce, R.M. Acheson and A.J. Ross, Heterocycles, 1983, 20, 489).

(36)   (37)

N-Methyl-2,1-benzisothiazoline-3-thione (38) reacts with DMAD to give the *o*-iminobenzoquinone methide (39). The latter then undergoes either dimerisation, resulting in the diazocine (40), or further reaction with DMAD to give the spiro compound (41) (D.M. McKinnon, A.S. Secco and K.A. Duncan, Canad. J. Chem., 1987, 65, 1247).

3-Phenyl-1,2-benzisothiazole (42) does not react photochemically with DMAD but undergoes a photochemical cycloaddition with ethoxypropyne to give the thiazabicycloheptadiene (45), probably *via* the benzothiazepine (43) (M. Sindler-Kulyk and D.C. Neckers, J. org. Chem., 1983, 48, 1275).

(42)   (43)   (44)   (45)

However, with ethoxyethene the benzisothiazole (42) gives (44), a dihydro derivative of (43) (M. Sindler-Kulyk, D.C. Neckers and J.R. Blant, Tetrahedron, 1981, 37, 3377).

Irradiation of 2-phenyl-1,2-benzisothiazol-3-(2H)-one yields a dibenzothiazepinone, the reaction proceeding *via* a biradical intermediate (N. Kamigata *et al.*, Bull. chem. Soc. Japan, 1985, 58, 313).

(d) *Reactions of saccharin*

Saccharin and its analogues have an extensive chemistry associated with them, much of which is involved with their conversion into 1,2-benzothiazine-1,1-dioxides. However, many other reactions have been reported.

Saccharin reacts with 1-(diethylamino)-propyne to yield the cyclobutenyl saccharinate (46). The latter on bromination affords the bromide (47). However, N-methylsaccharin with the same alkyne yields the stable spirooxete (48) (R.A. Abramovitch *et al.*, Chem. Comm., 1984, 1583).

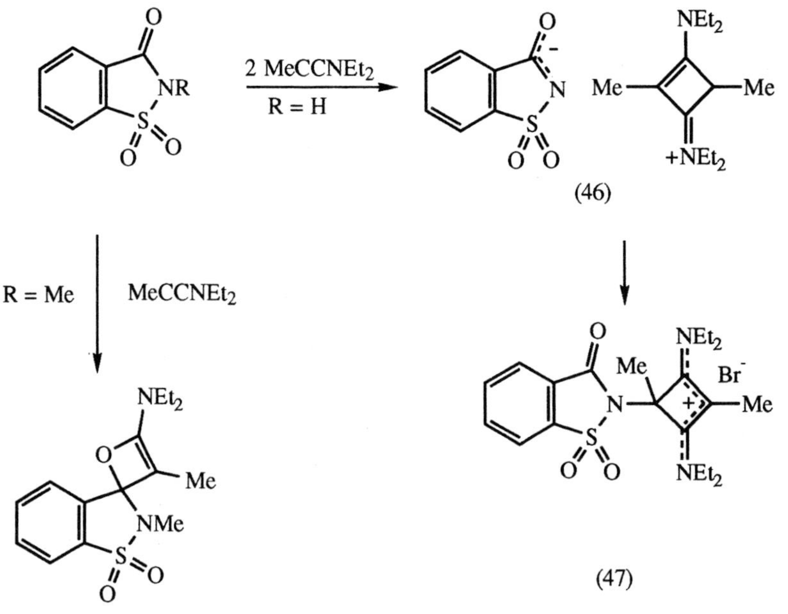

(46)

(47)

(48)

Since saccharin is a cyclic sulphonamide/carboxamide, it is possible for the heterocyclic ring to be cleaved. For example, a methanolic solution of the N-acetic acid, methyl ester derivative (49) cleaves easily across the C-N bond yielding the 2-sulphonamidobenzoic acid (50) when treated with sodium methoxide at room temperature (K. Unverferth, Chem. Abs., 1988, 108, 150061).

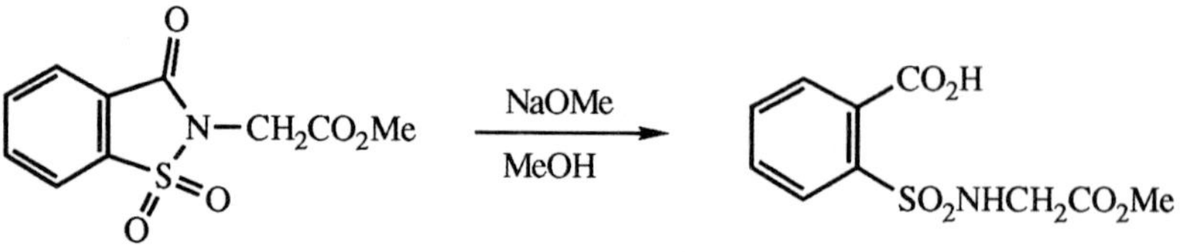

(49)   (50)

The reaction of the sodium salt of 1,2-benzisothiazole-3(2H)-thione-1,1-dioxide (thiosaccharin) (51) with an alkyl halide affords an alkyl thioether in high yield. Thus benzyl chloride gives (52) which with piperidine is cleaved to benzyl thiol (K. Inomata, H. Yamada and H. Kotake, Chem. Letters, 1981, 1457).

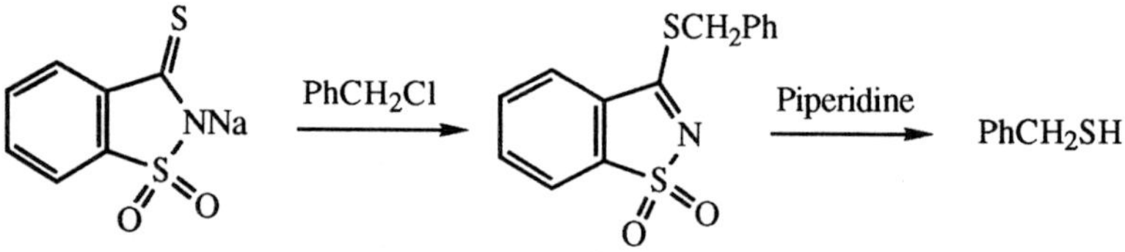

(51)   (52)

Thiosaccharin with diazomethane in ether-methanol yields the 3-methyl thioether (53). If ether alone is used as solvent, the N-methyl derivative (54) is formed which reacts further to give N-methylsaccharin (55) and the 3-methylene compound (56) (B. Unterhalt, F. Brunisch and J. Schweiger, Arch. Pharm., 1984, 317, 964).

(53)

(54)   (55)   +   (56)

The antiinflammatory drug Piroxicam [3-(2-pyridylamino)carboxy-2-methyl-4-hydroxy-2*H*-1,2-benzothiazine-1,1-dioxide] (61) is prepared from *N*-methylsaccharin or sodium saccharin. Thus, condensation of *N*-methyl saccharin with methyl chloroacetate gives the addition compound (57), which readily rearranges to the benzothiazine (60) Treatment with 2-aminopyridine yields (61) (P.D. Weeks *et al.*, J. org. Chem., 1983, 18, 3601).

Treatment of sodium saccharin (58) with methyl chloroacetate leaves the active group on C-2 (59). Drastic conditions are then required for ring expansion to (60) (J. He, Yiyao Gongye, 1987, 18, 531).

(See also: M. Potacek, Chem. Abs., 1989, 111, 232844; P. D. Weeks, Chem. Abs., 1983, 98, 198258, J. Svoboda, J. Palecek and V. Dedek, Coll. Czech. Chem. Comm., 1986, 51, 1133).

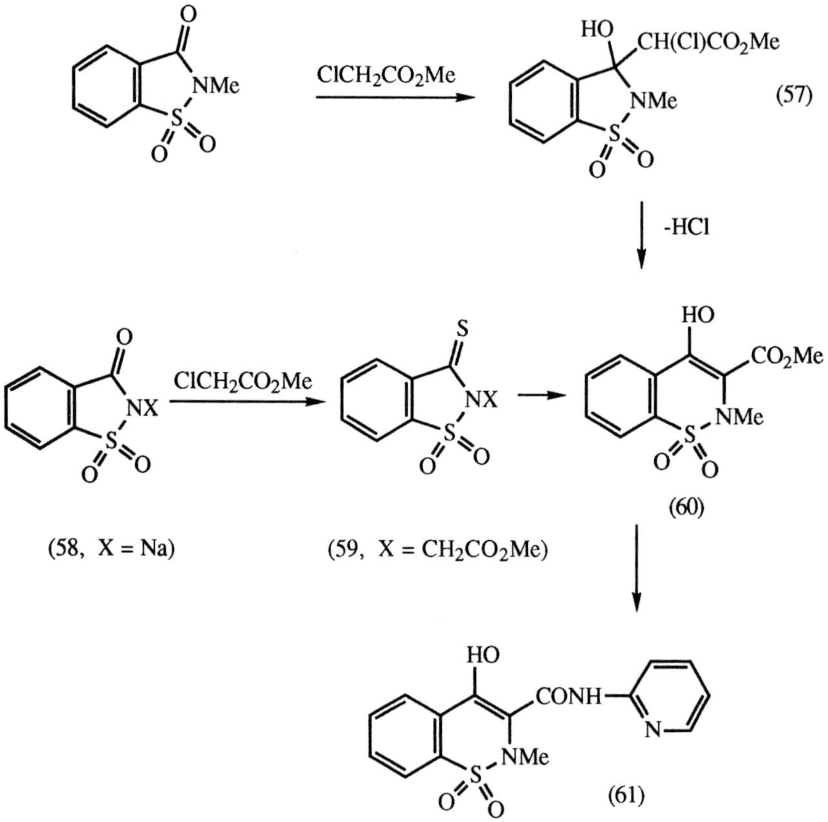

(57)

(58, X = Na)   (59, X = CH$_2$CO$_2$Me)

(60)

(61)

A related ring expansion to give the benzothiadiazine (62) occurs when the 3-thioxo-analogue (59) is treated with hydrazine. Reaction of (59) itself gives only the corresponding hydrazone (63) (U.V. Nakar, M.S. Mayadeo and K.D. Deodhar, Indian J. Chem., 1988, 27B, 109).

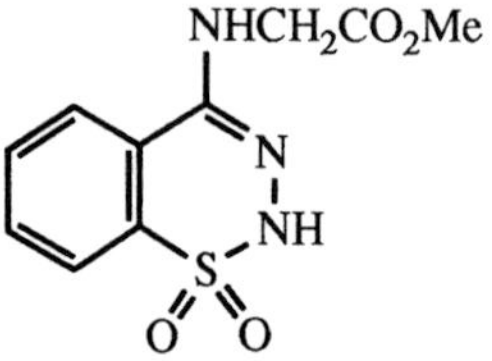

(62) 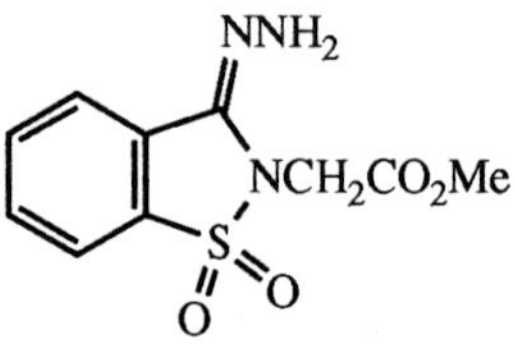   (63)

Saccharin and its derivatives may be used for the synthesis of β-lactams. Saccharin is acylated with, for example, phenoxyacetyl chloride and the N-acyl derivative (64) is treated with the N-phenyliminocinnamaldehyde to give (65), or with 3,4-dihydroisoquinoline to give β-lactam (66) (M. Miyake, N. Tokutake and M. Kirisawa, Synth. Comm., 1984, 14, 353).

7.   Thiazole and its Derivatives

Several reviews have been published in this area. These include the chemistry and biological activity of the 1,3-thiazolidinones (S.P. Singh *et al.*, Chem. Rev., 1981, 81, 175); the chemistry of the 1,3-thiazolinone system (G.C. Barrett, Tetrahedron, 1980, 36, 2023); synthetic methods in the preparation of condensed 4-thiazolidinones (H.K. Pujari, Adv. Heterocyclic Chem., 1990, 49, 3); the synthesis of 1β-carbapenems using C-4 chiral thiazolidine-2-thione derivatives (Y. Nagao, Kagaku (Kyoto), 1987, 42, 3) and the use of silylated thiazoles as heteroaryl anion equivalents in synthesis (A. Dandoni, Phosphorus and Sulfur, 1985, 24, 381).

(a) Thiazoles

(i) Synthesis

Further examples of the well-established route to thiazoles, N-C-S plus C-C, have been recorded. Thus the reaction of thiourea and ethyl 2-oximo-3-oxo-4-chlorobutanoate (1) yields the 2-aminothiazolyl system

(2) which is present in the side chain of several commercial cephalosporins (G. Kilper, Chem. Abs., 1981, 94, 103392; R. Heymes, Chem. Abs., 1981, 94, 208851).

$ClCH_2COC(NOH)CO_2Et$ $\xrightarrow{(H_2N)_2CS}$

(1)   (2)

Variations on this type of synthesis include the use of 2-imidazolidenethione as a cyclic thiourea to give a thiazoloimidazoline (3) (E. Campaigne and T. Selby, J. heterocyclic Chem., 1980, 17, 1255).

$\xrightarrow{ClCH_2COCH_2CO_2Et}$

(3)

5-Bromoalkyl-2-iminothiazolidin-4-ones (4) are obtained from *N,N*-disubstituted thioureas and 1,4-dibromoalkanoyl chlorides and may be converted into spiro-compounds (5) under phase transfer conditions (T. Okawara *et al.*, Chem. pharm. Bull., 1986, 34, 380).

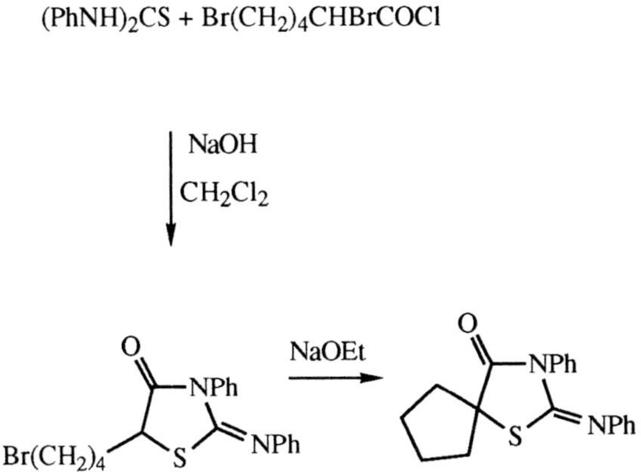
(PhNH)2CS + Br(CH2)4CHBrCOCl
NaOH
CH2Cl2
NaOEt
O
NPh
NPh
Br(CH2)4
S
O
NPh
S
NPh

(4) (5)

3-Arylprop-2-enethioamides with ethyl bromopyruvate yield a 2-styrylthiazole (6). Such compounds have analgesic properties (F. Bonina *et al.*, Farmaco Ed. Sci., 1985, 40, 875).

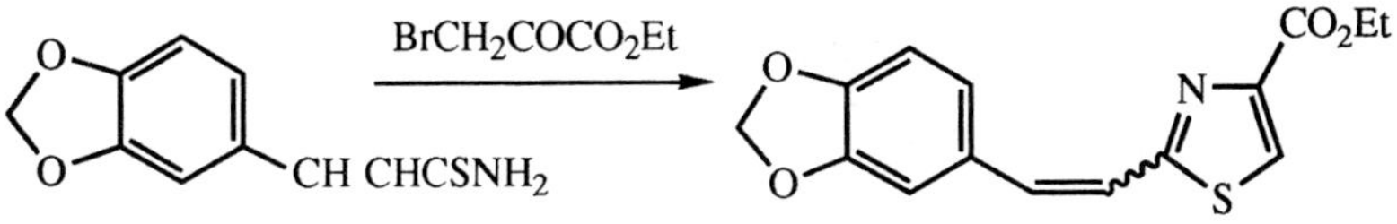
BrCH2COCO2Et
O
O
CH CHCSNH2
O
O
CO2Et
N
S

(6)

The N-C-S unit required for thiazole synthesis may be part of a heterocycle. Thus the thioimidazopyridine (7) reacts with phenacyl bromide yielding the thiazoloimidazopyridine (8) (A.M. Abdel Alim, M.A. Abdelkader and G.S. Alkaramany, Arch. Pharm., 1984, 12, 715). Thiazolo[3,2-*f*]xanthines (9) are similarly prepared (S.H. Gormash *et al.*, Khim. geterosikl. Soedin, 1987, 1534), as are thiazolo[4,3-*a*]isoquinolinium salts (Y. Kovtun and N.N. Romanov, *ibid.*, 1989, 553).

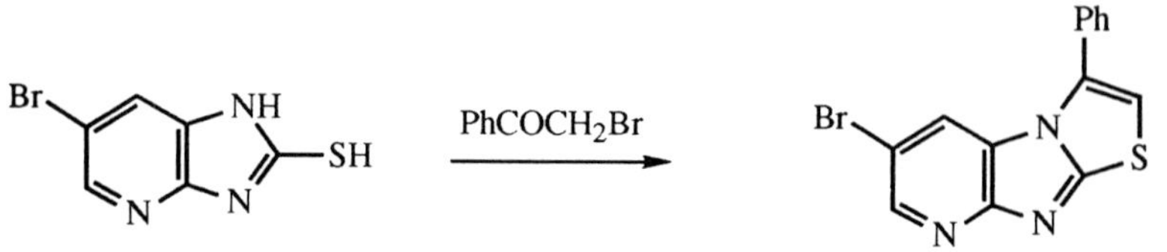

(7)   (8)

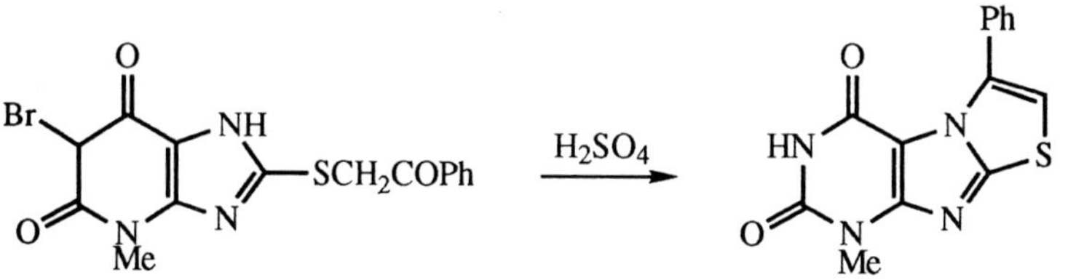

(9)

The reaction of 1-(α-chlorobenzyl)isoquinoline hydrochloride with thioacetamide affords the fused isoquinolinium (10) which is used in the preparation of cyanine dyes (Y.P. Kovtun, Khim. geterosikl. Soedin, 1989, 553).

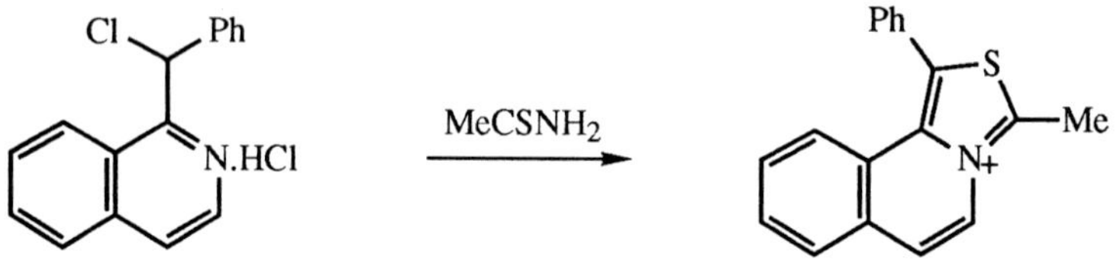

(10)

Treatment of methyl dithiocarbamate with chloroacetaldehyde gives the expected 2-methylthiothiazole (11). This may be demethylsulphenylated to thiazole by lithium in liquid ammonia (L. Brandsma, R.L.P. De Jong and H.D. Ver Kruijsse, Synthesis, 1985, 948).

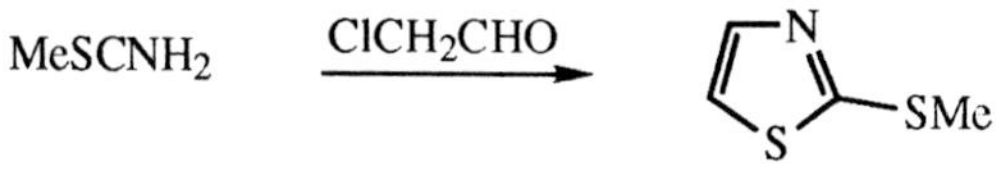

(11)

Several new thiazole syntheses utilise compounds with an *N*-substituted imino moiety. Thus, cyclisation of a thiocyanomethylketone with hydroxylamine hydrochloride yields a 2-aminothiazole *N*-oxide (12). The same compound arises from the oxidation of a 2-aminothiazole derivative with peracids (E. Perrone *et al.*, J. heterocyclic Chem., 1984, 21, 1097).

NCSCH₂COCH₂CO₂Et

+

HONH₂.HCl

(12)

1-Ethyl-1*H*-pyrrole-2-carboxaldoxime is dehydrated in acetic anhydride and subsequent treatment with hydrogen sulphide and cyclisation with an arylacetophenone gives the pyrrol-2-yl-thiazole (13) (K. Yoshino, Chem. Abs., 1986, 104, 129898).

(Ar = 4-MeOC₆H₄-)

ArCOCH(Br)Ar

(13)

Along similar lines, enehydrazines can be converted into thiazolones. Thus (14) reacts with chlorosulfphenylcarbonyl chloride to give the thiazol-2-one (15) which contains a single N-N bond (K. Grohe and H. Heitzer, Ann., 1982, 894).

(14)   ClSCOCl   (15)

Pyrimidinethiones contain the necessary groupings to furnish thiazolones and thiazolidinediones. Thus the reaction of (17) with a 2-haloketone in ethanol containing hydrochloric acid gives a thiazolone (16), whilst treatment of (17) with 2-chloroalkanoic acid yields a thiazolidine-2,4-dione (18) (H. Singh, P. Singh and K. Deep, Tetrahedron, 1983, 39, 1655).

(16)   (17)   (18)

3-Aryl-5-benzoyl-2,3-dihydro-2-imino-1,3,4-thiadiazole reacts with DMAD to give the sulphurane (20) by an intramolecular 1,3-dipolar addition involving the zwitterion (19), and thence the thiazole (21) by elimination of benzoyl cyanide (Y. Yamamoto et al., Bull. chem. Soc. Japan., 1989, 62, 211).

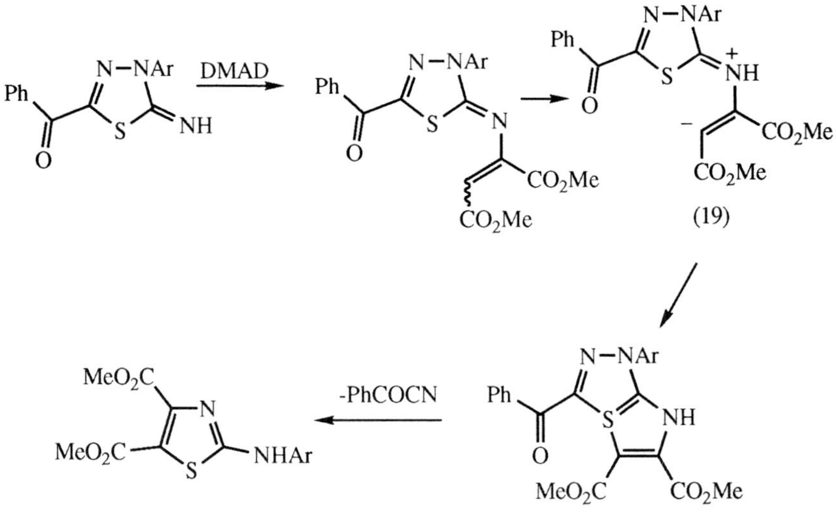

(19)

(21)   (20)

2,2-Dimethyl-$2H,5H$-thiazoline (22) results from the reaction of chloroacetaldehyde with ammonia, acetone and sodium bisulphide in aqueous acetone (J. Martens, H Offermans and P. Scherberich, Angew. Chem., 1981, 93, 680). The CN bond is susceptible to nucleophilic attack and so (22) on treatment with hydrogen cyanide and hydrolysis of the resulting thiazolidine (23) gives racemic cysteine.

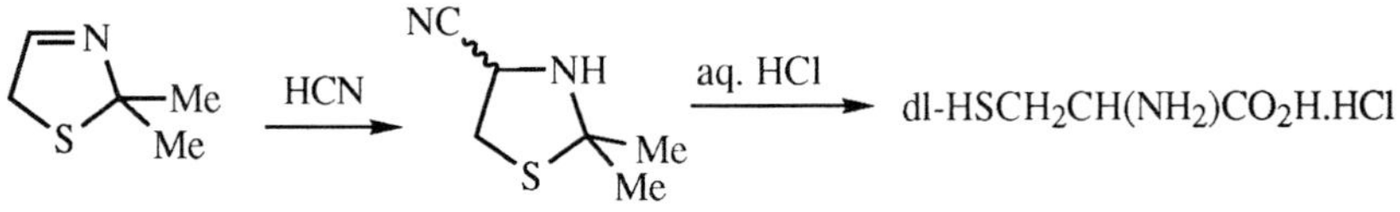

(22)   (23)

Another route to 3-thiazolines (26) involves the reaction of an ethyl 2-aryldiazo-3-aminobutenoate (24), prepared as shown below, with the dithiane (25) (J. Casteiger and V. Strauss, Ber., 1981, 114, 2336).

(24) (25) (26)

4,5-Dihydrothiazoles (28) are prepared from imidates and cysteamine. Thus, ethyl cyanoacetate is converted into the imidate (27) with ethanolic hydrogen chloride, and then heated with cysteamine. Compound (28) on reduction, hydrolysis and cyclisation with Mukaiyame-Ohno's reagent gives the β-lactam (29) (T. Chiba *et al.*, Chem. pharm. Bull., 1985, 33, 3046).

$$NCCH_2CO_2Et \xrightarrow[\text{EtOH}]{\text{HCl}} EtOC(NOH)CH_2CO_2Et \xrightarrow{HS(CH_2)_2NH_2}$$

(27)

(28)

NaBCNH$_3$
HCl / MeOH

(29)

Penicilloic acids, known β-lactamase inhibitors, are obtainable by ethanolysis of penicillanic acid or by cyclocondensation of D-penicillamine with ethyl 3-oxopropanoate. The acid (31) may also be synthesised from *t*-butyl 3-benzylaminoprop-2-enoate and D-penicillamine (30) (P.J. Claes, P. Herdewijn and H. Vanderhaeghe, Nouv. J. Chim., 1982, 6, 273).

$$Me_2C(SH)CH(NH_2)CO_2H \xrightarrow{PhCH_2NHCH:CHCO_2CMe_3}$$

(30) (31)

Treatment of benzyl penicillinate dioxide (32) with trifluoroacetic acid in deuteriated chloroform causes ring cleavage to yield the thiazolidine-1,1-dioxide (33) (P.H. Crackett and R.J. Stoodley, Tetrahedron Letters, 1984, 1295).

(32)   (33)

β-Lactam cleavage also occurs with diborane to give the corresponding amino alcohol (34) and the thiazolidine carboxylate (35) (P.G. Sammes, S. Smith and G.T. Wooley, J. chem. Soc., Perkin I, 1984, 2603).

(34)

(35)

The first synthesis of the basic skeleton of penicillins has been effected in three steps from cysteamine and ethyl propiolate. Cyclisation of thiazolidine-2-acetic acid to the β-lactam (36) is achieved using triphenylphosphine-2,2'-dipyridyl disulphide in acetonitrile (T. Chiba et al., Chem. Letters, 1985, 659).

$$HSCH_2CH_2NH_2 \xrightarrow{HC\equiv CCO_2Et}$$

(36)

A convenient preparation of penam (40) starts with the cyclisation of 2-iodoethyl isothiocyanate with di-*tert*-butyl sodiomalonate to afford di-*tert*-butyl 2-(thiazolidin-2-ylene)malonate (37). Hydrolysis of the latter and reduction of the monoester (38) gives (39) which is cyclised to (40) on treatment with carbodiimide (122) (R.C. Cambie *et al.*, Austral. J. Chem., 1985, 38, 745).

$$\text{(I-NCS)} \xrightarrow{NaCH(CO_2Bu^t)_2} \text{(37)} \xrightarrow{F_3CCO_2H} \text{(38)}$$

(37)

(38)

(40)

(39)

The 1-(alkenylthio)cyclopropyl azide (41) decomposes smoothly at 70 °C with evolution of nitrogen. The resultant nitrene interacts not only with the
3-membered ring to give thiazolidine (42), but also with the double bond yielding the thiazoline (43) (R. Jurritsma, H. Steinberg and J.J. De Boer, Recl. Neth. Chem. Soc., 1981, 100, 307).

(41)   (42)   (43)

Thiazolidinones and thiazolidinethiones are prepared by cyclocondensation of N-(2-thioethyl)hexylamine with carbon disulphide. The resultant thione (44) on treatment with phosgene and then sodium carbonate affords 3-hexylthiazolidinone (Patent, Chem. Abs., 1981, 95, 187239).

Spirocompounds (46) and (47), of interest for photochromic applications, may be prepared from diiminofurans (45) and thioacetic acid. The reaction is believed to proceed by 2,3-cyclocondensation at the β-amino group (L. Capuano and P. Moersdorf, Ann., 1982, 2178).

(45)   (46)   (47)

(ii) Reactions

Thiazole reacts with three equivalents of sulphur trioxide in 1,2-dichloroethane to give thiazole-5-sulphonic acid (49). Use of equimolar amounts of thiazole and sulphur trioxide yields only (48) (T.P. Bochkareva *et al.*, Khim geterosikl. Soedin, 1987, 1353).

4-Methyl-2-methylaminothiazole and dimethylaminothiocarbamoyl chloride gives a mixture of four products (50-53). It is suggested that the 2-methylimino-3-thiocarbamoylthiazoline (50) undergoes a proton-induced ring transformation to yield the 2-thiazolylidene derivative (53) (Y. Yamamoto *et al.*, Chem. pharm. Bull., 1984, *32*, 4292).

Thiazolyl hydrazines are useful synthetic intermediates often used for the synthesis of thiazolyl heterocycles. Thus 2-hydrazino-4-phenylthiazole reacts with pentan-2,4-dione to afford the 2-(dimethylpyrazolyl)-4-phenylthiazole (54) (S.P. Singh *et al.*, Indian J. Chem., 1986, 25B, 1054).

(54)

Similarly, 2-hydrazinyl-4,5-diphenylthiazole hydrochloride reacts with crotononitrile in ethanol containing sodium ethoxide to yield (55) (J. Uhlendorf, Chem. Abs., 1987, 107, 154328).

(55)

2-(Trimethylsilyl)thiazole (57) reacts with acyl chlorides and aldehydes. With the former, *ipso*-substitution of the silyl group occurs, yielding a 2-acylthiazole, (56), whilst aldehydes add to the C-Si bond, effectively giving an insertion product (58) (A. Medici *et al.*, Tetrahedron Letters, 1983, 2901).

Carbodesilylation of 2-(trimethylsilyl)thiazole by treatment with the half ester acid chloride of succinic acid, followed by bromination and cyclisation of the 2-acyl adduct (59) yields the (2,4'-bithiazol-5-yl)ethanoic acid (60) (A. Dondoni *et al.*, Gazz., 1986, 116, 133).

2-(Trimethylstannyl)thiazoles are prepared by quenching 2-lithiothiazoles with chlorotrimethylstannane. The 4- or 5-isomer may be prepared by blocking the 2-position with the TMS group. The reaction of 5-(trimethylstannyl)-thiazole (61) with iodine in THF at room temperature results in rapid iododestannylation to give 2-iodothiazole in 92% yield. Other halothiazoles may be similarly prepared. Treatment of 2-(trimethylstannyl)thiazole first with iodine in chloroform and subsequently with bromine affords 5-bromo-2-iodothiazole. Reversing the order of halogenation gives the 2-bromo-5-iodo isomer (A. Dondoni et al., Synthesis, 1986, 757).

(61)

2-(Trimethylstannyl)thiazole on heating with 2-bromothiazole and a 5% equivalent of tetrakis(triphenylphosphino)palladium (IV) in refluxing benzene yields the thiazole dimer (62). 2,5-Dibromothiazole yields the corresponding thiazole trimer (63) (A. Dondoni *et al.*, *Synthesis*, 1987, 185).

(62)   (63)

Thiazolium salts include thiamine, vitamin B₁ (64). The various reactions and ring transformations of the thiazolium system, and its use as an *in vitro* catalyst are of considerable interest.

(64)   (65)

Thiazolium salts for catalytic use are normally synthesised polymer bound from monomers such as (65) (K. Yamashita, H. Tokuda and K.

Tsuda, J. polymer Sci., 1989, 27, 1333; H.J. Van de Berg, G. Challa and U.K. Pandit, J. molec. Catal., 1989, 51, 13).

A novel approach to a thiazolium salt catalyst is to use the macrobicyclic thiazolium host (66e). This is prepared by diacylating the bisaminomethyl compound (66a) with a diester to give (66b). Debenzylation gives (66c) which with bromoacetyl bromide forms (66d). The latter with
4-methylthiazole gives (66e) (H.D. Lutter and F. Diederich, Angew. Chem., 1986, 98, 1125).

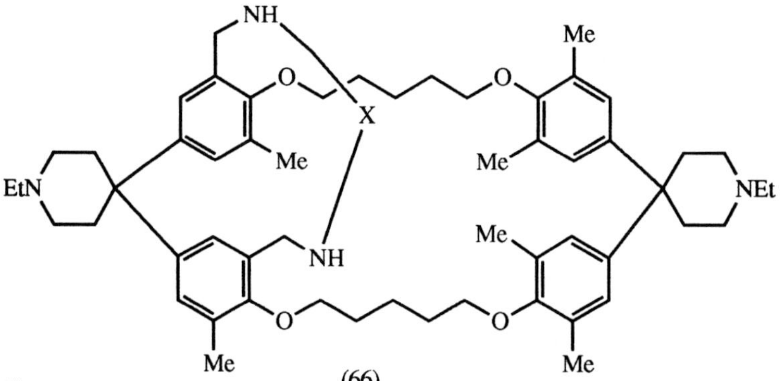
NH
Me
O
O
X
Me
Me
EtN
NEt
Me
NH
O
O
Me
Me

(66)

X =

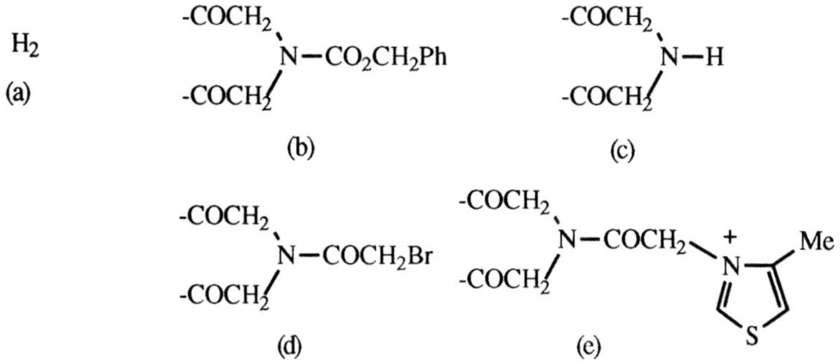
H2
(a)
-COCH2
N−CO2CH2Ph
-COCH2
(b)
-COCH2
N−H
-COCH2
(c)
-COCH2
N−COCH2Br
-COCH2
(d)
-COCH2
N−COCH2
+
N
Me
S
-COCH2
(e)

Thiamine itself is oxidised by cyanogen bromide under alkaline conditions to give the fluorescent compound shown below (T. Ogawa *et al.*, Bitamin, 1981, 55, 453).

Base-induced rearrangement of the thiazolium nucleus is possible. Hydroxide attack at the electrophilic C-2 of (67) yields (68). Rearrangement gives the thiolate and thence the thietane (69) (H.J. Federsel and G. Mereniji, J. org. Chem., 1981, 46, 4724).

Reaction of the novel 5,5'-bridged bis(thiazole) (70) with DMAD results in elimination of sulphur and formation of the phenylenebis(pyridone) derivative (71) (H. Gotthardt and W. Pflaumbaum, Ber., 1987, 120, 1017).

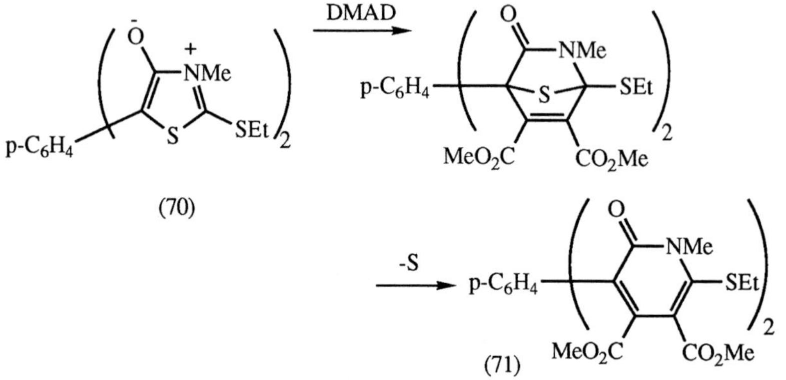

(70)

(71)

(b) Thiazolinethiones

The chemistry of 4-aminothiazolinethiones has been explored (K. Gewald *et al.*, Monatsh., 1981, 112, 1393). The thione group reacts with nucleophiles in a similar manner to the carbonyl group in a ketone. Thus, hydrazones, malononitriles and cyanamides can be formed which may undergo further reactions. The thiazolinylacetate (72) reacts with dimethyl sulphate and malononitrile to give the cyclised diaminopyrrolothiazole (73).

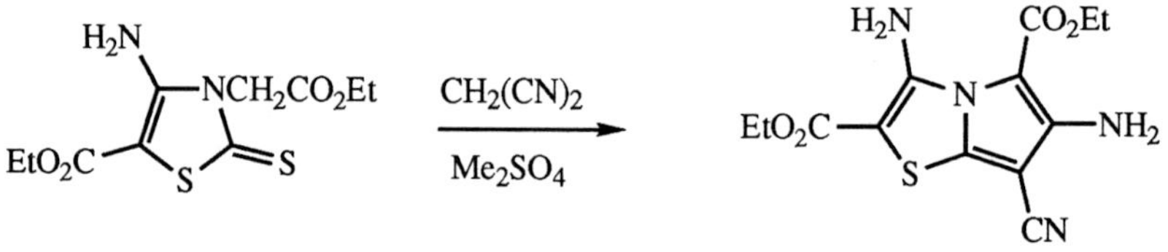

(72)   (73)

3,4-Disubstituted thiazoline-2(3*H*)-thiones may be metallated at C-5 as an intermediate step in their conversion into the 3,4,5-trisubstituted compound (74) (A.R. Katritzky, D. Winwood and N.E. Grzeskaviak, Synthesis, 1980, 800).

(74)

Thiazoline-5(4H)-thiones (76) react with organocuprates *via* carbophilic addition to the C–S bond to give dihydrothiazolethiols (75). However, if lithium di(*tert*-butyl) cuprate is used, the dihydrothiazolethiol (77) results (C. Jenny, P. Wipf and H. Heimgartner, Helv., 1986, 69, 773 and 1837). Addition reactions with thiazolethiones (76) may also be followed by cyclisation, leading to spiro-compound formation. The azadithiaspiro(4,4)nonadiene (78) arises from the reaction of (76) with diphenylcyclopropenone, and the diazadithiaspiro compound (79) is formed from (76) and 2,2-dimethyl-3-phenyl-2H-azirine (C. Jenny, D. Obrecht and H. Heimgartner, Tetrahedron Letters, 1982, 3059).

(75)   (76)   (77)

(78)   (79)

Other additions include that of DMAD, which with the 2-arylthiazole-5(4H)-thiones (76), yields the methylidenedithiole (80). Use of an excess of DMAD leads to formation of the thiophenethione (81). Addition of an ynamine to (76) yields the condensed thiazolylidene thioamides (83), presumably *via* the initial [2+2]-cycloaddition to form the intermediate thiete (82) (C. Jenny and H. Heimgartner, Helv., 1986, <u>69</u>, 174 and 419).

(80)   (76)   (81)

(76)   (82)   (83)

A ring transformation of the 2-thiazolethione (84) occurs under basic conditions to give the imidazole (85) (G. L'Abbe, W. Meatermans and M. Bruynseels, Bull. Soc. chim. Belg., 1986, <u>95</u>, 1129).

(84)   (85)

(c) Thiazolidines

Thiazolidines are useful precursors of aliphatic thiols. Simple ring opening with diborane yields a 2-(thioethyl)methylamine (86) (C. Melchiore, D. Giardina and P. Angeli, J. heterocyclic Chem., 1980, $\underline{17}$, 1215).

$$\text{(thiazolidine with -N(CH}_2)_6\text{NH}_2\text{)} \xrightarrow{\text{B}_2\text{H}_6} \text{HSCH}_2\text{CH}_2\text{NMe(CH}_2)_6\text{NH}_2$$

(86)

The 2-(alkylamino)ethanethiol (88) with a tertiary group on nitrogen is formed by the reaction of an allylic Grignard reagent with the thiazolidine (87). Diallylated (89) and triallylated (90) products may be prepared similarly from 2-alkylthiazolines and 2-(methylthio)thiazolines respectively (J. Laduranty, F. Barbot and L. Migimiac, Canad. J. Chem., 1987, $\underline{65}$, 859).

$$\text{(87)} \xrightarrow{\text{CH}_2\text{:CHCH}_2\text{MgBr}} \text{CH}_2\text{:CHCH}_2\text{CMePrNHCH}_2\text{CH}_2\text{SH}$$

(87)   (88)

(CH$_2$:CHCH$_2$)$_2$CMeNHCH$_2$CH$_2$SH   (CH$_2$:CHCH$_2$)$_3$CNHCH$_2$CH$_2$SH

(89)   (90)

The thiazolidine system may be involved in ring transformations, as in the decarboxylative ring expansion caused by heating thiazolidine-4-carboxylic acid with formaldehyde and methyl propiolate. Typically formed are eight-membered azadienes (91) (H. Ardill et al., Chem. Comm., 1987, 1296).

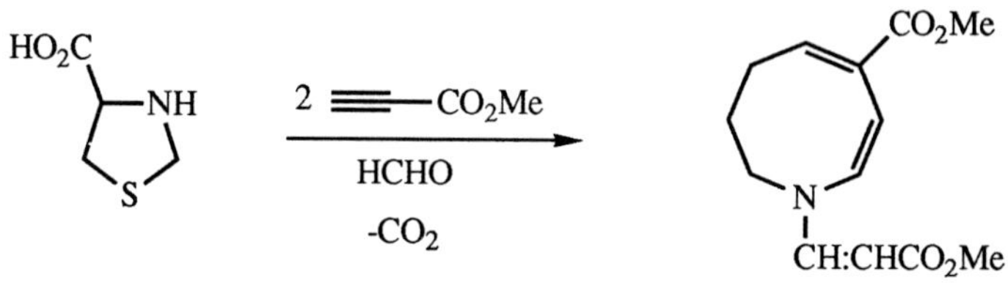

(91)

The racemic azathiabicyclohexane (92) is derived from 4-(hydroxymethyl)
thiazolidine on treatment with triphenyl phosphine and triethylamine (T. Takata *et al.*, Chem. Letters, 1985, 939).

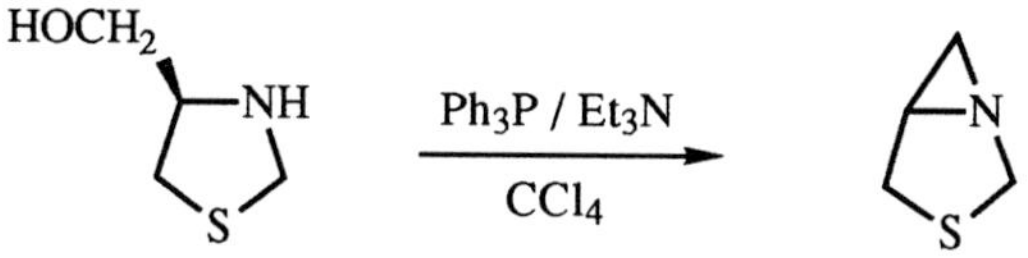

(92)

Thiazolidinediones may be transformed into furandiones. Thus the 4,5-thiazolidinedione (93) with piperidine yields the pyrrolinethione (94) which with alkali gives the 2,3-furandione (95) (B. Zaleska, Monatsh., 1986, 117, 671).

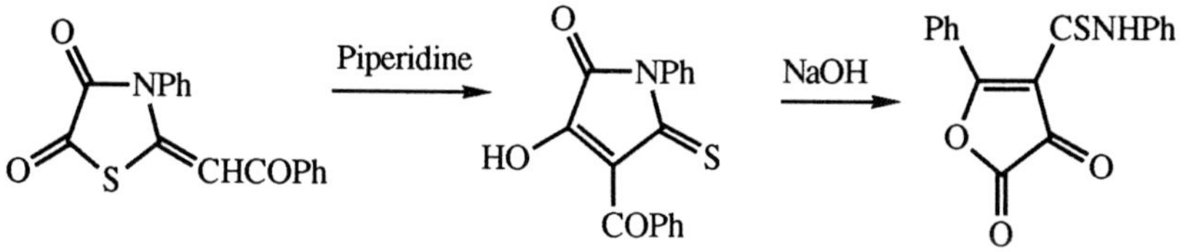

(93)   (94)   (95)

The benzylidenethiazolidinedione (96) on treatment with phenylhydrazine rearranges *via* the addition product (97) to give the pyrazolone (98) (M.T. Omar and F.A. Sherif, Synthesis, 1981, 742).

(96)   (97)   (98)

(d) Reactions of the thiazole system yielding fused heterocycles.

(i) Oxygen- and sulphur-containing fused thiazoloheterocycles.

Thiazoles will add to unsaturated compounds *via* either cycloaddition or Michael type reactions.

2,4-Dimethylthiazole reacts with benzophenone when irradiated to give the thiazolo-oxetane (99) (T. Nakano *et al.*, J. heterocyclic Chem., 1980, 17, 1777). Michael-addition products occur in the reactions of thiazoles with ketenes. For example, 2-(isopropyl)thiazole forms a 2:1 adduct with dichloroketene, the product of the reaction being the thiazolopyranone (100) (A. Medici *et al.*, J. org. Chem., 1984, 49, 590).

(99)   (100)

Similarly, ethyl benzylidenecyanoacetate reacts with the thiazolone (101) to yield the pyranothiazole (102) (M.K.A. Ibrahim *et al.*, Indian J. Chem., 1987, 27B, 216).

(101)   (102)

A similar reaction occurs with the oxothioxothiazole (103) and the nitrile (104) yielding the thiopyranothiazole (105) (F.H. Abdelrazek *et al.*, Synthesis, 1985, 432).

(103)   (105)

Thiopyrano-derivatives (107) result from the [4+2]-cycloaddition of acrylonitrile to 5-ethoxymethylene-2-oxo-4-thioxothiazolidine (106) (H.A. Ead *et al.*, Arch. Pharm., 1987, <u>320</u>, 1227).

(106)   (107)

(e) Nitrogen-containing fused thiazoloheterocycles.

(i) 5-Membered fused rings containing one nitrogen atom.

Pyrrolothiazoles may be prepared utilising the 2-methyl group of thiazoles and for example bromoacetone. Intramolecular cyclisation and dehydration gives the pyrrolo[2,1-*b*]thiazole (108) (J.C. Brindley, D.G. Gillon and G.D. Meakins, J. chem. Soc., Perkin I, 1986, 1255).

MeCOCH₂Br

(108)

4-Methylthiazole undergoes cycloadditions with DMAD after treatment with (trimethylsilyl)methyl trifluoromethanesulphonate to afford the pyrrolino[2,1-*b*]thiazole (109) (A. Padwa, U. Chiacchio and M.K. Venhatraman, Chem. Comm., 1985, 1108).

(109)

Pyrrolo[1,2-*c*]thiazolidines (110) are prepared by cyclo-condensation of 3-formyl-2-heteroarylthiazolidine-4-carboxylic acids with ethyl 2,3-dichloropropionate (J.L. Fabre, Chem. Abs., 1988, 109, 128992).

(110)

(ii) 5-Membered fused rings containing two nitrogen atoms

The imidazolo[2,1-c]thiazole (113) is produced by the base catalysed cyclisation of the cyanimine derivative (112) which in turn is obtained by the reaction of the thiazolidine (111) with 2-(chloroethyl)phenylketone (Patent, Chem. Abs., 1985, 103, 104967).

(111)   (112)   (113)

Alternatively N-4 may be furnished from 2-aminothiazole by treatment with phenacyl bromide. The latter attacks the nitrogen heteroatom of the thiazole, the reaction proceeding *via* (114) (G.D. Meakins *et al.*, J. chem. Soc., Perkin I, 1989, 643).

(114)

The tetrahydroimidazole analogues, the imidazolidino[4,1-c]thiazoles (115) may be prepared either by treatment of a thiazolidinecarboxylic acid with methyl isocyanate (O. Palla, Chem. Abs., 1985, 102, 71306) or by the reaction of a (dimethylamino)azirine and 1,3-thiazolidine-2-thione (S.M. Ametamey *et al.*, Helv., 1986, 69, 2013).

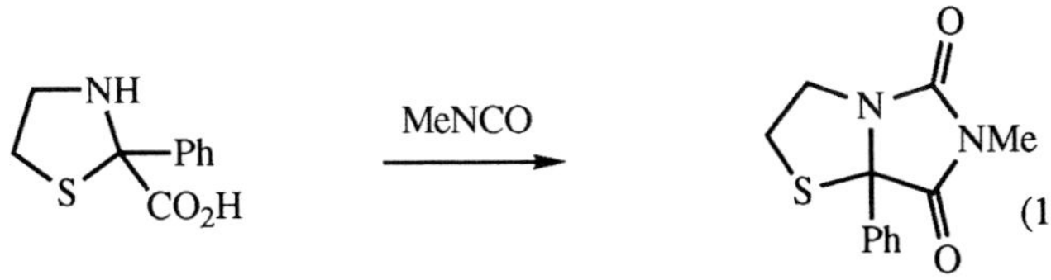

(115)

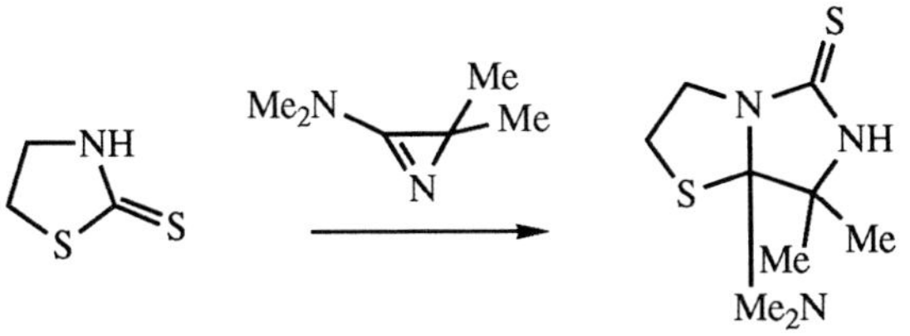

Pyrazolo[5,1-*b*]thiazoles (116) and (118) are prepared by the reaction of a 3-aminothiazoline-2-thione with phenacyl bromide and treatment of the thiazolo[2,3-*b*]thiadiazines so formed with triethylamine, or by the reaction of the thiazolium salt (117) with malononitrile (P. Molina *et al.*, Heterocycles, 1987, 26, 1323).

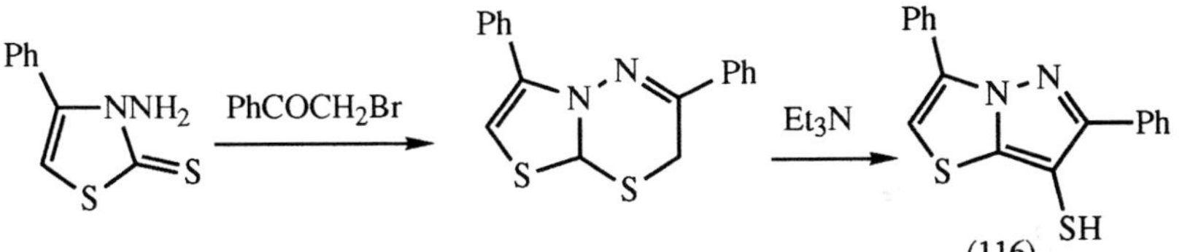

(116)

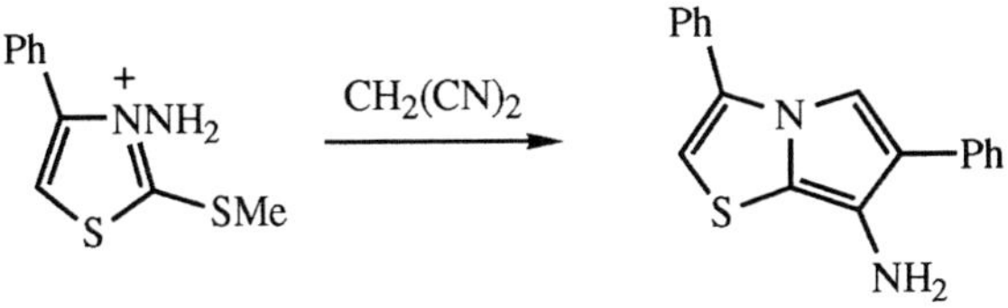

(117)

(118)

(iii) 5-Membered fused rings containing three nitrogen atoms

Thiazolo[2,3-*e*]triazoles (119) are usually produced by cyclisation of hydrazinylthiazoles with phosphorus oxychloride (M. Veverka, Chem. Abs., 1987, <u>106</u>, 138454).

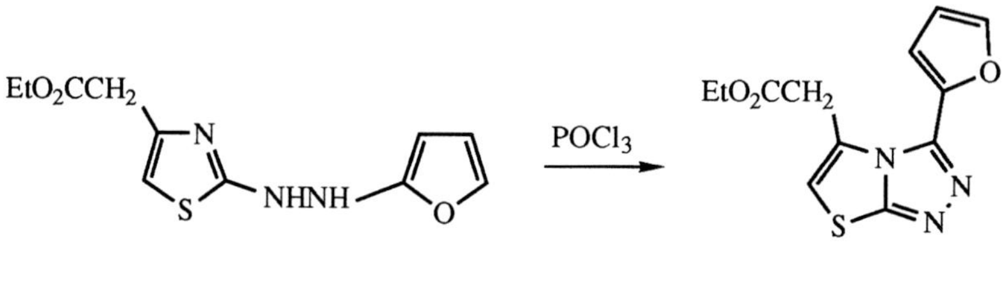

(119)

Thiazolinehydrazones (120) react with thioacetic acid to yield the thiazolidylaminothiazolinones (121), which may be cyclised in concentrated sulphuric acid to the bisthiazolotriazolones (122) (H. Singh, L.D. Yadav and A.K. Singh, J. Indian chem. Soc., 1985, <u>62</u>, 147).

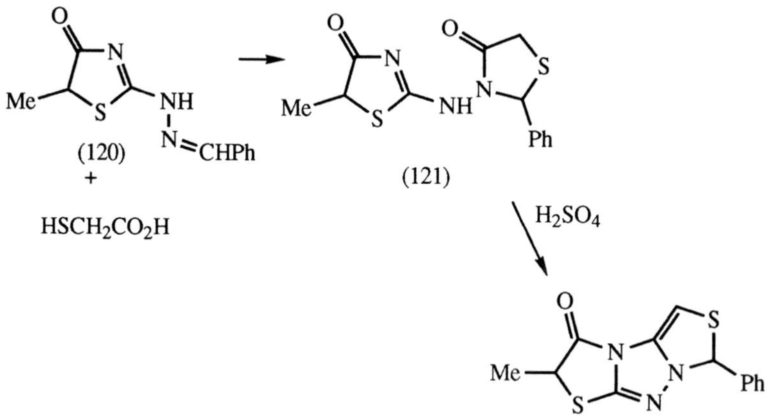

(122)

To produce the thiazolo[3,4-*b*]triazoles (123) it is necessary to have a 3-aminothiazolidine with a reactive group (e.g. carbonyl) at C-4. Cyclisation may then be achieved with a chloroimine (E.K. Mikitenko and N.N. Romanov, Khim. geterosikl. Soedin, 1982, 331).

(123)

(iv) 6-Membered fused rings containing one nitrogen atom

2-Cyanomethylthiazoles (124) undergo a Michael reaction with ethyl propenoate. The adduct may then cyclise to yield a thiazolopyridone (125) (F.A. Khalifa, B.Y. Riad and F.H. Hafez, Heterocycles, 1983, 20, 1021).

(124)

+

H₂C:CHCO₂Et

(125)

Alternatively, the ester group may be replaced by a cyano function in which case a dihydropyridine analogue of (125) is formed. (K.U. Sadek et al., Z. Naturforsch, 1984, 39B, 824).

Thiazolium salts having a reactive methylene group at C-2 and at C-3 condense readily in the presence of acid with 1,3-diketones. Thus the thiazolium salt (126) and pentan-2,4-dione yields the thiazolo[3,2-a]pyridinium salt (127) (N.A. Chuiguk and K.V. Fodotov Ukr. Khim. Zh., 1980, 46, 1306; C. Galera et al., J. heterocyclic Chem., 1986, 23, 1889).

(126)   (127)

(v) 6-Membered fused rings containing two nitrogen atoms

As with other fused pyrimidines, compounds derived from the cyclisation of thiazole and a two nitrogen, six membered heterocycle are of interest as potential physiologically active compounds.

To furnish the necessary nitrogen atoms, aminothiazole derivatives are generally used. Thus, cyclisation of 5-acetylamino-2-methylthiazole-4-carboxamide with aniline using phosphorus pentoxide as dehydrating agent yields the 7-aminothiazolo[5,4-*d*]pyrimidine (128) (K.E. Andersen, M. Hammad and E.B. Pedersen, Ann., 1986, 1255).

(128)

Condensation of 2-aminothiazoles with 1,3-dicarbonyl compounds is also useful, as in the reaction of 2-aminothiazole and diethyl 2-(4-chlorophenyl)malonate to give the thiazolopyrimidine (129) (Patent, Chem. Abs., 1983, 98, 72124 and 98, 215611; J.K. Sahu *et al*, Indian J. Chem., 1984, 24B, 117; F. Ye, B. Chen and X. Huang, Synthesis, 1989, 317).

Benzenoid analogues, the thiazoloquinazolones (130) may be prepared from oxothioxothiazolidines and anthranilic acid (H.A. Dabouna nd M.A. Abdel Aziz, Arch. Pharm., 1983, 316, 394).

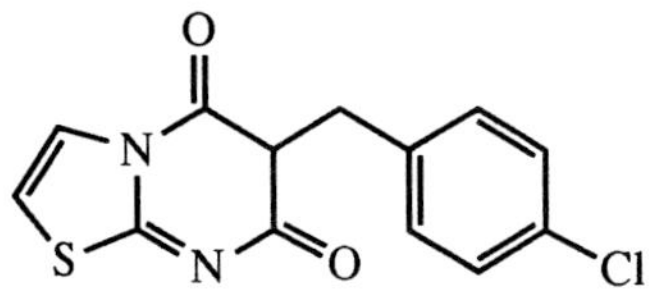

(129)

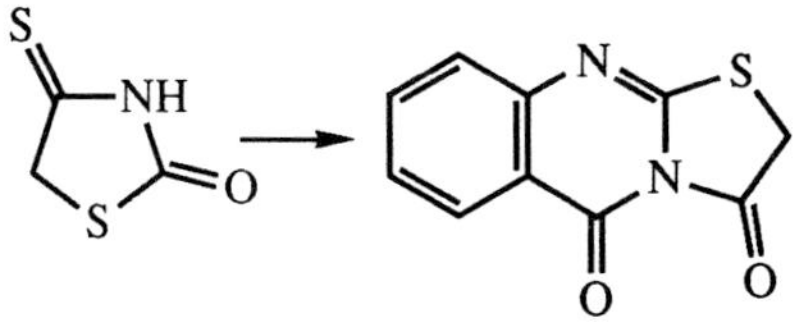

(130)

The reactivity of the halogen in 2-chlorothiazole allows 5-chloro-anthranilamide to be used in a similar reaction to give (131) (G. Hamprecht, Chem. Abs., 1988, 109, 73466).

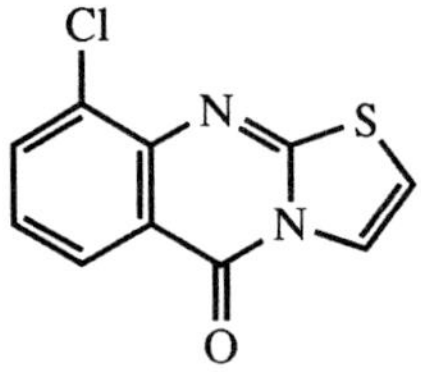

(131)

(vi) 6-Membered fused rings containing three nitrogen atoms

Formation of a fused triazine ring can be effected when the thiazole contains a hydrazine type substituent.   Thus treatment of 2-(phenylhydrazono)-4-thiazolidinone with phenacyl bromide gives an adduct (132) which readily cyclises to the thiazolotriazinedione (133) (S.O. Abd Allah, M.R. Elmoghayar and H.A. Ead, Ann. Quim., 1984, 80C, 232).

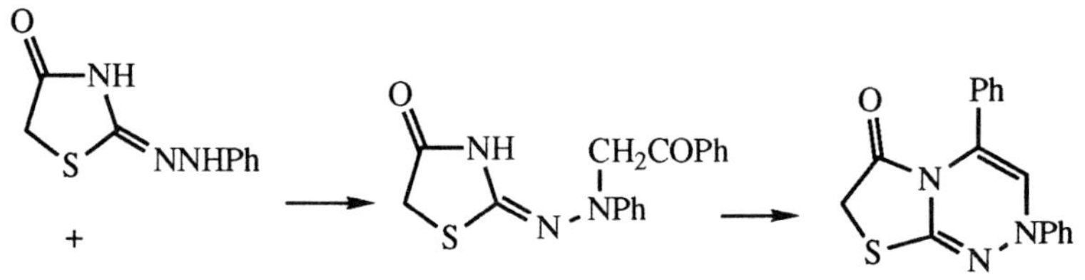

(132)

(133)

Diazotisation of the amino group adjacent to an amide group in the thiazole (134) yields the thiazolo[4,3-*e*]triazine (135) (W. Ried and D. Kuhnt, Ann., 1986, 780).

(134)   (135)

(vii) Larger nitrogen-containing fused rings

Hydrazino groups on thiazoles can react with 1,3-dicarbonyl compounds to yield triazepines (136) (B.V. Alaka, D. Patnails and M.K. Rout, J. Indian chem. Soc., 1982, 59, 1168).

(136)

Larger ring systems may be constructed using the bridged dithiazolamine (137) which with triethyl orthoformate gives the dithiazolodiazocine (138) (L.K.A. Rahman, J. heterocyclic Chem., 1986, 23, 1435).

(137)   (138)

8.   Benzothiazoles

Benzothiazoles remain prominent in the area of cyanine dyes for use in the photographic industry. The bioluminescent compounds such as Luciferin are finding increasing use at the biotechnology interface, for example as aids to immunoassay.

(a) Synthesis

Many of the routes to benzothiazoles involve 2-aminothiophenol which reacts with aldehydes to give a 2-substituted benzothiazole. This reaction has been adapted by using 2-aminothiophenol and pyridine-2,6-dicarboxaldehyde to give bis-2,6-(2-thiazolyl)pyridine (139), a bidentate ligand (A.S. Salamah *et al.*, Polyhedron, 1982, 1, 543).

2-Aminothiophenol and *tert*-butyl 3-oxo-2,2-dimethylbutanoate yield a 2-substituted benzothiazoline (140) which undergoes further cyclisation to give the tricyclic β-lactam (141) (G. Santar *et al.*, Z. Naturforsch, 1988, 43, 758).

(139)

(140)

CHMeCMe$_2$CO$_2$CMe$_3$

(141)

Benzothiazole (143) itself may be prepared by irradiation of 2-bromo-carbamoylthiobenzene (142) in methanol (R. Paramisivam and V.T. Ramakrishnan, Indian J. Chem., 1987, 26B, 930).

(142) $\xrightarrow[\text{MeOH}]{h\nu}$ (143) (144)

Improved yields of 2-aminobenzothiazole have been achieved using a mixture of an arylthiourea, sulphonyl chloride and magnesium oxide (H. Rentel, Chem. Abs., 1984, 101, 23461). 6-Trifluoromethyl-2,4-dinitrochlorobenzene reacts with thiourea at 100 °C in sulfolane to yield 2-amino-7-trifluoromethyl-5-nitrobenzothiazole (144) (R. Hamprecht, Chem. Abs., 1984, 101, 211131).

A 2-methylaminobenzothiazole (147) is formed by oxidative cyclisation of the o-phenylenediamine thiourea derivative (145). The reaction is thought to proceed via the quinone imine (146) (S. Rajappa and R. Sreenivasan, Tetrahedron, 1980, 36, 3087).

(145) $\xrightarrow[\text{HCl / EtOH}]{[O]}$ (146)

(147)

A new catalyst for thiourea cyclo-condensations such as that with 2-iodoaniline is tetrakis(triethylphosphino)nickel (IV) in DMF. This enables the formation of 2-aminobenzothiazole to be effected at 60 °C in 87% yield (K. Takagi, Chem. Abs., 1988, 109, 110420).

6-Methoxy-2-aminothiazole is the starting point for a new synthesis of D-luciferin (148) (A.G. Batz, Chem. Abs., 1982, 95, 169172).

(148)

Reagents: i) Diazotization   ii)CuCl   iii) Me₃SiI / aq. MeOH

A new synthesis of 2-cyanomethylbenzothiazole starting from 2-aminothiophenol uses malononitrile as the carbon source (G.E.H. Elgemeie *et al.*, Heterocycles, 1986, 24, 349).

Aryliminoderivatives of 2-aminothiophenol (149) have been oxidatively cyclised to the 2-arylbenzothiazoles (150) using barium manganate (VII) (R.G. Srivastava and P.S. Venkataraman, Synth. Comm., 1988, 18, 1537).

(149)   (150)

The ethoxythioanilide (151) is cyclised under basic conditions with potassium ferricyanide to give the 2-ethoxybenzothiazole (152). This may be converted into the corresponding benzothiazolin-2(3H)-one (153) by treatment with ethanolic hydrogen chloride (W. Engel, Chem. Abs., 1982, <u>96</u>, 68981).

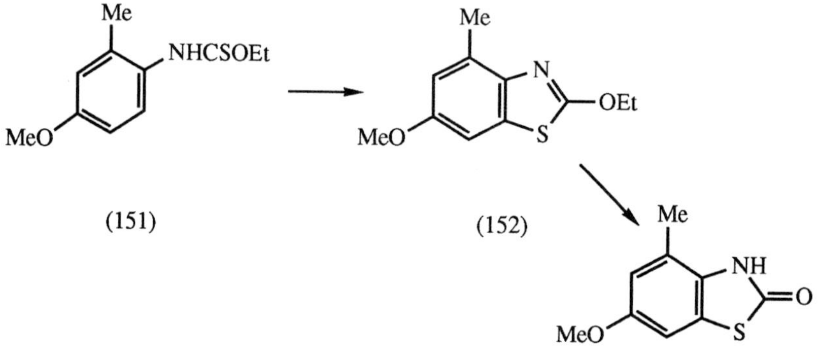

(151)   (152)

(153)

An alternative cyclising agent used in the conversion of 5-fluoro-2-aminothiophenol to 6-fluorobenzothiazol-2(3H)-one (154) is N,N'-carbonyldiimidazole (W. Engel, Chem. Abs., 1982, <u>96</u>, 40925).

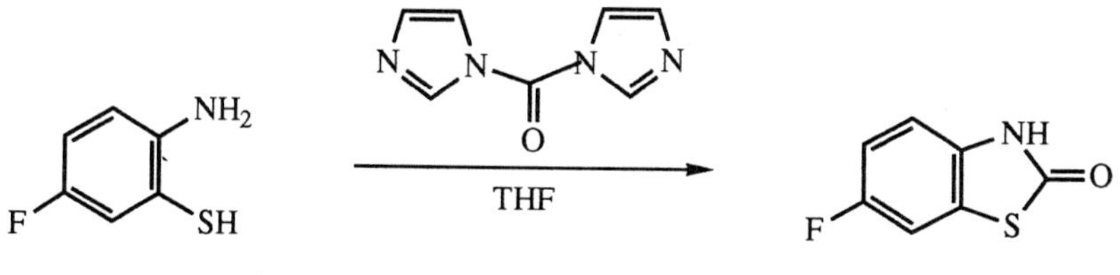

(154)

2,5-Dinitrophenylthiocarbamide (155) may be reductively cyclised to 6-nitro- benzothiazol-2(3H)-one (156) (J. Schulze, Chem. Abs., 1982, <u>97</u>, 162979).

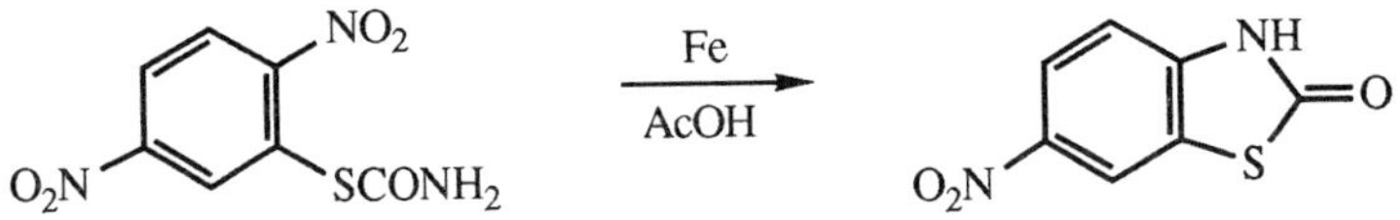

(155)   (156)

(b) Reactions

2-Methylbenzothiazole is the precursor to some cyanine dyes. The reaction of 3-ethyl-2-methylbenzothiazolium tosylate (157) with the pyran salts (158) gives a dye (159) which absorbs at around 1000 nm in dichloromethane (N.V. Monich et al., Khim. geterosikl. Soedin, 1981, 1631).

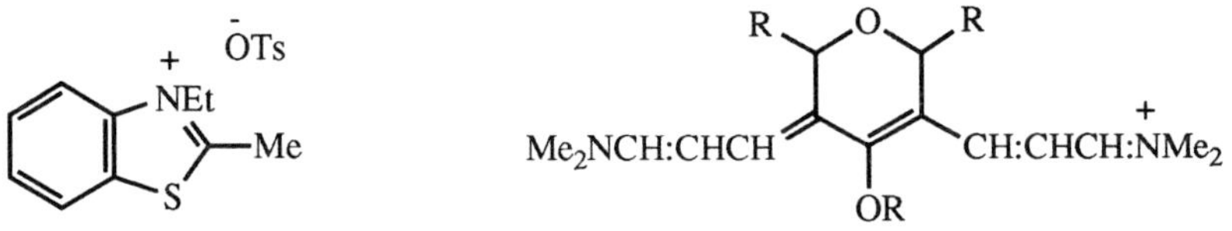

(157)   (158)

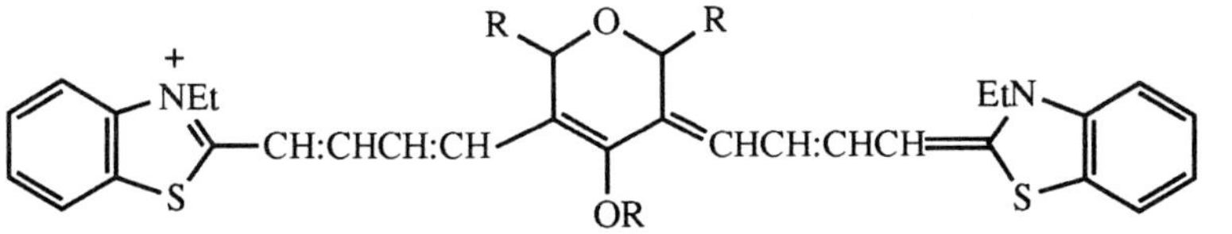

(159)

Condensation of the 2,3-dimethylbenzothiazolium iodide (160) with phenanthraquinone monoxime yields a photochromic spiro compound (161) (M. Reichenbaecher, Chem. Abs., 1982, 97, 164536).

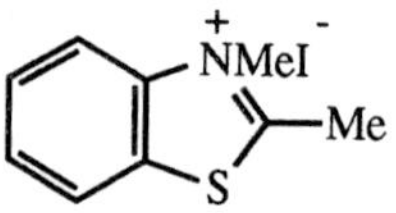

(160)

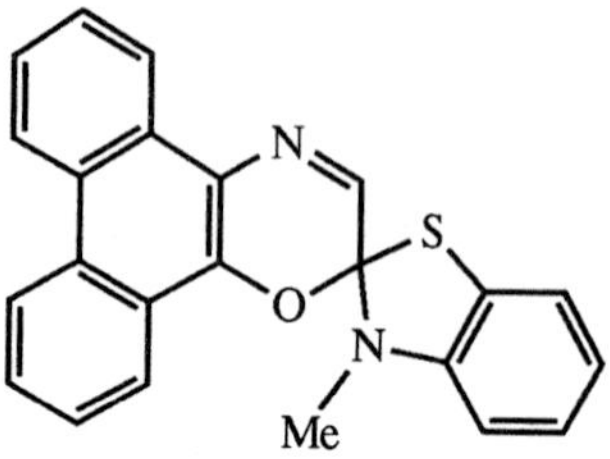

(161)

2-Methylbenzothiazole reacts with 2-chloro-5-nitrobenzaldehyde to give a styrene derivative (162) which on irradiation cyclises to the benzothiazolo[3,2-*a*] quinolinium salt (163) which has anti-tumour activity both *in vivo* and *in vitro via* DNA intercalation (O. Cox *et al.*, J. med. Chem., 1982, 25, 2378).

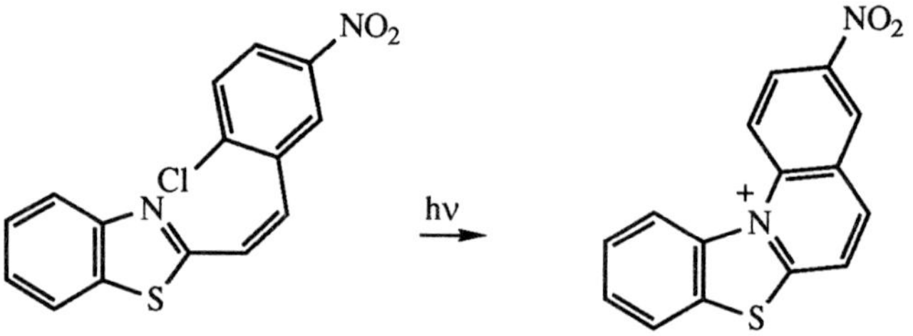

(162) hν → (163)

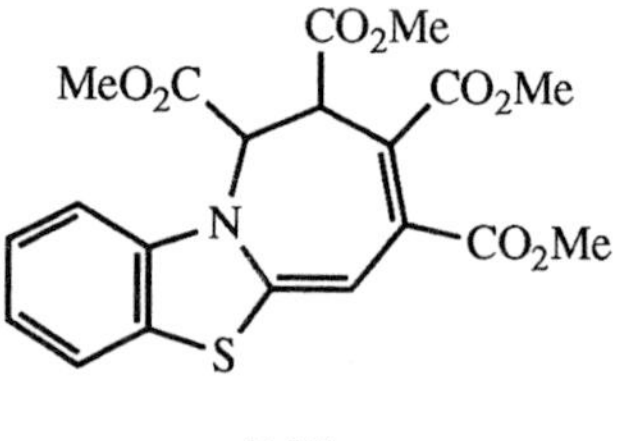

(164)

As in other areas of thiazole chemistry, photochemical reactions are important. 2-Methylbenzothiazole reacts with dimethyl acetylenedicarboxylate under irradiation to give the ring expanded azepine system (164) (T. Itoh *et al.*, Heterocycles, 1983, 20, 1321. See also R.M. Letcher, K.K. Cheung and D.W.M. Sin, J. chem. Res. (S), 1989, 115).

3-Chloroalkylbenzothiazolium salts (165) undergo ring expansion under basic conditions (K.J. Federel and J. Bergman, Tetrahedron Letters, 1980, 2429; G. Liro *et al.*, J. heterocyclic Chem., 1981, 19, 279).

(165)

Ring expansion of benzothiazoline-1-oxides (166) usually takes place in refluxing acetic anhydride. Ring expansion of benzothiazoles gives rise to the 1,4-benzothiazine derivative rather than the 1,2-benzothiazines observed with ring expansion of benzoisothiazoles (M. Hori *et al.*, Tetrahedron Letters, 1981, 1701; H. Shimizu *et al.*, Chem. pharm. Bull., 1984, 32, 2571).

(166)

2-Formylbenzothiazole gives a phosphorane adduct which undergoes a Wittig reaction with aldehydes (A. Dondoni *et al.*, Tetrahedron, 1988, 44, 2021).

2-Phenylbenzothiazole reacts regio- and stereo-specifically under photochemical conditions with ethoxyethene to afford the benzothiazepine (167) (M. Sindler-Kulyk and D.C. Neckers, Tetrahedron Letters, 1981, 2081).

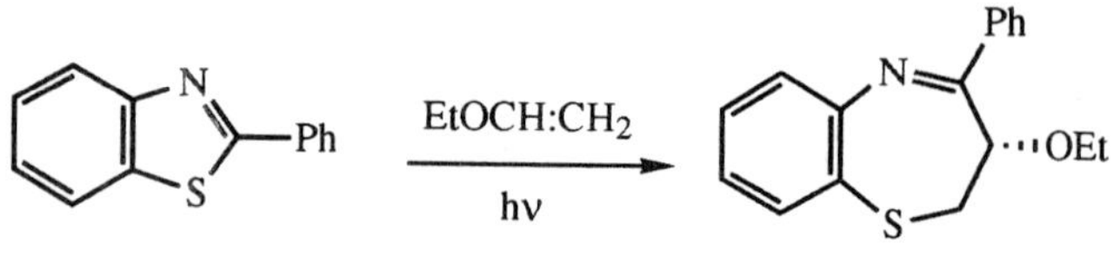

(167)

2-Aminobenzothiazole is a useful starting material for the preparation of fused heterocycles such as 4*H*-pyrimido[2,1-*b*]benzothiazol-4-ones (168) (J.R. Wade *et al.*, J. med. Chem., 1983, 26, 608). Similar pseudo-purines (169) and (170) are afforded by the reaction of 2-aminobenzothiazoles with DMAD in ethanol or with ethyl benzoylacetate and polyphosphoric acid (F. Russo *et al.*, J. heterocyclic Chem., 1988, 25, 949).

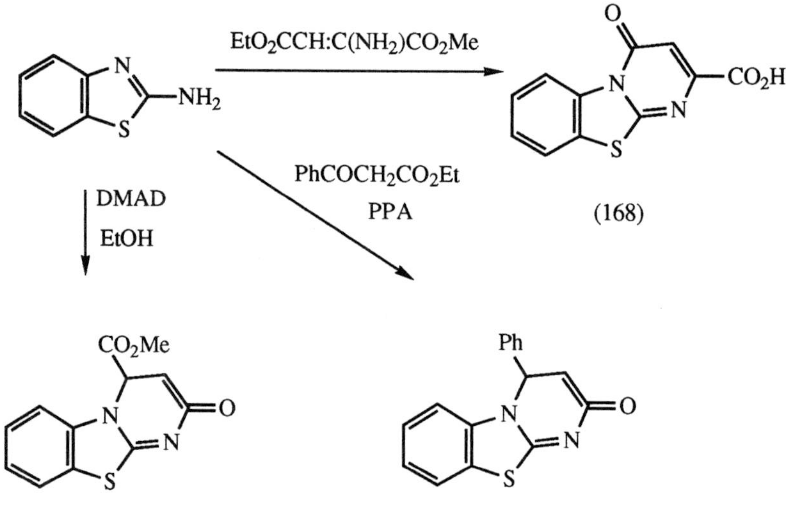

(168)

(169)

(170)

Fused triazoles (171) are produced from 2-hydrazinobenzothiazoles and formic acid (Patent, Chem. Abs., 1982, 97, 162998).

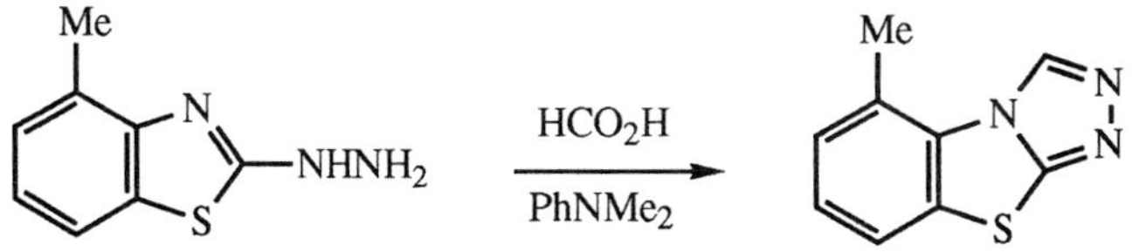

(171)

Other condensed benzothiazoles include the tetrahydro-derivative (172) formed by the reaction of 2-aminotetrahydrobenzothiazole with phenacylbromide (M.N. Balse and C.S. Mahajonshetti, Indian J. Chem., 1980, 19B, 263).

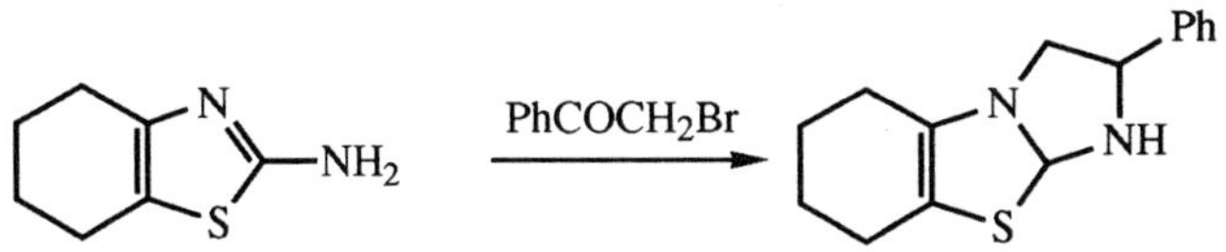

(172)

2-Substituted-amino-3-methylbenzothiazolium salts are a useful source of monomethyl secondary amines (A.R. Katritzky, M. Drewniak and J.M. Aurrecoechea, J. chem. Soc., Perkin I, 1987, 2539). Treatment of 3-methyl-2-methylthiobenzothiazolium iodide (173) with either an aromatic or an aliphatic primary amine followed by elimination yields the 2-imino compound (174). This is methylated with iodomethane to give (175) and finally the monomethylamine is liberated by treatment with butylamine.

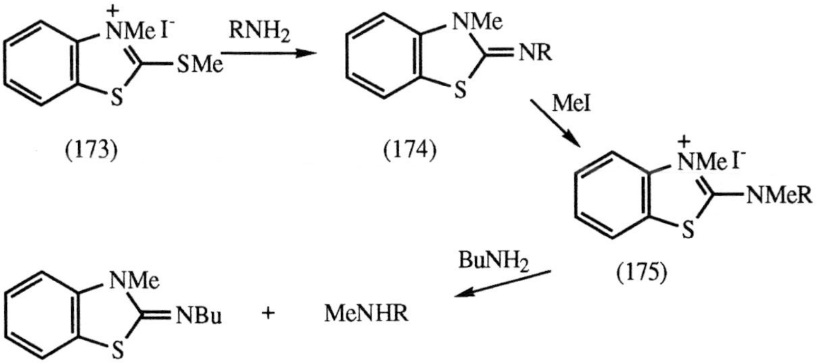

2-Chlorobenzothiazole reacts with Grignard reagents derived from 2-(bromoalkyl)benzothiazoles to afford a cross-coupled product (176) (F. Babudi *et al.*, Heterocycles, 1986, 24, 2215).

Fused heterocycles are also obtainable from 2-chlorobenzothiazole. The benzothiazolo[3,2-*a*]quinolone (177) results from its reaction with ethyl (3-fluoro-4,6-dichlorobenzoyl)acetate (D.T.W. Chu, P.B. Fernando and A.G. Pernot, J. med. Chem., 1986, 29, 1531).

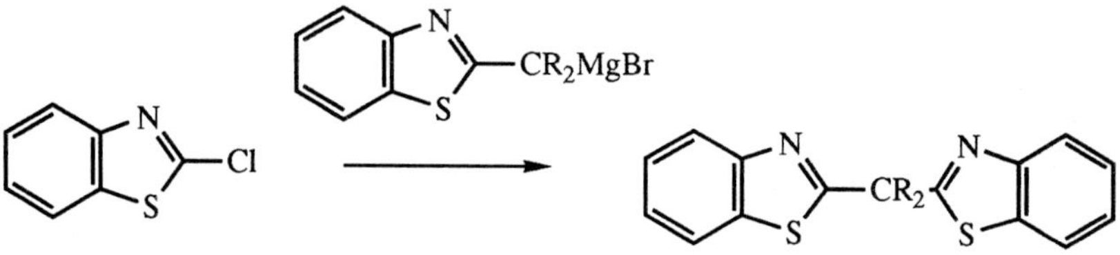

(176)

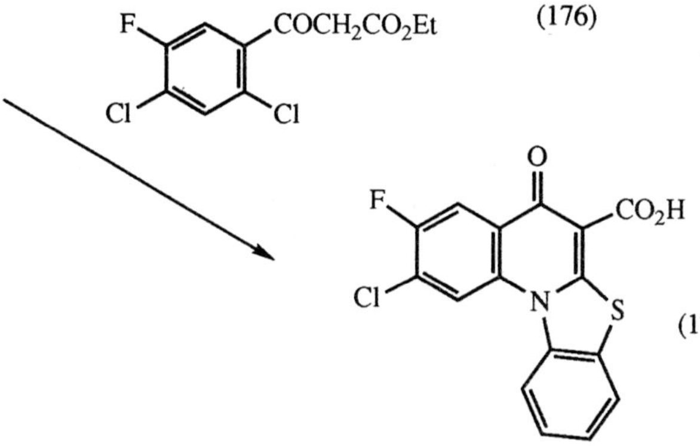

(177)

2-Alkyl- or aryl-benzothiazoles (178) may be ring opened with an allylic Grignard reagent to yield a 2-(bisalkenyl)aminothiophenol (179) as the disulphide (F. Babudi *et al.*, Tetrahedron letters 1984, 2047).

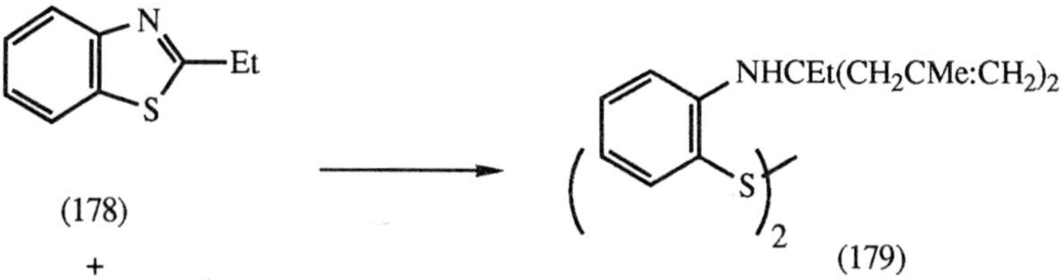

(178)

+

CH₂:CMeCH₂MgBr

(179)

Reissert compounds are well known in six membered *N*-heterocyclic chemistry. This is not so with the five membered species because of their facile ring opening. However, the use of trimethylsilylcyanide as the source of cyanide in a single phase system leads to the formation of Reissert compound (180) in good yield from benzothiazole itself. This gives a useful method of extending the benzothiazole nucleus, for example to the tricyclic pyrido-analogue (181) (B.C. Uff *et al.*, Chem. Comm., 1984, 1245).

Reagent; (a) Me₃SiCN, Cl(CH₂)₃COCl, AlCl₃ , CH₂Cl₂

Benzothiazole reacts with phenacyl bromide to give the *N*-phenacylide (182), which undergoes a cycloaddition with the endocyclic double bond of the methylene cyclopropene (183) to give the novel product (184) (O. Tsuge *et al.*, Chem. Letters, 1981, 1493).

The thiazole analogue (186) of benzomorphan has been prepared from the 4,5,6,7-tetrahydrobenzothiazole (185) (K. Katsuura *et al.*, Chem. pharm. Bull., 1983, 31, 1518).

(185)   (186)

9. Isoselenazoles

The redox chemistry of the seleno-cysteine systems has been studied using compounds (1) and (2) as models for the selenoenzyme glutathione-peroxidase (H. J. Reich and C. P. Jasperse, J. Amer. chem. Soc., 1987, 109, 5549).

(1)   (2)

Isoselenazole carboxylic acids (3) have been prepared by oxidation of the corresponding methyl derivative using selenium dioxide (F. Lucchesini, V. Bertini and A. De Munro, Heterocycles, 1985, 23, 127).

(3)   (4)

The isoselenazole nucleus is reactive towards electrophilic substitution. Nitration of 3-methylisoselenazole occurs at C-5 (F. Lucchesini et al, Heterocycles, 1988, 27, 2431).

In common with its sulphur analogue, isothiazole, isoselenazole has excellent anti-tumour properties when part of a fused pyrimidine system (4) (T. Ueda, H. Yoshida and J. Sakakibara, Synthesis, 1985, 695; Tetrahedron Letters, 1987, 4579). A similar compound (5) is active against the Ehrlich ascites tumour (H. Ito, J. Sakakibara and T. Ueda, Cancer Letters, 1985, 28, 61).

Pyrimido[5,6-c]isoselenazoles (5) are converted into pyrimido[5,6-b]pyridines (6) by treatment with active methylene compounds (loc. cit.).

$$\text{MeCOCH}_2\text{CO}_2\text{Et} \xrightarrow{\text{NaOEt}}$$

(5)   (6)

10.   Benzisoselenazoles

(a) Synthesis

Ebselen, 2-phenyl-3(2H)-benzisoselenazolone (7) and its analogues, e.g. (8), are anti-arterioschlerotic and general anti-inflammatory agents (A. Welter, Chem. Abs., 1984, 101, 38454).

(7)   (8)

Synthesis of benzisoselenazolones such as Ebselen uses 2-(chlorocarbonyl)phenylselenyl chloride and a primary amine (R. Weber and M. Renson, J. heterocyclic Chem., 1973, 10, 267; A. Welter Chem. Abs., 1982, 96, 199699; 1987, 106, 50198; C. Lambert et al., Bull. chim. Soc. Belg., 1987, 96, 383).

A novel synthesis of 3-substituted benzisoselenazoles involves the addition of benzyne to a selenadiazole (9) (M.R. Bryce *et al.*, J. chem. Soc., Perkin I, 1981, 607).

(9)

(b) Reactions

The reactions of the benzisoselenazole nucleus depend mainly on the weakness of the Se-N bond and the different oxidation states of selenium. Thus, ring opening of 2-(4'-methoxyphenyl)-3(2H)-benzisoselenazolone with thiophenol in dichloromethane at room temperature gives 94-99% of (10) (N. Kamigata *et al.*, Heterocycles, 1986, 24, 3027). Such ring-opened species exhibit glutathione-peroxidase-like activity and so are useful in the treatment of cell damage caused by active oxygen metabolites, for example in radiation sickness. (N. Dereu, Chem. Abs., 1986, 105, 114733; A. Welter, Chem. Abs., 1987, 106, 84181).

(10)

This antioxidant activity relies on the conversion of selenium in the organoselenium compound into the Se-oxide. Thus, Ebselen catalytically reduces hydroperoxides and is oxidised to 1-oxo-2-phenyl-3-benzisoselenazoline (11) (H. Fischer and N. Derek, Bull. Soc. chim. Belg., 1987, 96, 757).

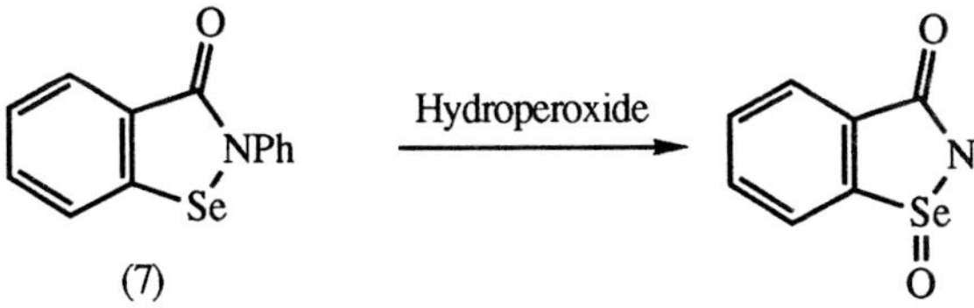

(7)   Hydroperoxide → (11)

Other species may be oxidised by "Ebselen oxide" (11) or its analogues as shown below.

PhSPh $\xrightarrow{(11)}$ Ph_2SO

Ph$_3$P $\xrightarrow{(11)}$ Ph_3PO

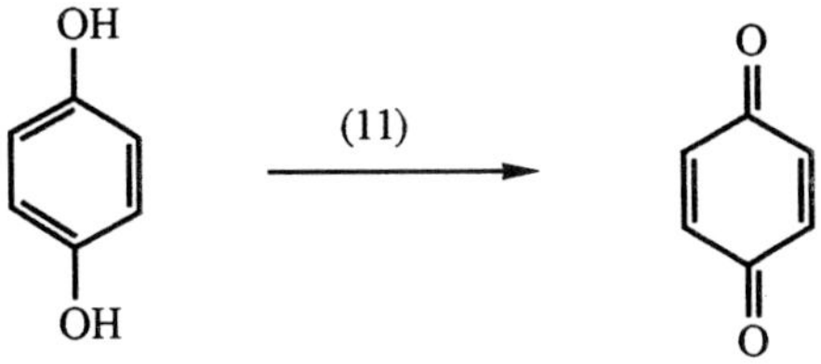

The Se-oxide is usually prepared using ozone or hydrogen peroxide (N. Kamigata *et al.*, Sulfur Letters, 1986, 5, 1).

3-Benzisoselenazolethiones (12) are prepared by sulphuration of the corresponding benzisoselenazolones with Lawesson's reagent or phosphorus pentasulphide (A. Welter, Chem. Abs., 1986, 104, 50881).

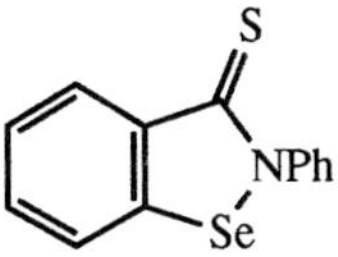

(12)

2-Aryl-1,2-benzisoselenazol-3(2H)-ones undergo a photochemical reaction in benzene to yield dibenzo[b,f][1,4]selenazepin-11(10H)-ones (N. Kamigata *et al.*, Bull. chem. Soc. Japan, 1986, 59, 2179).

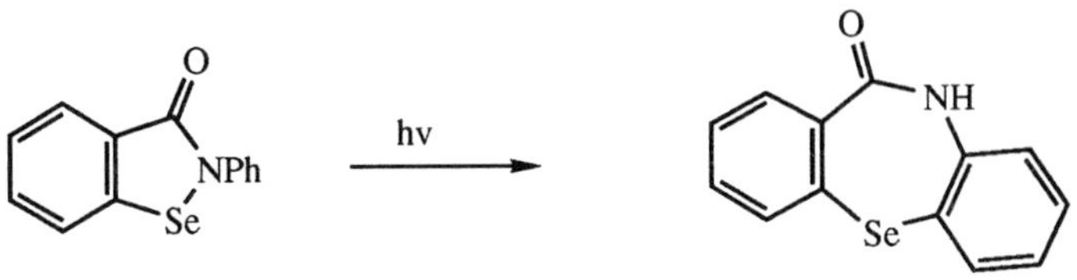

11. Isotellurazoles and Benzisotellurazoles

The tellurium analogue of isothiazole (13) was first prepared by the reaction of a disubstituted acetylene with potassium telluride and hydroxylamine-O-sulphonic acid (F. Lucchesini and V. Bertini, Synthesis, 1983, 824).

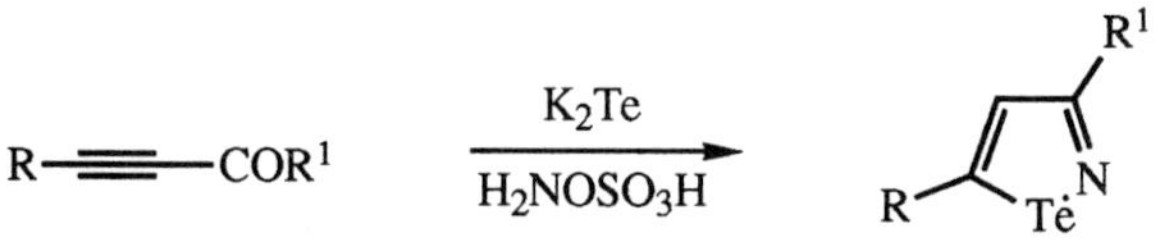

(13)

The same reagents can also be used to give vinyltellurazoles (14) using hex-5-en-3-yne-2-one to supply the carbon backbone (F. Lucchesini *et al.*, Heterocycles, 1987, <u>26</u>, 1587).

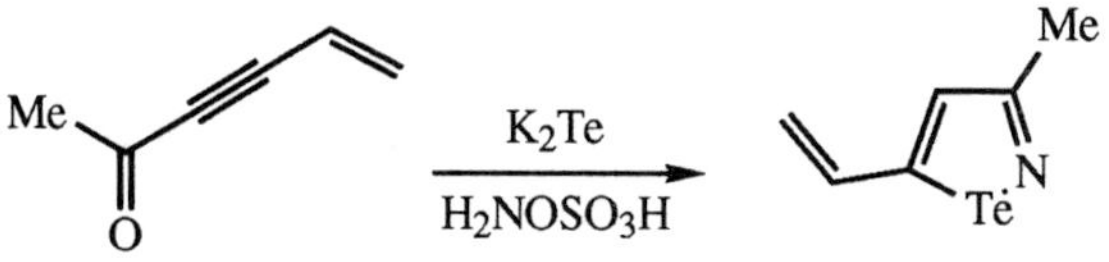

(14)

1,2-Benzisotellurazoles are prepared by the action of ammonia on o-(butyltelluro)benzaldehydes (R. Weber and M. Renson J. heterocyclic Chem., 1978, <u>15</u>, 865). This method has been extended by using a substituted arylimino derivative to give an N-substituted analogue (15) (A.A. Makseimenko *et al.*, Zh. Obshch. Khim., 1988, <u>58</u>, 1176).

(15)

12. Selenazoles

(a) Synthesis

A novel route to the selenazole nucleus uses a polyfunctional reagent such as the vinyl phosphonium salt (16), which with sodium hydrogen selenide gives the salt (17), cyclisation of which yields the phosphonium derivative (18) (V.S. Brovarets and B.S. Drach, Zh. Obsch. Khim., 1986, 56, 321).

Another route utilises the cyclisation of the selenosemicarbazide (19) with, for example, chloroacetone. The hydrazones so formed are tridentate ligands (J. Gospodarek and S. Bilinski, Chem. Anal., 1981, 26, 411).

A more conventional route employing selenourea and C-aroyl-N-aryl formohydrazidoyl bromide (20) yields an arylazoselenazole (21) (H.M. Hassaneen et al., Org. Prep. Proc. Int., 1988, 20, 505).

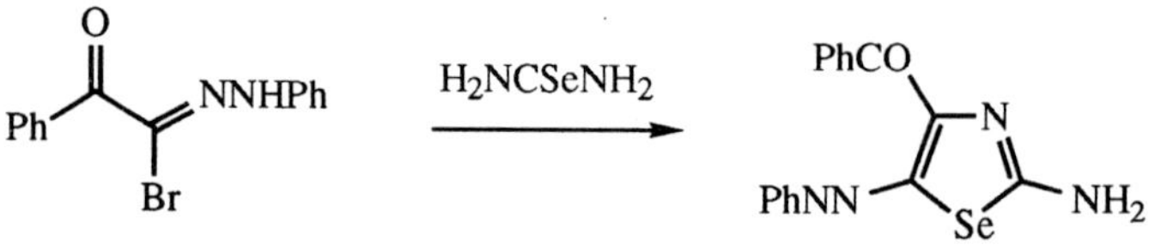

(20)   (21)

(b) Reactions

Alkylation takes place at N-3 on the treatment of either a 2-acetylamino-Δ^2-selenazolin-4-one or a 5-alkyl-selenazolidin-4-one with an alkyl halide apparently *via* the initial alkylation of the 2-acetylamino group in (22) (I.B. Levshin *et al.*, Zh. org. Khim., 1985, 21 641)

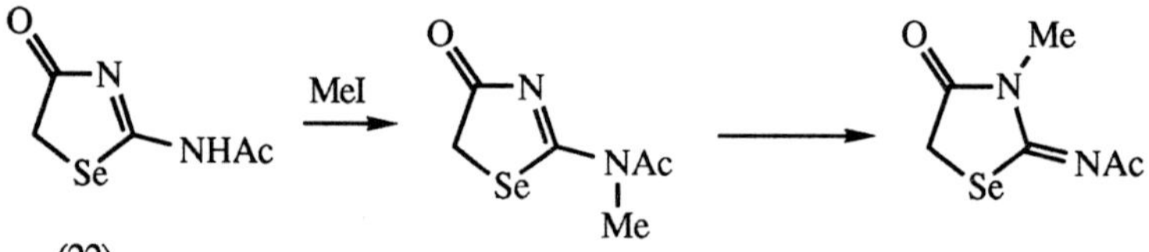

(22)

Selenazolium salts may be prepared by the treatment of the corresponding oxazolium salt (23) with sodium hydrogen selenide. The addition compound (24) ring opens to (25) which then cyclises to (26) The latter is useful in the synthesis of a variety of other heterocycles including imidazoles (27), benzimidazoles (28) and triazines (29) (O.P. Shvaika and V.F. Lipnitski, Zh. obshch. Khim., 1981, 51, 1842).

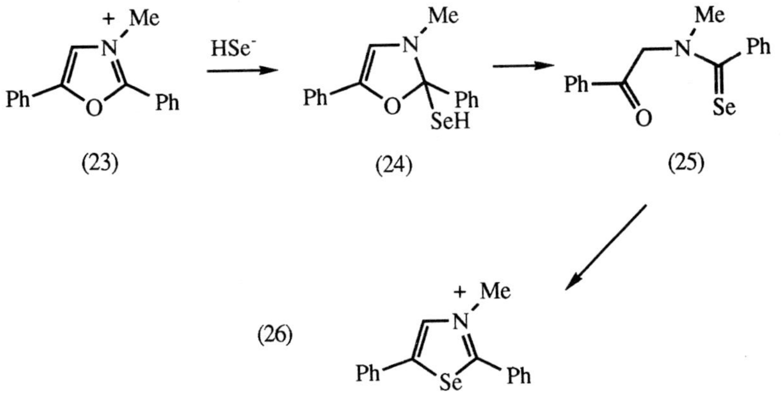

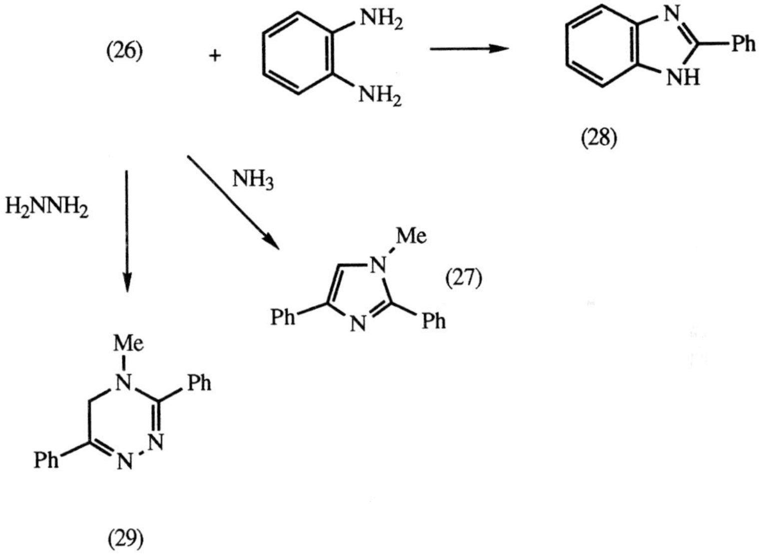

(23) (24) (25)

(26)

(26) + [structure with NH₂ groups] → (28)

H₂NNH₂

NH₃

(27)

(29)

(c) Fused Heterocyclic Selenazoles

2-(Allylseleno)pyridines (30) are the starting materials in the synthesis of 2,3-dihydroselenazolo[3,2-a]pyridinium salts (31). (A. M. Shestopalov *et al.*, Zh. obshch. Khim., 1988, 58, 840).

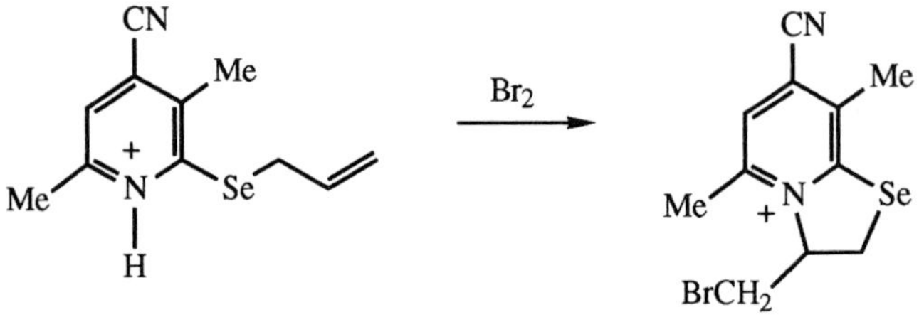

(30)   (31)

Similarly isoquinolinyl selenates (32) on treatment with chloroacetonitrile form condensed selenazoles such as the 3-aminoselenazolo[3,4-c]isoquinolinium amide inner salt (33) (H. Singh and N. Malhotra, Indian J. Chem., 1983, 22B, 328).

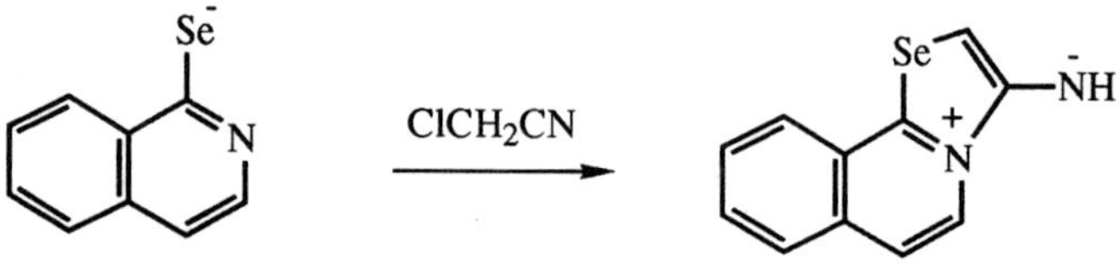

(32)   (33)

13. Benzoselenazoles

The selenadiazole (34) reacts with benzyne to yield the benzoselenazole (35) in low yield (M.R. Bryce et al., Chem. Comm., 1982, 299).

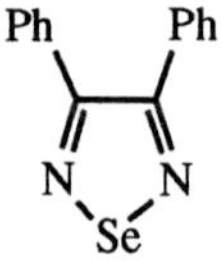

(34) 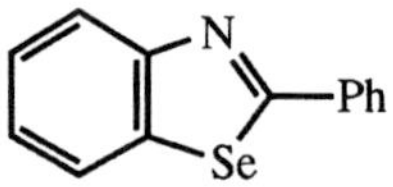   (35) 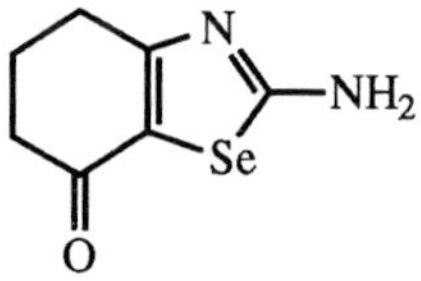   (36)

The tetrahydrobenzoselenazole (36) is produced by cyclocondensation of 2-bromo-1,3-cyclohexanedione with selenourea (V. Barkane et al., Chem. Abs., 1986, 106, 213843).

14. *Benzotellurazoles*

The benzotellurazole (38) is synthesised from the telluroanilide (37) by cyclisation with phosphorus oxychloride (M. Mbuyi, Tetrahedron Letters., 1983, 5873. See also: G.M. Abakarov *et al.,* Khim. geterosikl. Soedin, 1988, 276 and I.D. Sadekov *et al., ibid.,* 1989, 120)

(37)   (38)

An improved synthesis employs azobenzene and tellurium tetrachloride as starting materials. Borohydride reduction of the resulting 2-(phenylazo)phenyltellurium trichloride (39) and subsequent cyclisation with acetic anhydride yields 2-methylbenzotellurazole (40). The derivative (41) may be prepared similarly from the corresponding naphthalene derivative (T. Junk and K.J. Irgolic, Phosphorus and Sulfur, 1988, <u>38</u>, 121).

(39)   (40)

(41)

Chapter 18

FIVE-MEMBERED HETEROCYCLIC COMPOUNDS WITH THREE HETERO-ATOMS IN THE RING

DAVID J. ROWE

1. Introduction

Since the publication of Volume IVD in 1981, "Comprehensive Heterocyclic Chemistry" (ed. A.R. Katritzky and C.W. Rees Pergamon 1984) in eight volumes has been published. Volumes 5, "Five-membered rings with Two or More Nitrogen Atoms " and 6 "Five-membered Rings with Two or More Oxygen, Sulfur or Nitrogen Atoms", both edited by K.T. Potts, contain comprehensive reviews of the literature on these topics to 1981, with some mention of later publications. Volume 5, Chapters 4.01, 4.02 (A.R. Katritzky and J.M. Lagowski) and 4.03 (K.T. Potts) review (respectively) the structure, reactivity and synthesis of five-membered rings with two or more heteroatoms.

The two-volume work "1,3-Dipolar Cycloaddition Chemistry" (ed. A. Padwa, Wiley, 1984) provides information relevant to a number of the heterocycles included in this chapter.

2. 1,2,3-Triazoles

These are reviewed by H. Warnoff in Comprehensive Heterocyclic Chemistry, Vol. 5, Ch. 4, p. 669. 4-Amino-1,2,3-triazoles have the subject of a separate review (A. Albert, Adv. heterocyclic Chem., 1986, 40, 129). The electronic properties of triazoles have also been discussed (V.P. Mamaev, O.P. Shkuro and S.O. Baram, Adv. heterocyclic Chem., 1987, 42, 1).

(a) Synthesis of 1,2,3-triazoles

(i) 1,3-Dipolar addition reactions
The method most frequently used for the synthesis of 1,2,3-triazoles is the 1,3-dipolar addition of an azide to an alkyne. (See W. Livowski, Padwa, Vol. 1, Ch. 5, p. 559, and K. Turnbull, Chem. Reviews, 1988, 88 297). Only selected examples are given here.

N-Unsubstituted compounds (2) can be prepared using 4-methoxybenzyl azide (1) followed by solvolysis; cleavage occurs more readily than with the benzyl derivative (D.R. Buckle and C.J.M. Rockell, J. chem. Soc. Perkin 1, 1982, 627).

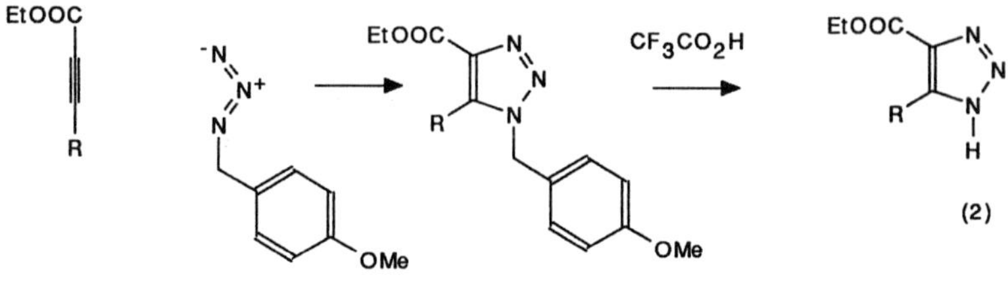

(1)

R=CN,H,Ph,OH,Cl,PhO,4-OMeC$_6$H$_4$S

1-Methyltriazoles (*e.g.* 5) have been prepared using (trimethylsilyl)methyl azide (3) [from (chloromethyl)trimethylsilane and sodium azide)], a synthetic equivalent of the highly explosive methyl azide; the cycloadduct (4) readily undergoes protiodesilylation (O. Tsuge, S. Kanematsu and K. Matsuda, Chem. Letters, 1983, 1131).

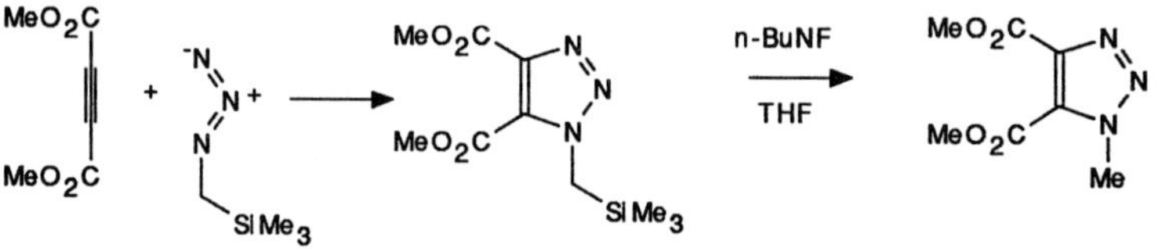

(3)　　　　(4)　　　　(5)

The propargylic azide (6) readily dimerises to the bis(triazole) (7), but reacts with DMAD to yield (8) (H. Prieve, Acta chem. Scand., 1984, B38, 623).

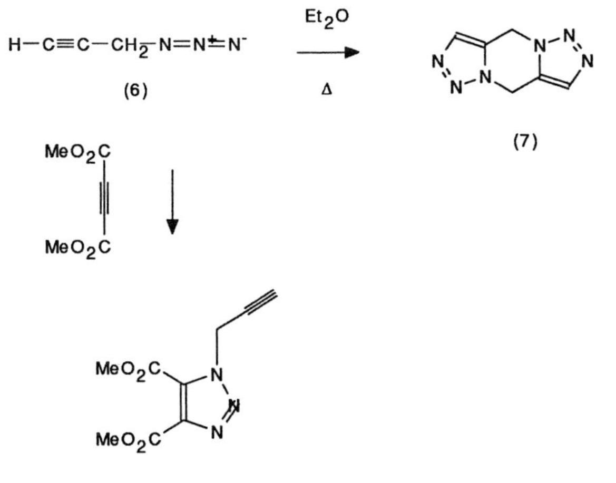

(6)

(7)

(8)

Both azide groups of (9) react with DMAD to give the di(triazolyl)naphthalene (10) (K. Handa *et al.*, Chem. Comm., 1984, 450).

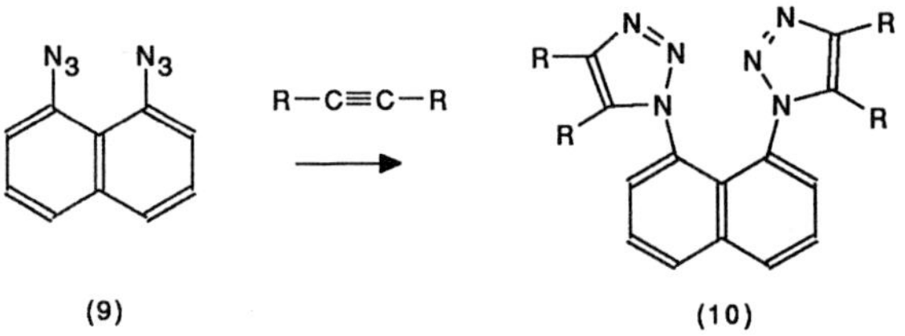

(9)

(10)

R=CO₂Me

A 1,3-dipolar addition is the first step in the synthesis of the benzodiazepine analogue (11) (F. Melane, *et al.*, J. heterocyclic Chem., 1989, 26, 1605).

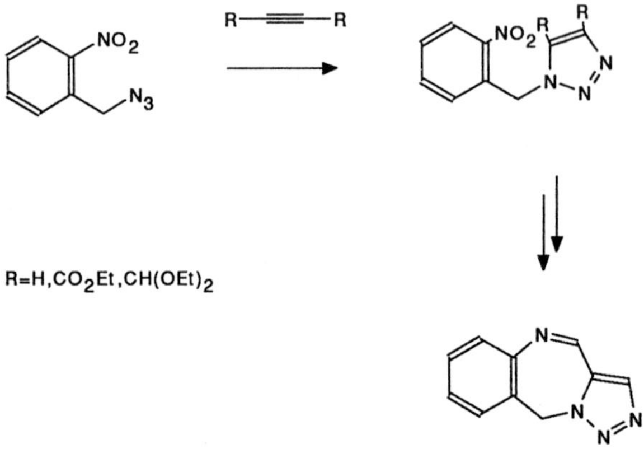

R=H,CO$_2$Et,CH(OEt)$_2$

(11)

An important azide in medicinal use is the "anti-AIDS" drug AZT(12) from which a large number of triazoles have been prepared (P. Wigerinck *et al.*, J. heterocyclic Chem., 1989, 26, 1635).

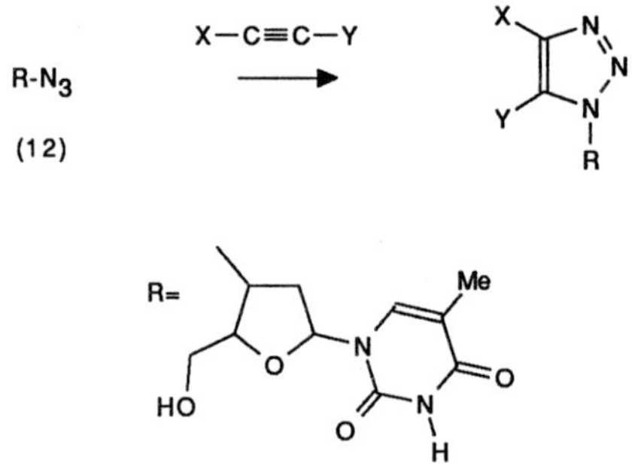

R-N$_3$

(12)

X,Y=H,SiMe$_3$,CH$_2$Hal,Me,Co$_2$Et,Ph,OEt,NO$_2$

The rate of the reaction between the cationic reagents (13) and (14) is increased by a factor of 5×10^4 in the presence of the encapsulating nonadecacyclic cage-compound, cucurbituril (W.L. Mock, T.A. Irra, J.P. Wepsiec, J. org. Chem., 1983, 48, 3619).

$H_3\overset{+}{N}-CH_2-C\equiv CH$

(13)

$N_3-CH_2-CH_2-\overset{+}{N}H_3$

(14)

The reactions of conjugated enynes (e.g. 15), with phenyl azide occur preferentially at the triple bond (V.A. Galishev, I.A. Maretina and A.A. Petrov, J. org. Chem. U.S.S.R. 1984, 19, 404; 1985, 20, 1446).

(15)

The enol tautomer of a 1,3-dicarbonyl compound may act as 1,3-dipolarophile. Thus the azide (16) reacts with pentane-1,3-dione to give (17) after loss of water (C. Carral et al., J. heterocyclic Chem., 1987, 24, 1301).

$-H_2O$

(16)

(17)

1,2,3-Triazoles are also formed from the 1,3-dipolar addition of a diazoalkane to a nitrile (M. Regitz and H. Heydt in Padwa, Vol. 1, Ch. 4, p. 462). For example diazomethane adds to the nitrile (18) to give a mixture of methylated 1,2,3-triazoles presumably formed by methylation of the triazole (19) (G. Mouysset et al., J. heterocyclic Chem., 1988, 25, 1167).

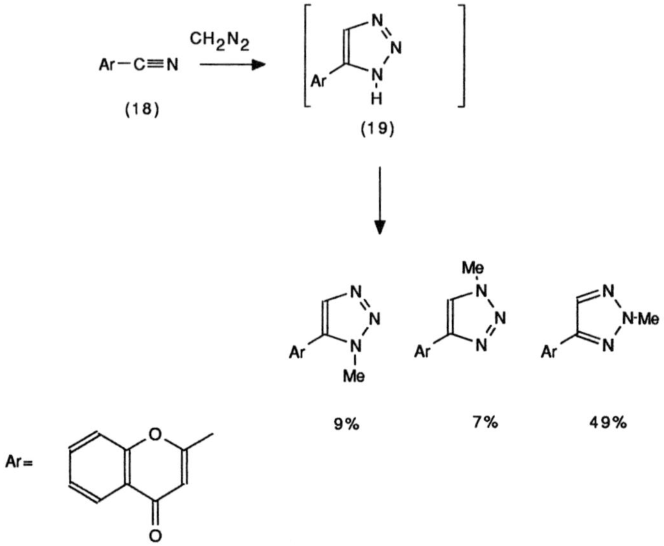

Ar=

(b) Oxidative synthesis

Oxidative cyclisation of bis(hydrazones) ("osazones") is the oldest established method of 1,2,3-triazole synthesis. The mechanism of this process has been discussed in a review of the oxidation of nitrogen compounds with lead (IV), thallium (III) and mercury (II) acetates (R.N. Butler, Chem. Reviews, 1984, 84, 249). This process and its relevance to other cyclisations in heterocyclic chemistry is discussed in the section on Rearrangements(p.352).

An example of an oxidative synthesis in the preparation of an 1-aroyloxy-4,5-dimethyl- 1,2,3-triazole (20) from an α-hydroxamino aroylhydrazone of butane-2,3-dione (21); the structure of (20c) has been confirmed by X-ray crystallography. (A.B. Theocharis et al., J. chem. Soc. Perkin 1, 1989, 619)

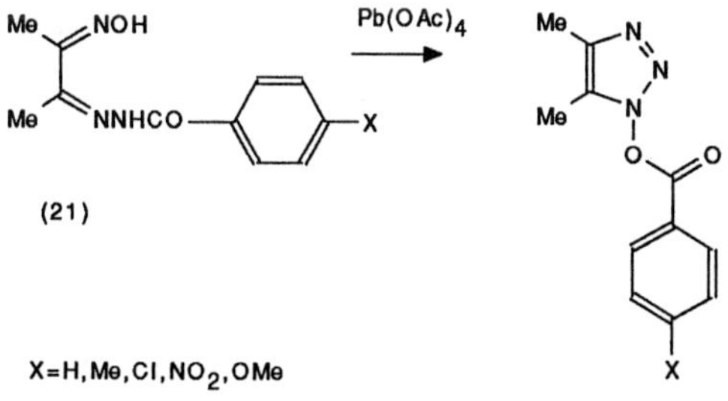

$X = H, Me, Cl, NO_2, OMe$

(c) Reactions and Uses of 1,2,3-triazoles

1-Hydroxy benzotriazole (22) (HOBT) is extensively used as a "coupling agent" in peptide synthesis, where it functions as both a catalyst and a racemisation suppressant ("The Practice of Peptide Synthesis", M. Bodanszky and A. Bodanszky and "Principles of Peptide Synthesis, M. Bodanszky, both Springer-Verlag, N.Y. 1984).

The mode of action of (22), and in particular the participation of *O*-(23) and *N*-acylated (24) derivatives has been studied.

(22)

(23) (24)

The *O*-acyl compound (25) rapidly rearranges to (26) in moist acetone and equilibration occurs even on attempted column chromatography; the existence of a ketene intermediate (28) follows from the formation of (29) when (27) rearranges in methanol (S. Nagarajan, S.R. Wilson, K.L. Rinehart Jnr., J. org. Chem., 1985, 50, 2174.

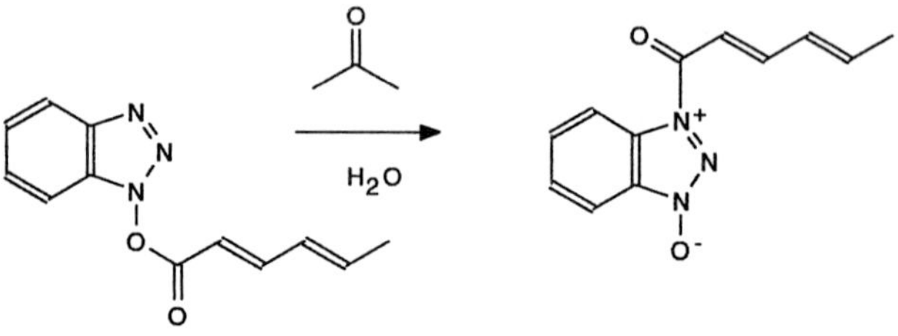

(25)   (26)

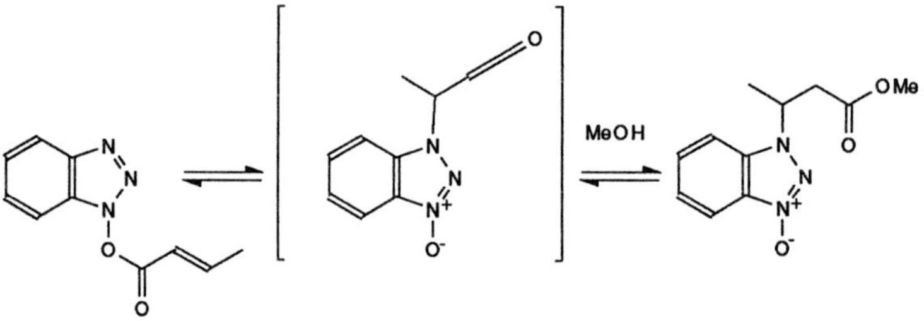

(27)   (28)   (29)

Acylation of HOBT with *N*-tritylmethionine occurs on nitrogen to give (30), the structure of which has been confirmed by X-ray crystallography; again equilibration with the *O*-acylated species (31) occurs in solution (K. Barlos, D. Papaisannou and S. Voliotes, J. org. Chem., 1985, 50, 696).

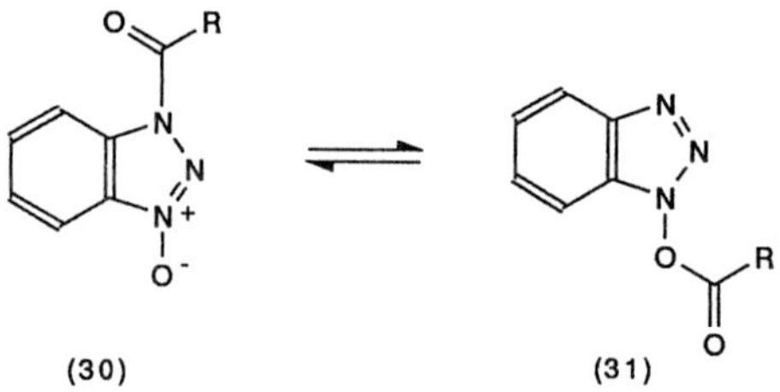

(30)   (31)

$$R=CH(CH_2SCH_3)NHCPh_3$$

Due to the facile interconversion of *O*- and *N*-acylated species, structural assignments based on solution data such as nmr spectroscopy are ambiguous. An investigation using X-ray crystallography (J. Singh *et al.*, J. org. Chem. 1988, <u>53</u>, 205) has shown that the *O*-acylated strucutres assigned to (32a) (S. Kim and H. Chang, Chem. Comm., 1983, 1357) (32b) (*idem*, Bull. Korean Chem. Soc., 1986, <u>7</u>, 70), and (32 c) (H. Kinoshita *et al.*, Chem. Letters, 1985, 515) are incorrect and has confirmed the structures of (32d) and (32e).

R

a	$(CH_3)_3CO$
b	$PhCH_2O$
c	$PhCH=CHCH_2O$
d	CH_3SO_2
e	(see structure)

(32)

The above results casts doubt on the structures of (33a) (A. Paquet, Canad. J. chem., 1982, <u>60</u>, 676) (33b) (R.E. Shute and D.H. Rich, Synthesis, 1987, 346) and (34) (K. Takeda, *et al, ibid.*, 1987, 557), which were not determined by X-ray crystallography.

(33)   (34)

R

a

b $Me_3SiCH_2CH_2O$

Acyl derivatives of monocyclic 1,2,3-triazoles also rearrange. Oxidation of (35) yields (37); this may proceed *via* the *N*-aroyl compound (36), with rearrangement occurring *via* an intramolecular [1,3]aroyl migration or an intermolecular aroyl transfer. Support for the latter mechanism comes from identification of (38) as a by-product in this reaction, assumed to be formed by aroylation of the starting material by (36) or (37). Treatment of (35) with (37) gives a near-quantitative yield of(38)(A.B. Theocharis *et al.*, J.chem.Soc.Perkin 1, 1989, 619).

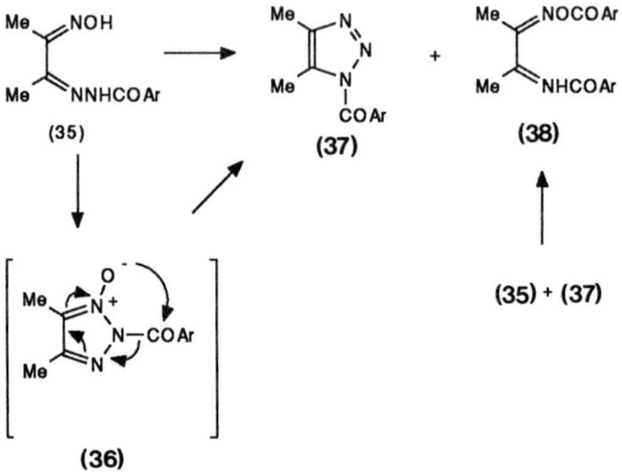

The mechanism of the rearrangement of acylated benzotriazoles is unclear; it has been suggested that *O*-acylation predominates under kinetic control, with the thermodynamically stable *N*-acylated product formed *via* an intermolecular transfer between two components of an ion pair, as shown below (J. Singh, *et al.*, J. org. Chem., 1988, 53, 205).

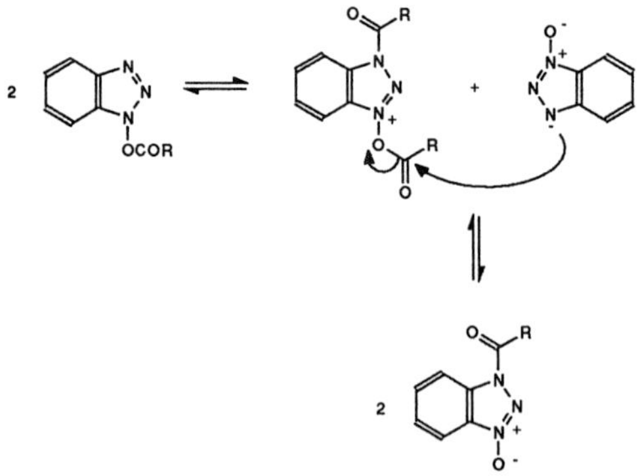

(Alkylamino)triazoles also rearrange. The "bis-triazole" (39), shown to have this structure in the solid phase by X-ray analysis, rearranges in solution, probably *via* species such as (40) and (41) (A.R. Katritzky *et al.*, J. chem. Soc. Perkin 1, 1987, 2673).

(d) *Extrusion of nitrogen from benzotriazoles*

The synthesis of carbazoles by thermal or photochemical extrusion of nitrogen from benzotriazoles (the Graebe-Ullman synthesis) is a possible route to 4aH-carbazoles (45). For example, thermolysis of (42) gives, *inter alia*, the carbazoles (43) and (44), whose formation has been rationalised as arising *via* rearrangement of (45) (J.J. Kulagavski, C.J. Moody and C.W. Rees, J. chem. Soc. Perkin 1, 1985, 2725).

The intermediacy of (45) is also proposed in the formation of the cyclopentaquinoline (46) in the photolysis of (42), with (45) undergoing an aza-di-π-methane rearrangement (J.J. Kulagavski, C.J. Moody and C.W. Rees, *loc. cit.*).

(46)

The triazole (47), the benzo-analogue of (42), is stable, being obtained in 24-30% yield on photolysis of (48), together with the by-products (49) and (50) (*idem, ibid,* 2733).

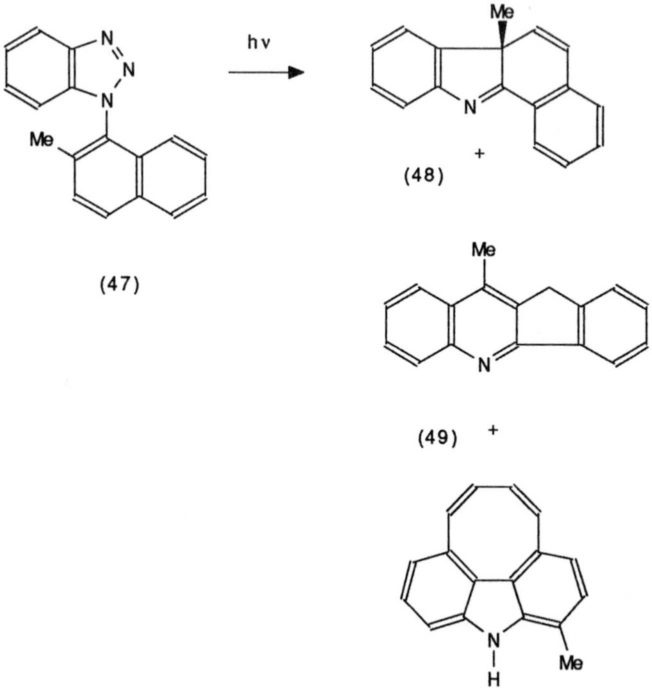

(47)

hv

(48)

+

(49)   +

(50)

Photolysis of an 1-alkenylbenzotriazole yields an indole.   For example 1-chlorobenzotriazole (51) adds rapidly to cyclohexene to give (52) which is dehydrohalogenated to (53); photolysis of the latter gives the tetrahydrocarbazole (54). Benztriazole (55) adds to ethyl propiolate in refluxing toluene to give a mixture of the *E* and the *Z*-isomer of (56). Photolysis of this mixture gives (57) (P.A. Wender and C.R. Cooper, Tetrahedron, 1986, 42, 2985).

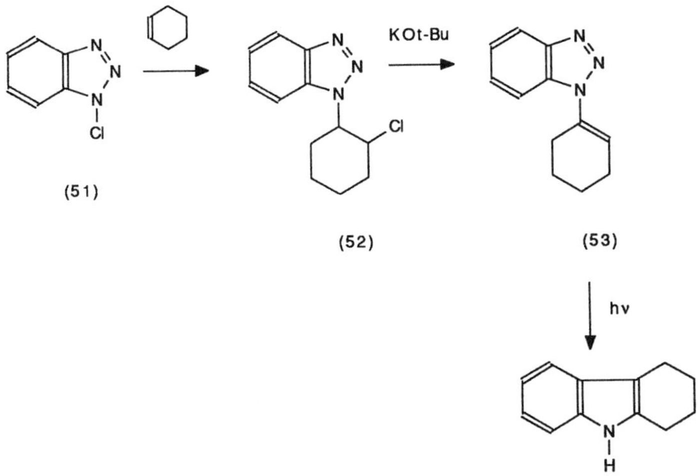

(51) (52) (53)

(54)

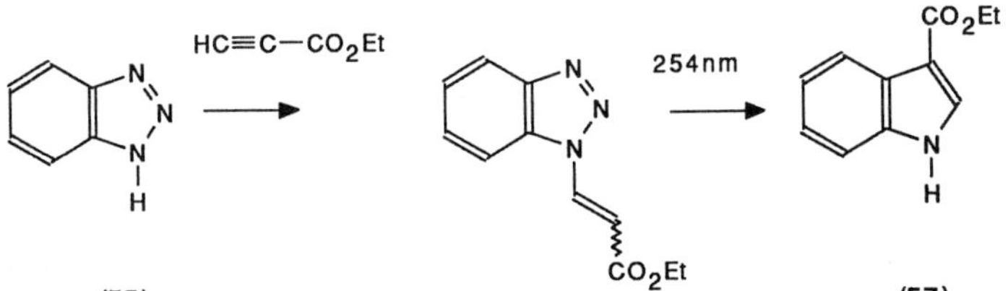

(55) (56) (57)

Insight into the mechanism of nitrogren extrusion is provided by the observed formation of phenylketenimine (59) by thermolysis of 1-vinylbenzotriazole (58); at a higher temperature this isomerises to indole (60) (A. Maquestiau, *et al.*, J. chem. Soc. Perkin 2, 1988, 1071).

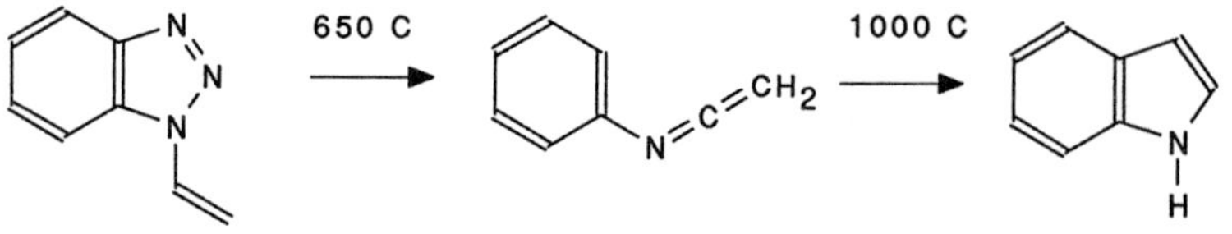

(59) (60) (61)

Benzotriazole (61) is a powerful nucleophile, adding to aldehydes to give alcohols (62a) which precipitate from non-polar solvents; in solution, (62a) is in equilibrium with the strongly hydrogen bonded ion pair (62b).

The alcohol (62a) is highly reactive, yielding aminomethyl compounds (63) with amines. Formation of (63) can be carried out as a "one-pot" synthesis from benzotriazole (56) (A.R. Katritzky, S. Rachival, B. Rachival, J. chem. Soc. Perkin 1, 1987, 791, 799).

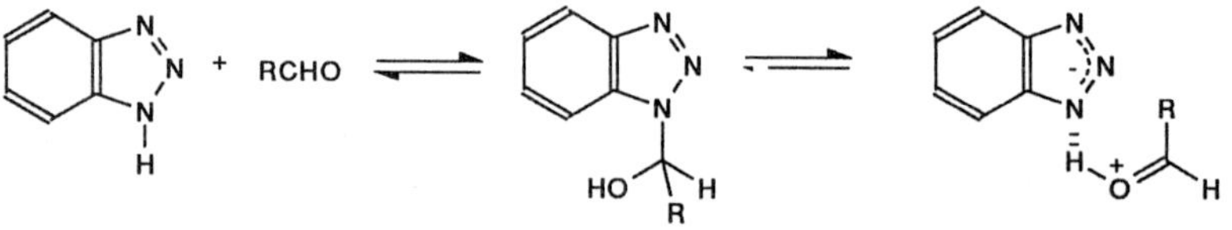

(61) (62a) (62b)

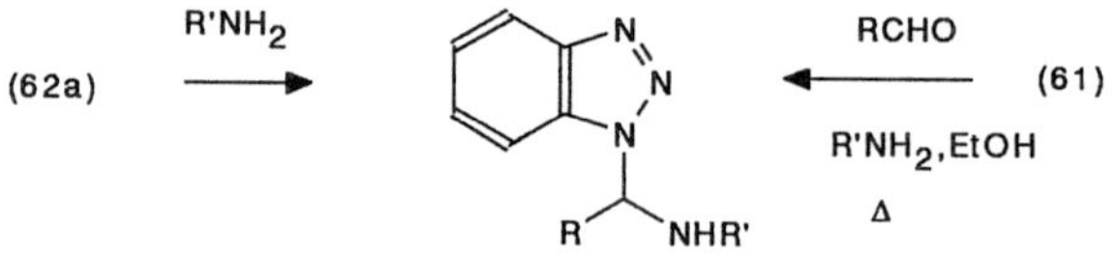

(63)

Aliphatic amines with benzotriazole tend to give bis(triazoles) (64) (A.R. Katritzky, S.Rachival and B.Rachival, *loc cit*).

(64)

Aromatic derivatives of (63) are cleaved by hydride reductants or Grignard reagents, making this reaction sequence an excellent method form monoalkylation of aromatic and heteroaromatic amines (A.R. Katritzky, *et al.*, J. chem. Soc, Perkin 1, 1987, 805).

The adducts (65) yield tertiary amines (66) and (67) upon reaction with Grignard reagents and sodium borohydride respectively. The "dimer" (68) behaves similarly, as does the "trimer" (70) formed from a formaldehyde adduct (69) with formamide (A.R. Katritzky *et al.*, J. chem. Soc. Perkin 1, 1989, 225).

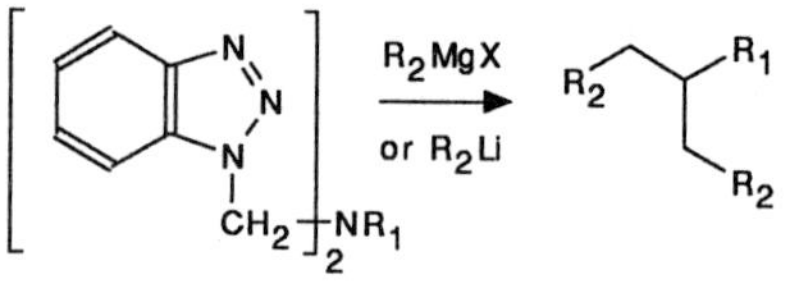

(68)

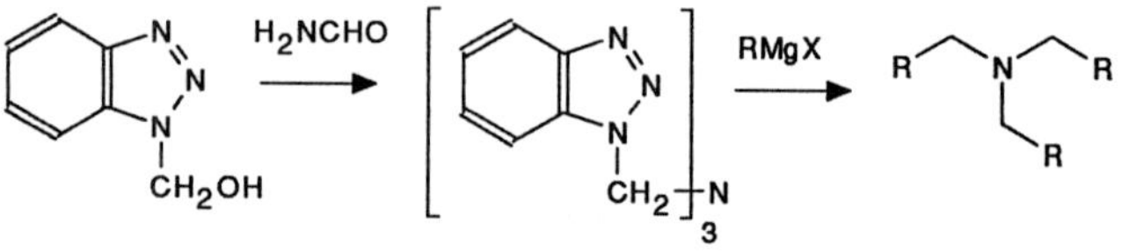

(69)　　　　　(70)

A similar reaction sequence can be carried out using amides instead of amines; the adduct can be cleaved to either an amide or an amine (A.R. Katritzky and M. Drewniak, J. chem. Soc. Perkin 1, 1988, 2339):

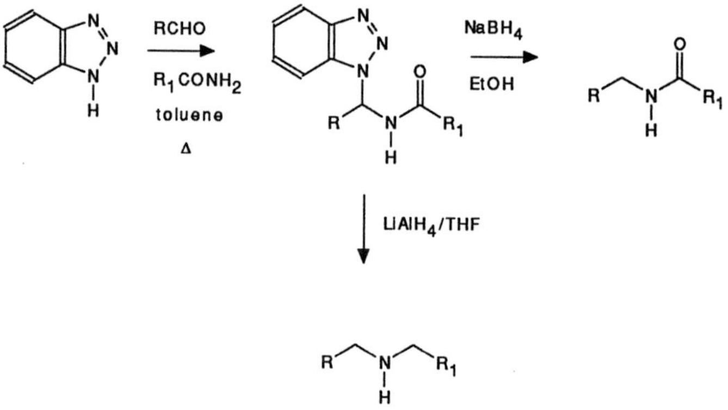

In the presence of thionyl chloride, the benzotriazole/aldehyde adducts form chloroalkyl derivatives (71) which react readily with nucleophiles (A.R. Katrizky, *et al.*, J. chem. Soc. Perkin 1, 1987, 781, 811).

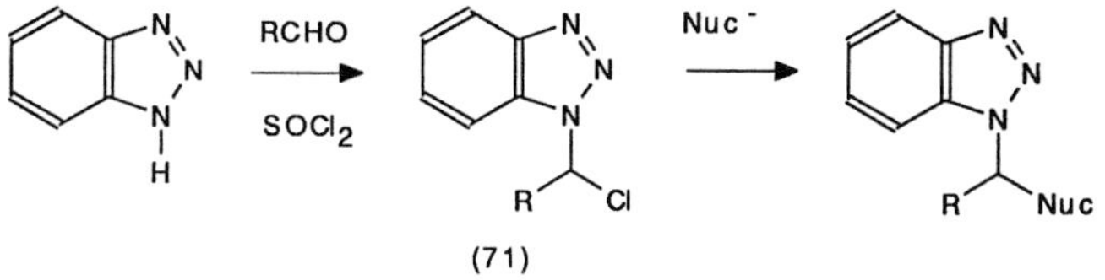

(71)

Nuc=H,OR,SR,N$_3$,CN,R'(via R'MgX)

The bis-compound (72), obtained from an aromatic aldehyde and benzotriazole, is readily metallated, the resulting compound (73) being highly nucleophilic. The compound (73) is the synthetic equivalent of the ArCO unit as illustrated below in the preparation of the ketones (74) and (75) (A.R. Katritzky and W. Kuzmierkiewicz, J. chem. Soc. Perkin 1, 1987, 819).

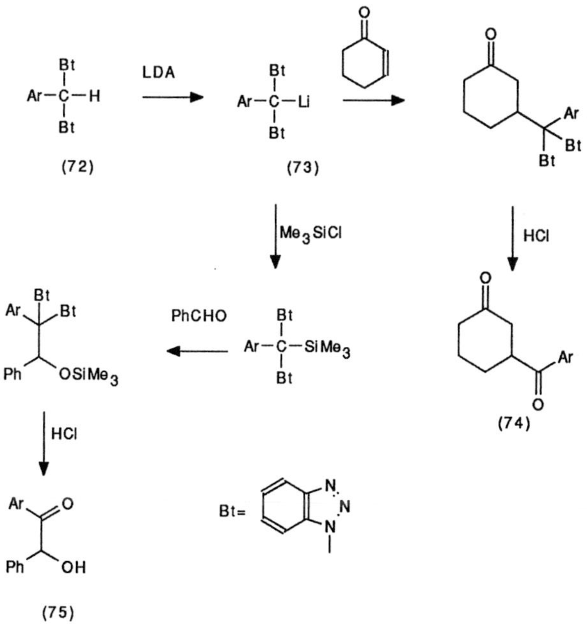

The triazolium imides (76) are powerful 1,3-dipoles reacting with methyl cyanodithioformate (77) to yield the 8π 1,3,4,5-thiatriazine system (78) *via* the rearrangement shown (R.N. Butler, *et al.*, J. chem. Soc Perkin 1, 1989, 371).

$Ar = Ph, 4-BrC_6H_4, 4-NO_2C_6H_4$

(78)

3. Triazolines

Δ^2-1,2,3-Triazolines (79), usually simply termed "triazolines" (P.K. Kadava, B. Stabonik and M. Toler, Adv. heterocyclic Chem., 1984, 37, 217) and the less common Δ^3- and Δ^4-1,2,3-triazolines (80), (81) (P.K. Kadaba, *ibid.*, 351) have been reviewed.

(79)   (80)   (81)

(a) Synthesis of triazolines

(i) 1,3-Dipolar addition of azides to alkenes
For general reviews, see the synthesis of 1,2,3-triazoles from azides (p.259).
Aryl azides add both regiospecifically and stereospecifically to the allene (82) to yield
(83) (D. Baraclough *et al.*, J. chem. Res. Synop., 1984, 102).

$Ar = Ph, 4-NO_2$ and $2-OMe-C_6H_4$

Spirotriazoles are formed when aromatic azides add to methylenefluorene (84) (K. Hirakawa and Y. Tanakiki, J. org. Chem., 1982, <u>47</u>, 280) and the polycycle (85) (L. Fitjer, Ber., 1982, <u>115</u>, 1047).

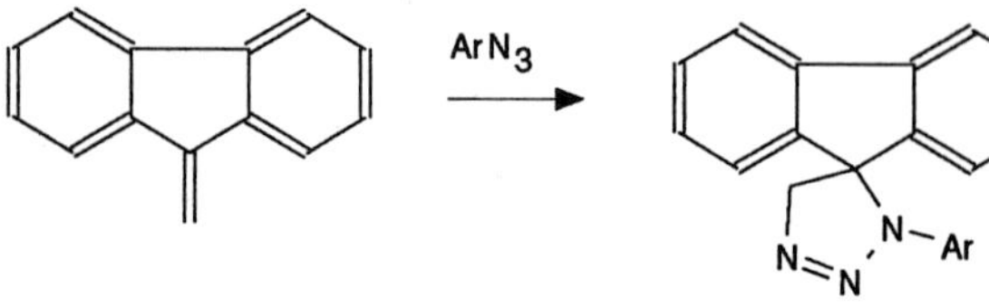

(84)

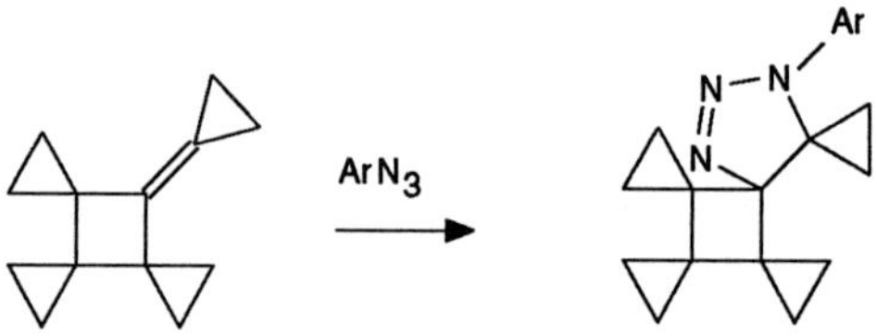

(85)

Intramolecular cycloaddition may occur as in the formation of (87) from the azide (86) (M.J. Pearson, Chem. Comm., 1981, 947).

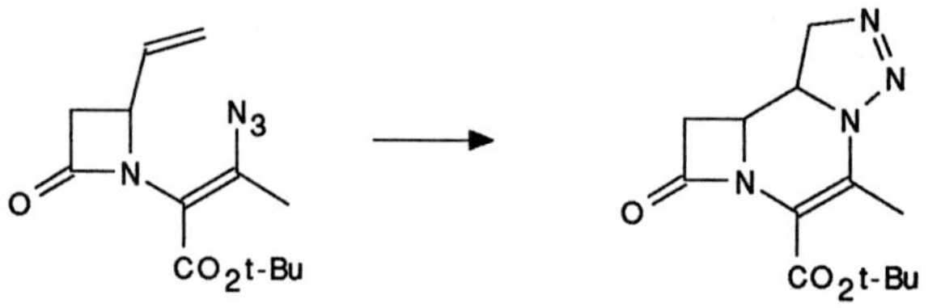

(86)   (87)

The normally sluggish addition of azides to dimethyl maleate is accelerated by complexation of the azide to a metal ion (T. Kammerich *et al.*, Inorg. Chem., 1982, 21, 1226).

The 4,5-diacyltriazolines formed by addition of aryl azides to *trans*-1,2-dibenzoyl- and to diacetyl-ethylene are unstable and decompose to a variety of products. *trans*-1,2-Diacetylethylene (88) reacts with the azide (89) to give the enamine (91) as the major product, formed *via* ring-opening of the triazoline (90) (L. Benati, P.C., Monevecchi and P. Spagnola, J. chem. Soc. Perkin 1, 1989, 2235). However the triazoline is formed in 35% yield from *trans*-1,2-dibenzoylethylene (92) and the azide (89).

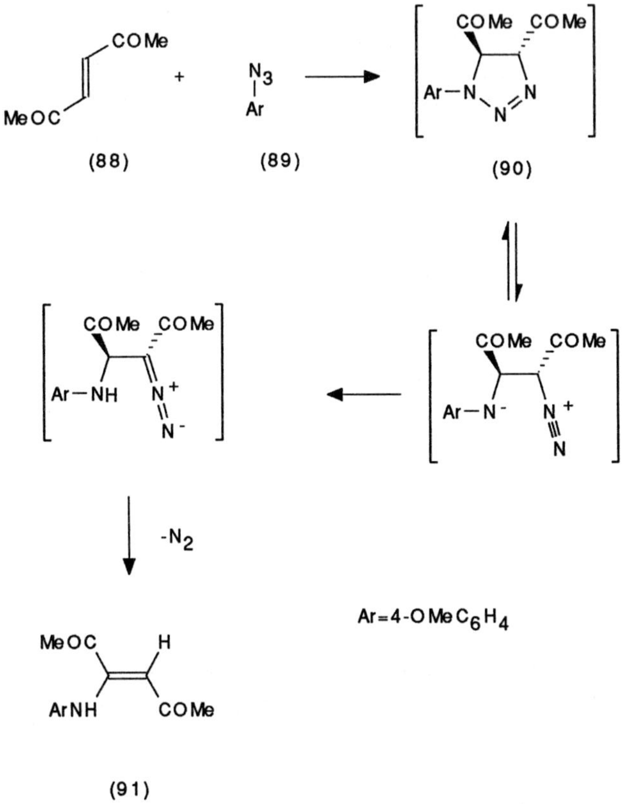

(88) + (89) ⟶ (90)

(91)

Ar=4-OMeC$_6$H$_4$

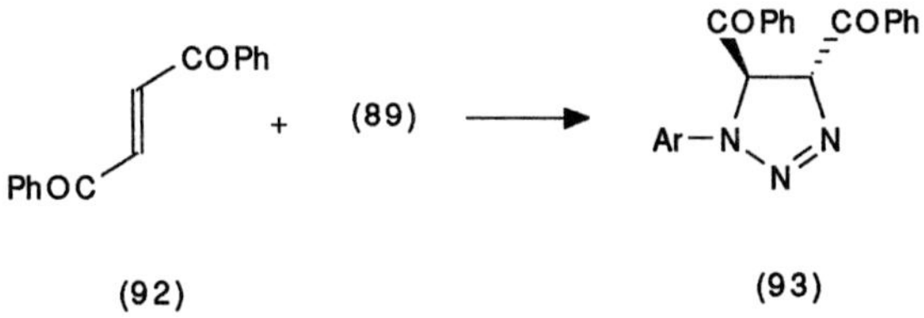

(92) + (89) ⟶ (93)

Triazolines are valuable synthetic intermediates, which may be used without actual isolation. The azide (94) undergoes an intramolecular cycloaddition, forming (95) which spontaneously eliminates nitrogen to yield the aziridine (96); this in turn undergoes ring fission to yield, *inter alia*, the pyrrolizidine alkaloid supiridine (97) (T. Hudlicky, J.D. Frazier and L.D. Kwart, Tetrahedron Letters, 1985, 26, 3523).

The aziridinone (101) is formed by photolysis of the triazoline (100) obtained by addition of azide (98) to the lithium enolate (99) (H. Quast and B. Seiferling, Tetrahedron Letters, 1982, 23, 4681).

Triazolines are formed by 1,3-dipolar addition of diazomethane to the carbon-nitrogen bond of an imine (for a review see M. Regitz and H. Haydt in Padwa, Vol. 1, Ch. 4, p. 462). The imine (102) and diazomethane yield (103) (A Hammam, Ind. J. Chem., 1982, 21B, 348).

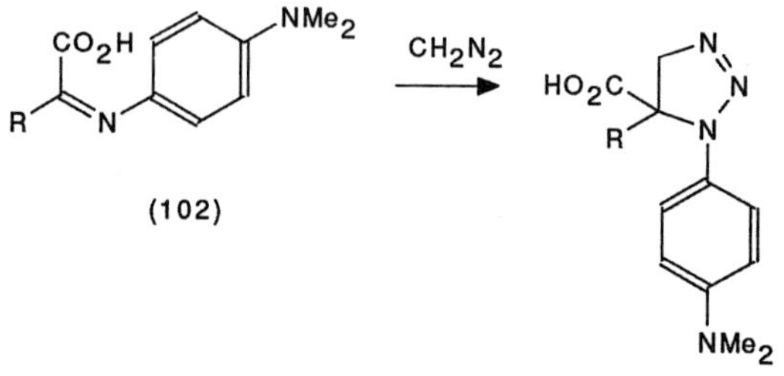

(102)

(103)

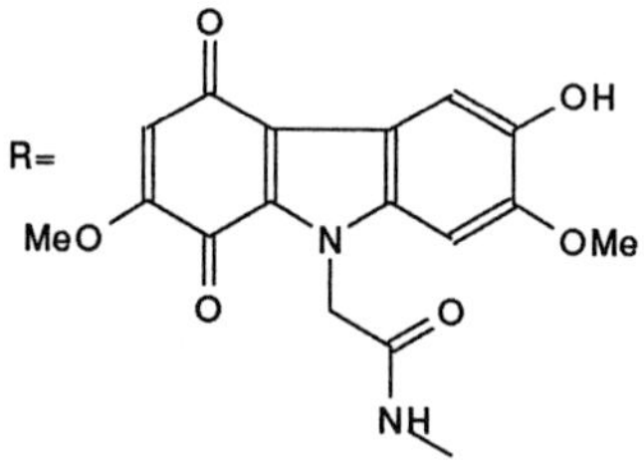

R=

Some triazolines can be oxidized to triazoles; potassium permanganate under phase transfer conditions has been recommended for this reaction (P.K. Kadaba, J. pr. Chem., 1982, 324, 857).

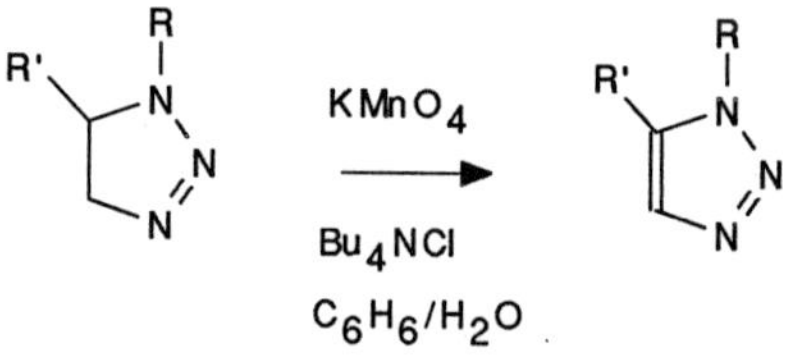

R=aryl;R'=aryl,heteroaryl

4. 1,2,4-Triazoles

For a review see J.B. Polya in Comprehensive Heterocyclic Chemistry, Vol. 5., Ch. 4.12, p. 733.

(a) Synthesis

The most used synthesis of 1,2,4-triazoles is the cyclisation of hydrazine derivatives; differences occur in the method of introducing the "additional" nitrogen atom. Isothiocyanates are frequently used, as illustrated by the synthesis of (104) and (105) (F. Kurzer and J.L. Secker, J. heterocyclic Chem., 1989, 26, 355).

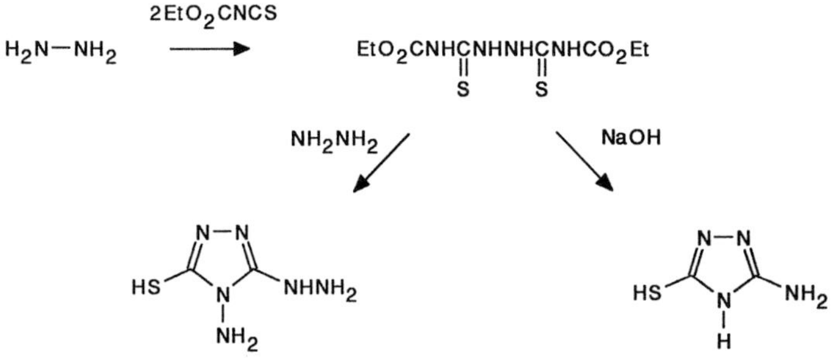

(104) (105)

Frequently, acid hydrazides are used instead of hydrazine itself, for example (106) (B. Rigo, J. heterocyclic chem., 1989, 26, 1723) and (107) (Y. Kurasawa *et al., ibid.,* 1986, 23, 1391):

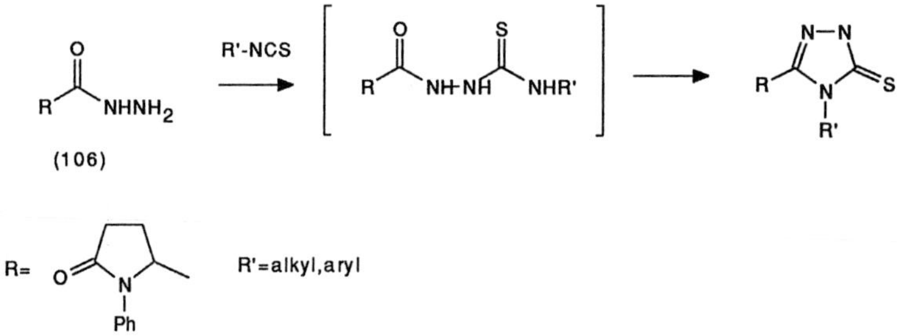

(106)

R=   R'=alkyl,aryl

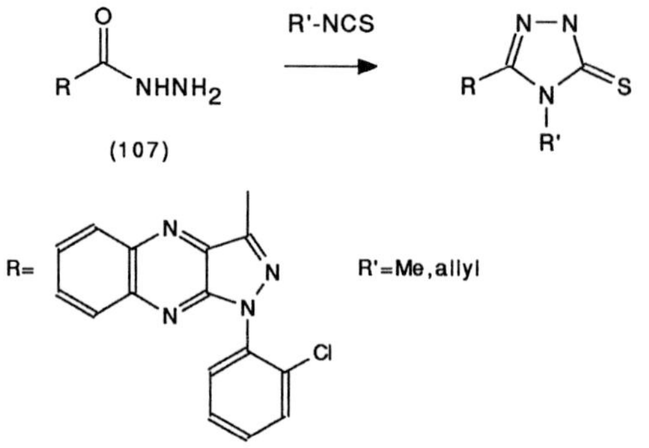

(107)

R'-NCS

R= (quinoxaline-pyrazole system with 2-Cl-phenyl)

R'=Me,allyl

The bis(hydrazides) (108) yield the bis(1,2,4-triazol-5-thiones) (109) (V.J. Ram *et al.*, heterocyclic Chem., 1988, <u>25</u>, 253; 1989, <u>26</u>, 625).

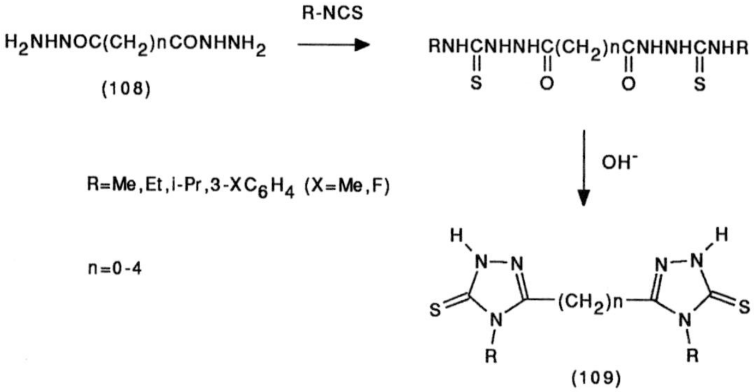

$H_2NHNOC(CH_2)nCONHNH_2$

(108)

R-NCS

$RNHCNHNHC(CH_2)nCNHNHCNHR$

S O O S

R=Me,Et,i-Pr,3-XC_6H_4 (X=Me,F)

n=0-4

OH^-

(109)

The 1-acylbithioureas (110 a-c) give 1,2,4-triazole-5-thiones(111a-c) when treated with sodium hydroxide under phase-transfer conditions; other derivatives yield 1,3,4-thiadazoles(p.334)(T. Okawara *et al.*, J. heterocyclic Chem., 1988, <u>25</u>, 1071).

$$
\underset{(110)}{\text{PhNHCNR}_1\text{NHCNHCR}_2}
$$

(110) → (111)

	R_1	R_2
a	H	CH_2Ph
b	Me	CH_2Ph
c	Me	Ph

Ethyl hydrazinecarboxylate (113) reacts with (112) to give the adduct (114), which readily cyclises with hydrazine (F. Malbec *et al.*, J. heterocyclic Chem., 1984, 21, 1769).

1,3,4-Oxadiazoles can act as hydrazine precursors. The fluoroalkyl derivatives (115) are cleaved by amines to give (116), which cyclise to the 1,2,4-triazoles (115) (D.B. Reitz and M.J. Finkes, J. heterocyclic Chem., 1989, 26, 225). Reaction of (115) with hydrazine proceeds in a similar manner to yield *N*-amino-1,2,4-triazoles (117), which can be deaminated with nitrous acid *(idem*, J. org. Chem., 1989, 54, 1760).

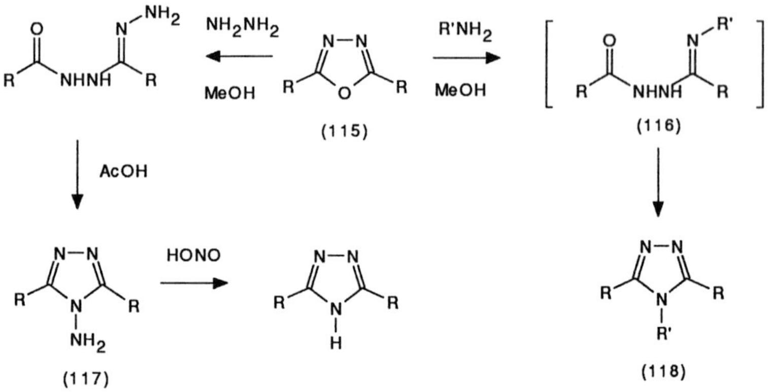

288

(115)

(116)

(117)

(118)

R=CF$_3$,C$_2$F$_5$,n-C$_3$F$_7$

R'=alkyl,aryl

The 1,3,4-oxadiazol-5-thione (119) behaves similarly (B. Rigo, J. heterocyclic Chem., 1989, 26, 1723).

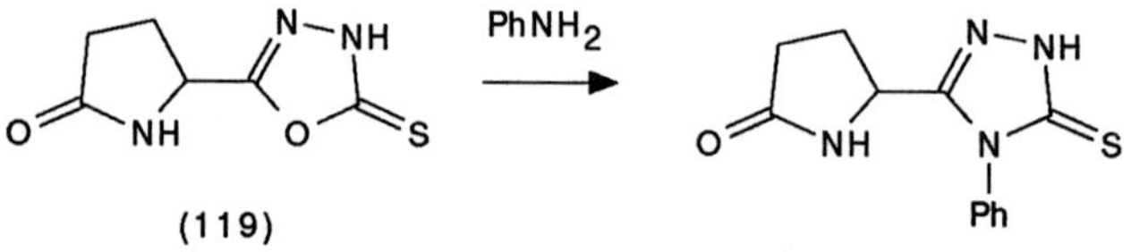

(119)

Diphenylcyano-carbonimidate (120), a reactive precursor, reacts with amines to yield the compounds (121) which cyclise readily with hydrazine (R.L. Webb *et al.*, J. heterocyclic Chem., 1982, 19, 1205; 1987, 24, 275).

R=Ph, CH$_2$Ph, CH$_2$CH$_2$Ph,

Lead dioxide oxidation of an aldehyde semicarbazone gives a high yield of a 2,4-dihydro-1,2,4-triazol-3-one, (122); a free radical mechanism has been proposed (T. H. Nguyen, R. Milcent and G. Barvier, J. heterocyclic Chem., 1985, 22, 1383).

A similar cyclisation occurs when aromatic aldehyde semicarbazones and thiosemicarbazones are treated with sulphur monochloride; the latter can also produce 1,3,4-thiadiazoles (p.333) (R. Milcent and T.-H. Nguyen, J. heterocyclic Chem., 1986, 23, 881).

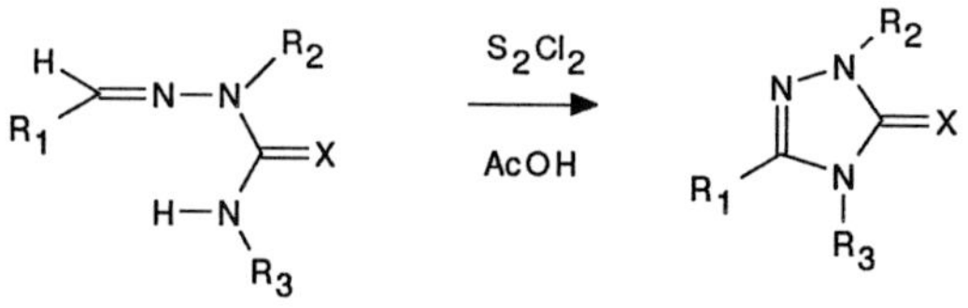

$X=O,S$

$R_1=Ph,4-OMe-,4-Cl-C_6H_4$

$R_2=H,Ph$

$R_3=H,Et,n-Pr,Ph,3-NO_2C_6H_4,2,5-Cl_2C_6H_3$

The complex hydrazone (125) is proposed as an intermediate in the reaction between the acyl hydrazide (124) and the sulphonylhydrazone (123) (H.M. Hassaneen et al., J. heterocyclic Chem., 1984, 21, 797).

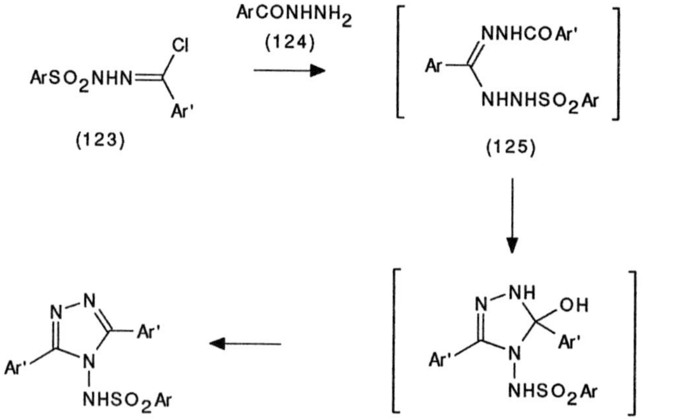

Ar=2-Naphthyl

$Ar'=4-X-C_6H_4;X=H,CH_3,OCH_3,NO_2$

The amidrazone (126) reacts with either ethyl chloroformate, ethyl chlorothioformate or phenyl isocyanate to yield the 1,2,4-triazolenone (127). The latter is formed from phenyl isocyanate *via* the isolable urea (123) (I.T. Barnish *et al.*, J. heterocyclic Chem., 1986, 23, 417).

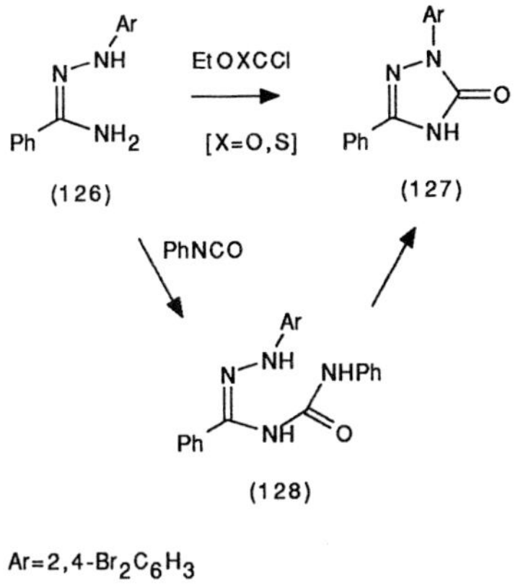
EtOXCCl
[X=O,S]
PhNCO
(126)
(127)
(128)
Ar=2,4-Br$_2$C$_6$H$_3$

The hydrazone unit of (130), generated by Curtius Rearrangement of the azide (129), undergoes intramolecular addition to the isocyanate function (H.S. Kim, Y. Kurasawa and A. Takada, J. heterocyclic Chem., 1989, 26, 1129).

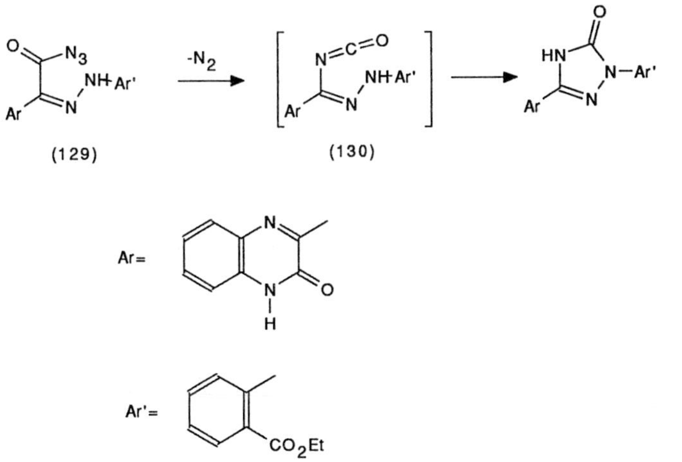
-N$_2$
(129)
(130)
Ar=
Ar'=

The complex thiosemicarbazone (131) when heated in pyridine gives a mixture of the triazoles (132) and (133); however when (131) is first alkylated to yield the isothiosemicarbazone (134), cyclisation proceeds cleanly to give (135) (H. Emilsson and H. Selander, J. heterocyclic Chem., 1988, 25, 565).

Ar=2,6-Cl$_2$C$_6$H$_3$

Cyclisation of isothiosemicarbazones is of great synthetic utility. For example (136) reacts with the vinyl ether (137) to yield (138), which cyclises in acetic acid to give the N-alkenyl-1,2,4-triazole (139). Use of the nitrile (140) instead of (137) yields a triazolopyrimidine such as (141) via an additional reaction at the nitrile function (C. Yamazaki et al., J.org. Chem., 1985, 50, 5513).

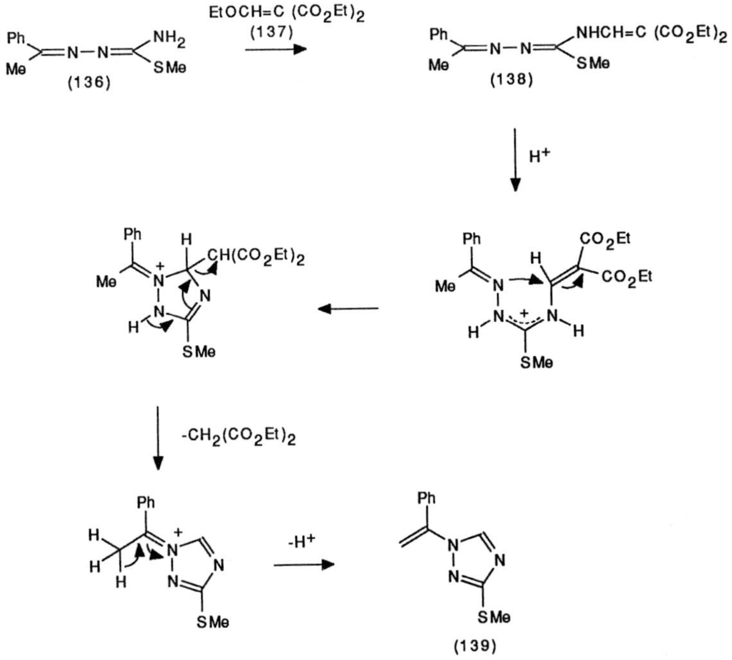

EtOCH=C (CO2Et)2
(137)
(136)
(138)
H+
-CH2(CO2Et)2
-H+
(139)

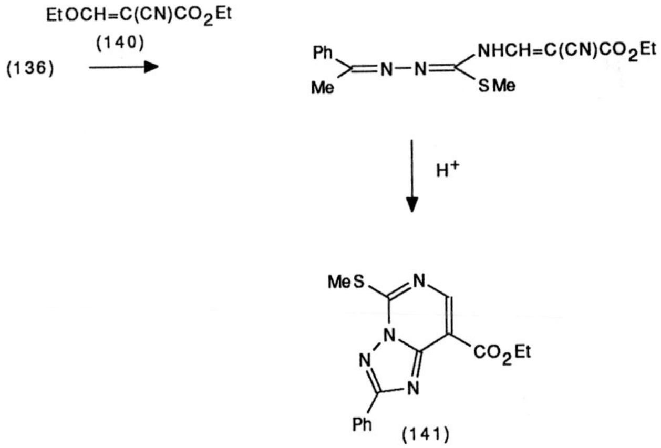

EtOCH=C(CN)CO2Et
(140)
(136)
H+
(141)

When isothiosemicarbazones are treated with the nitroacrylate (142) 1,2,4-triazoles are formed directly without isolation of an intermediate; when the isothiosemicarbazone is unsymmetrical the terminal "anti-Saytzeff" alkene predominates in the product. Thus (143) yields only (144) (C. Yamazaki *et al.*, J.chem. Soc. Perkin 1, 1987, 1567).

Bis (1,2,4-triazoyl) alkanes can be prepared using this approach. The isothiosemicarbazone (145) reacts with (146) in aqueous formic acid to yield (149); a mechanism involving addition of the adduct (147), analogous to (138), to an intermediate iminium ion (148), has been suggested (C. Yamazaki, T. Takahashi and K. Hata, J. chem. Soc, Perkin 1, 1988, 1897).

Nitrogen analogues of isothiosemicarbazones also undergo cyclisation. The diaminomethylene- hydrazone (150) react with the dinitrile (151) to yield (152) which undergoes acid-catalysed cyclisation to (153). When the dinitrile (151) is replaced by the cyanoacrylate (154), pyrimidine formation is suppressed and a N-alkenyl-1,2,4-triazole (155) is formed (Y. Miyamoto, Chem. pharm. Bull, 1985, 33, 2678; Y. Miyamoto and C. Yamazaki, J. heterocyclic Chem., 1989, 26, 327, 763).

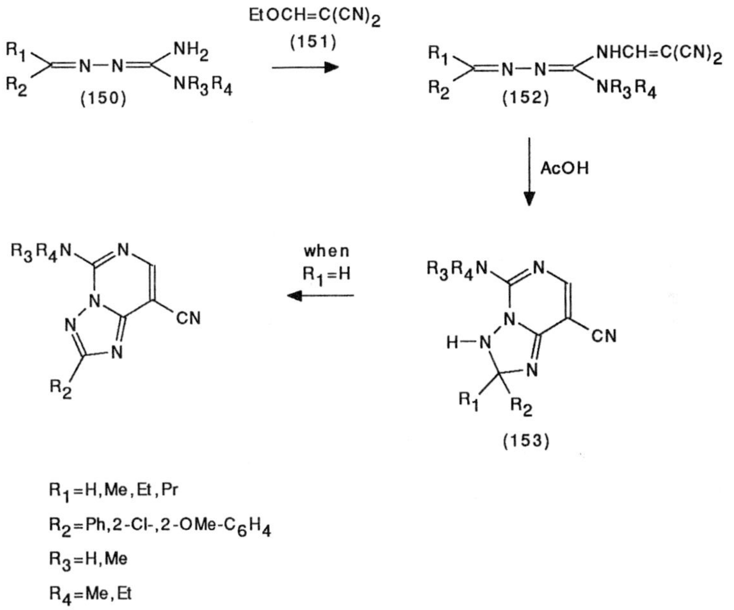

$R_1 = H, Me, Et, Pr$

$R_2 = Ph, 2-Cl-, 2-OMe-C_6H_4$

$R_3 = H, Me$

$R_4 = Me, Et$

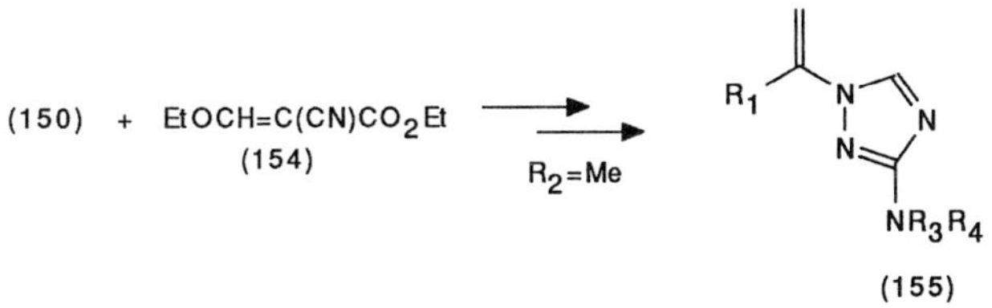

(i) By cycloaddition

The nitrilimine (157), generated from the hydrazonoyl halide (156) in the presence of base, undergoes 1,3-dipolar cycloaddition with the carbon-nitrogen triple bond of the isothiocyanate (158) to yield the 1,2,4-triazole (159) (D.J. Grieg, *et al.*, J. chem. Soc. Perkin, 1987, 607).

The nitrilimine (160) adds to the carbon-nitrogen double bond of the cyanate ion and to dicyclohexylcarbodiimide to form (161) and (162) respectively (K. Tanaka *et al.*,J.heterocyclic Chem., 1987, 24, 1391).

This reaction is discussed by P. Caramella and P. Grunanger in Padwa, Vol. 1, Ch. 3, p. 361.

(b) Reactions

In DMSO-d$_6$ thione (163) exists in equilibrium with its thiol tautomer (164) as indicated by ^{13}C nmr spectroscopy (T. Okawara, *et al.*, J. heterocyclic Chem., 1988, 25, 1071). Alkylation of (165) proceeds cleanly at sulphur to yield sulphide (166) which can be oxidized to the sulphone (167) (V.J. Ram *et al.*, *ibid.*, 1989, 26, 625).

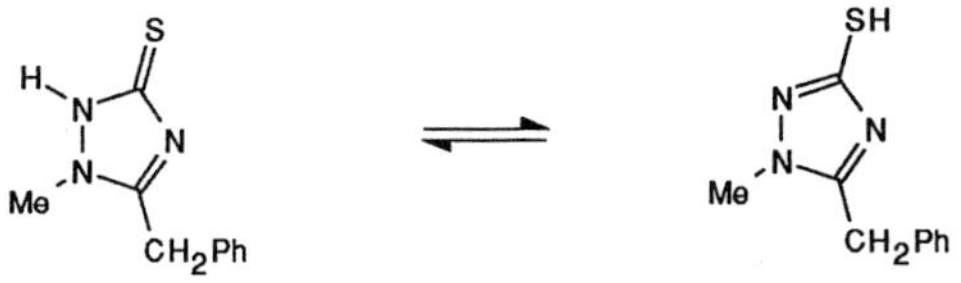

(163)   (164)

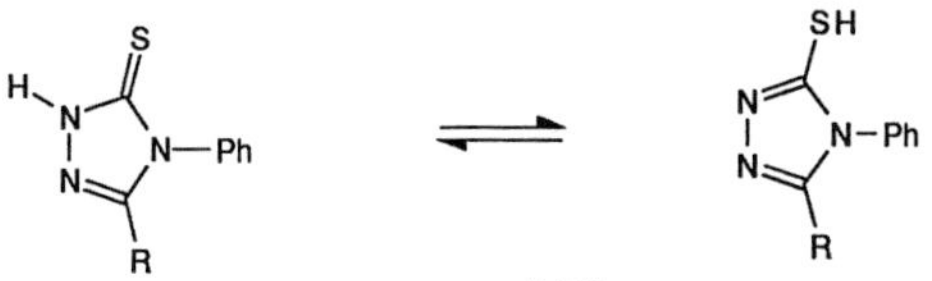

(165)

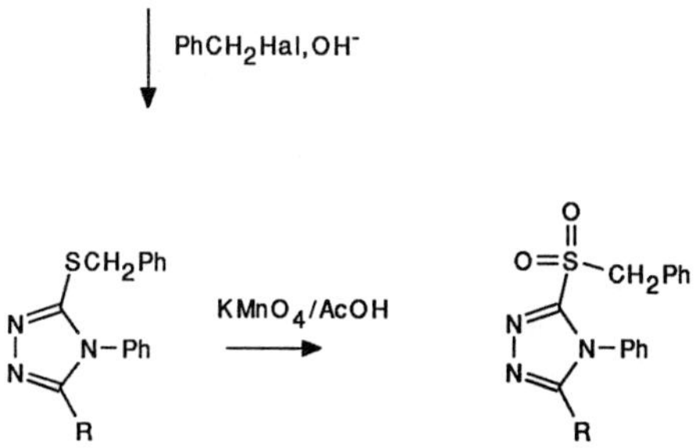

PhCH$_2$Hal,OH$^-$

(166)   (167)

KMnO$_4$/AcOH

R=

(CH$_2$)$_8$

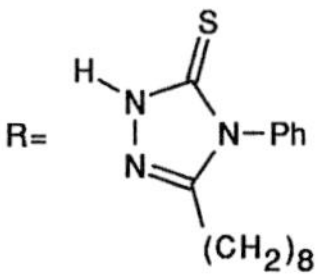

5. 1,2,3-Oxadiazoles

The chemistry of this ring system has been reviewed by L.B. Clapp in Comprehensive Heterocyclic Chemistry, Vol. 6., Ch. 4.21., p. 365.

The parent ring system remains unknown although its existence as an intermediate has been considered in the following reactions.

(a) The photochemistry of azo-oxides, for example (168) (A. Albini and N. Alpegiani, Chem. Reviews, 1984, 84, 43).

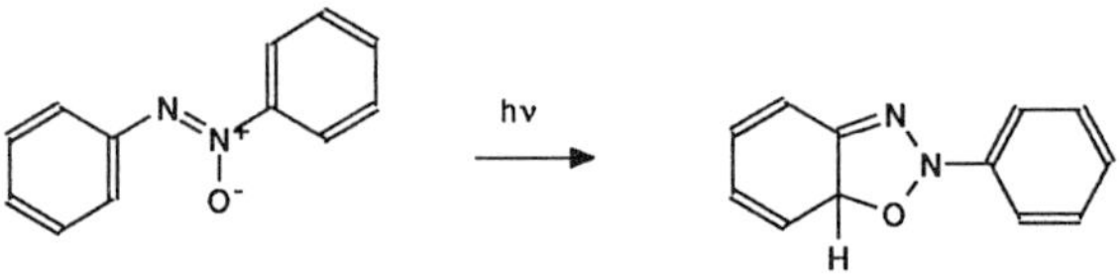

(b) The reactions of diazoalkanes with carbonyl compounds (M. Regitz and H. Heydt in Padwa, Vol. 1., Ch. 4, p. 467).

(c) The 1,3-dipolar addition of azo-oxides with alkenes (R.C. Storr in Padwa, Vol. 2, Ch. 10, p.187).

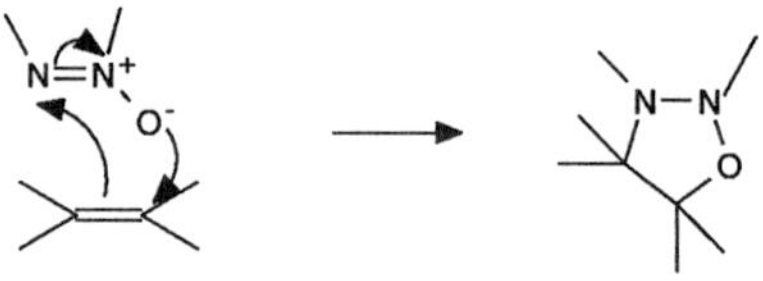

The ring is stable at its mesoionic form, e.g. (169). This compound is cleaved to the hydrazine (170) by acid (M.M. El-Bakoush and J. Pamick, J. heterocyclic Chem., 1988, 25, 1055).

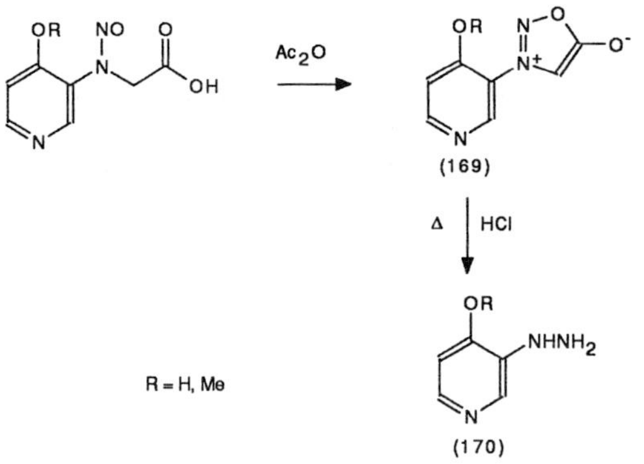

(169)

Δ | HCl

(170)

R = H, Me

6. 1,2,4-Oxadiazoles

These have been reviewed by L.B. Clapp in Comprehensive Heterocyclic Chemistry, Vol. 6., Ch. 4.21, p. 365.

(a) Synthesis

(b) 1,3-Dipolar cycloaddition methods

The highly regiospecific addition of nitrile oxides to nitriles has been extensively studied and has been reviewed by P. Caramella and P. Grunanger in Padwa, Vol. 1, Ch. 3., p. 361.

For example the nitrile oxides (171) add to malononitrile to give (172); the residual nitrile function can be hydrolysed to the corresponding acid or amide (J.J. Tegeler and C.J. Diamond, J. heterocyclic Chem. 1987, 24, 697)

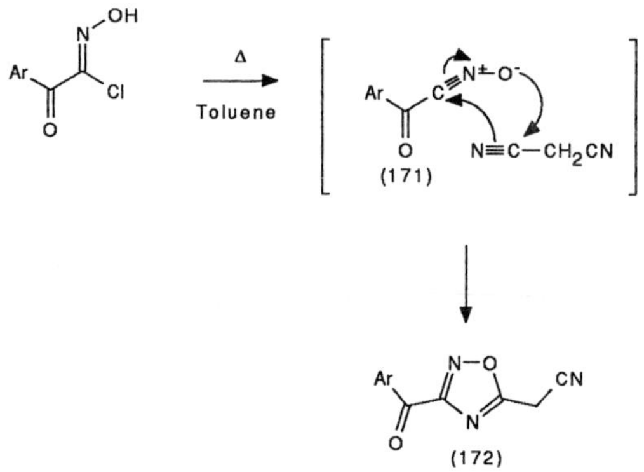

(171)

(172)

Where a 1,3-dipolarophile has both carbon-carbon and carbon-nitrogen multiple bonds, isoxazoline formation may compete with oxadiazole formation. The reaction of mesitylcarbonitrile oxide (173) with isopropylidene malononitrile gives a mixture of (174), (175) and (176) (P.S. Lianis, N.A. Radio and N.E. Alexandrou, J. heterocyclic Chem., 1989, 26, 1683).

Hydrogen bonding is a major factor in influencing product specificity. Addition of (177) to the nitrile-enamine (178) gives predominantly the 1,2,4-oxadiazole (179) if either R_1 or R_2 is hydrogen and the nitrile oxide can hydrogen bond to the substrate. This effect is pronounced if the solvent, e.g. benzene, cannot hydrogen bond. Isoxazoles (180) dominate if R_1 and R_2 are alkyl and the solvent is polar, e.g. dimethyl sulphoxide (A. Corsaro, et al., J. heterocyclic Chem.,1985, 22, 797).

Further evidence for the importance of hydrogen bonding comes from the predominant formation of 1,2,4-oxadiazoles from *o*-substituted benzonitriles (181) when the substituent can act as a hydrogen bond donor (A. Corsaro, *et al.*, J. heterocyclic Chem., 1984, 21, 949).

X = NHAc, NHCOPh, NHTs

Steric factors also play a part, as the fluorene derivatives (182), (183), give 1,2,4-oxadiazoles exclusively (P.S. Lianis, N.A. Radios and N.E. Alexandrou, J. heterocyclic Chem., 1988, 25, 1099).

$Ar = 2,4,6\text{-}Me_3\text{-}C_6H_2$

This process is an excellent method of preparing 3-halo-1,2,4-oxadiazoles, using dihaloformaldoximes (184) as a source of halocyanogen oxide; the halogen in the product is readily displaced by a nucleophile (G.R. Humphrey and S.H.B. Wright, J. heterocyclic Chem., 1989, 26, 23).

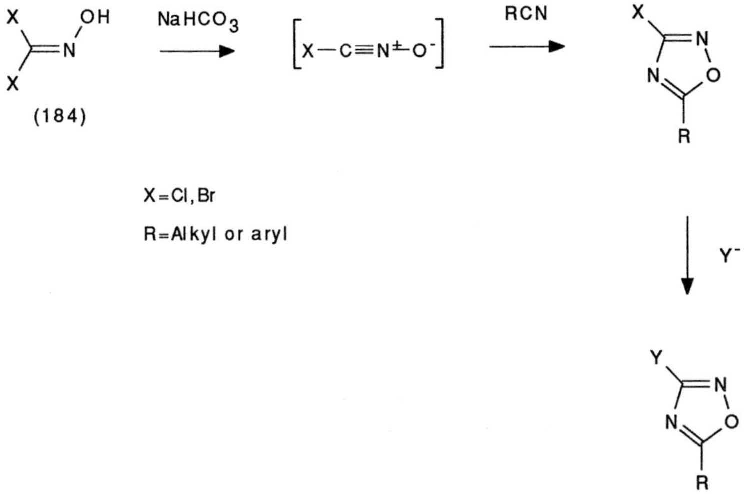
NaHCO3
RCN
(184)
X=Cl, Br
R=Alkyl or aryl
Y⁻
Y=OH,NH2,CN,N3 etc.

1,2,4-Oxadiazoles are minor products from nitrile oxides and base even in the absence of a nitrile; a mechanism involving 1,3-dipolar cycloaddition between the nitrile oxide and a derivative of the starting material has been proposed (P. Caramella, *et al.*, Tetrahedron Letters, 1983, 4377).

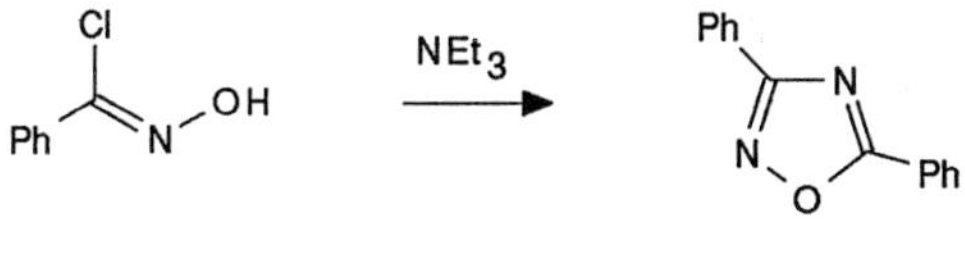
NEt3

Nitrile oxides add to the carbon-nitrogen triple bond of aryl thiocyanates and selenocyanates (D.J. Grieg, *et al.*, J. chem. Soc., Perkin 1, 1987, 607).

R = Ph, 4-X-C$_6$H$_4$[X=NO$_2$, OMe, Cl]

X = S, Se

Ar = X-C$_6$H$_4$[X=Cl, NO$_2$], 1-Naphthyl

The addition of nitrile oxides to nitriles has been reviewed by P. Caramella and P. Grunanger. (Padwa, Vol. 1., Ch. 3, p. 361), and in "Nitrile Oxides, Nitrones, and Nitronates; Novel Strategies in Organic Synthesis", K.B.G. Torsell, VCH, New York, 1988.

In an analogous reaction, nitrones (185) (the *N*-oxides of Schiff's bases) add nitriles to give Δ⁴-1,2,4-oxadiazolines. The nitrone (185) reacts quantitatively with trichloroacetonitrile to give (186) (P.H. Hermkens, *et al.*, Tetrahedron, 1988, <u>44</u>, 6491).

Nitrones such as (187) react with unsaturated nitriles (*e.g.* 188) at the carbon-nitrogen triple bond (cf. nitrile oxides, p.302). The formation of (189) demonstrates the stereoselectivity of such reactions, P.H. Hermkens *et al.*, *loc. cit.*; R. Plate *et al.*, J. org. Chem., 1987, <u>52</u>, 1047).

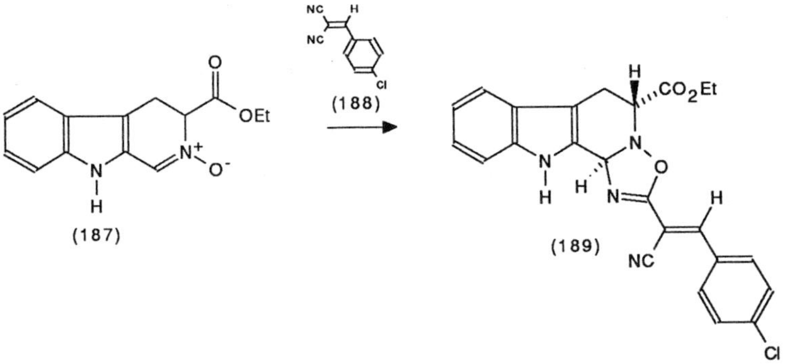

(187) (188) (189)

The cycloaddition reactions of nitrones are discussed by J.J. Tufariello (Padwa, Vol. 2, Ch. 9, p. 130) and in Torsell's monograph.

Δ^4-Oxadiazolines are produced in the reaction between nitrosobenzene and a Δ^2-oxazolin-5-on (190). Such reactions are believed to involve the oxadiazoline (190) in its tautomeric mesionic form (191) acting as a 1,3-dipole (H. Rodriguez, *et al.*, Tetrahedron, 1983, 39, 23):

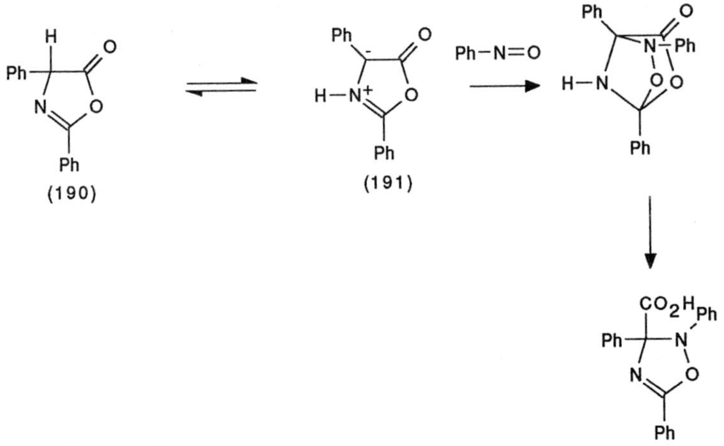

(190) (191)

The 1,3-dipolar cycloaddition reactions of mesionic compounds have been reviewed (K.T. Potts, Padwa, Vol. 2, Ch. 8, p.1).

(c) Other methods

Amidoximes such as (192) react with diketene (193) to give (194), which can be thermolysed to the oxadiazole (195) (B. Kuebel, Chem. Abs., 1982, 96, 20105) or reduced to the oxadiazole (196) (K. Tabei, *et al.*, Chem. pharm. Bull., 1982, 30, 336). The reaction of (192) with the diketene derivative (197) gives the oxadiazole (198) (T. Takada, *et al.*, Chem. pharm. Bull., 1982, 30, 3987).

The amidoxime (192) also reacts with *N,N*-dicyclohexylcarbodiimide (DCC) (199) to form the oxadiazole (200); when the reaction is carried out in the presence of acetic acid, (200) is not formed but a low yield of the oxadiazole (201) is obtained (E. Kawashima and K. Tabei, J. heterocyclic Chem., 1986, 23, 1657):

Aryl amidoximes(202) react with succinic anhydride in a "one-pot" synthesis of the oxadiazoles (204) presumably *via* (203) (R. M. Serivastava, M.B.A.B. Viana and L. Bieber, J. heterocyclic Chem., 1984, 21, 1193) and with aldehydes to form oxadiazolines (205) which readily aromatise to the oxadiazoles (206) (J.M.S. Srivastava, *et al.*, J. heterocyclic Chem., 1987, 24, 101).

R,R'=H or COCH₂CH₂CO₂H

Y=Ph,Me,Et

A potentially versatile route to oxadiazoles utilises diphenylcyanocarbonimidate (207) which reacts with aniline to give (208). The latter with hydroxylamine forms the oxadiazole (209a) in high yield together with a trace of its isomer (209b) (R.L. Webb, *et al.*, J. heterocyclic Chem., 1987, 24, 275).

(d) Reactions

Catalytic hydrogenation of 1,2,4-oxadiazoles leads to reductive ring cleavage, a valuable step in the construction of the pyrimidine ring of quinazolines (210), the triaza- and tetraza-napthalenes (211), (212) and the purine analogues (213) and (214) (D. Korbonitz, *et al.*, Ber., 1984, 117, 3183):

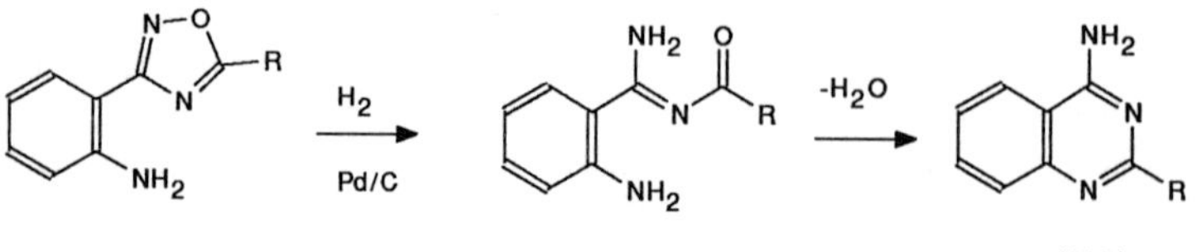

308

(210)

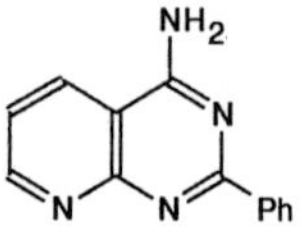

(211)

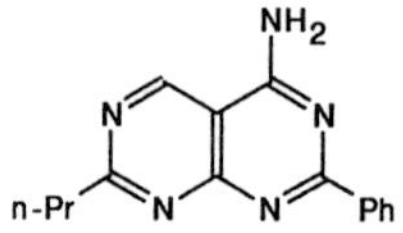

(212)

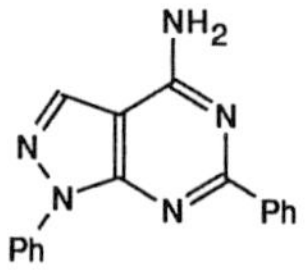

(213)

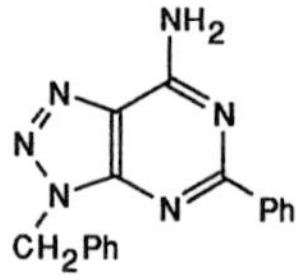

(214)

1,2,4-Oxadiazolines also undergo ring-opening to form amidines as illustrated by the formation of (216) from (215) (after initial decarboxylation) (H. Rodriquez, *et al.,* Tetrahedron, 1983, 39, 23) and (218) from (217) (R. Plate, *et al., J.* org. Chem., 1987, 52, 1047).

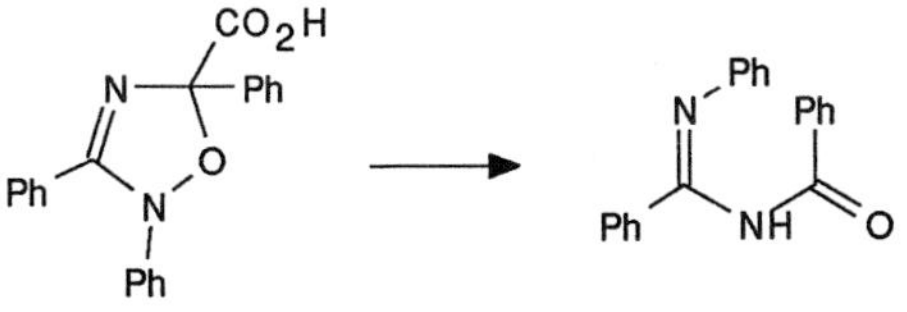

(215)

(216)

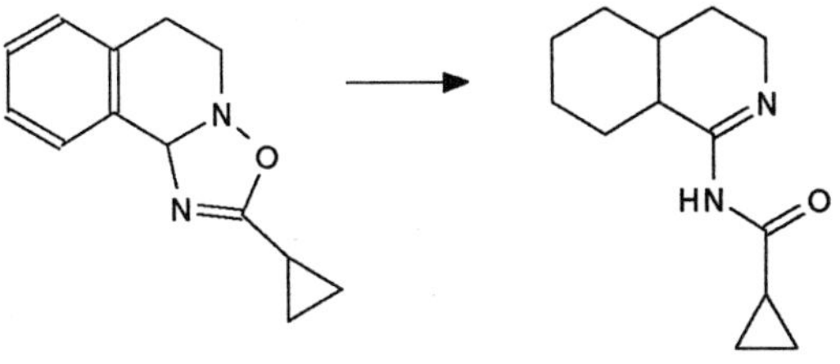

(217)

(218)

7. 1,3,4-Oxadiazoles

These have been reviewed by J. Hill in Comprehensive Heterocyclic Chemistry, Volume 6, Ch. 4. 23, p. 427.

(a) Synthesis

(b) From acyl hydrazides

This is a widely used approach, and reactions mainly differ in the source of the final ring carbon. Examples of the reagents used are given below.

(i) Orthoesters. The synthesis of (219) from (220) (N.P. Peet and S. Sunder, J. heterocyclic Chem., 1984, 21, 1807; Y. Kuvasawa et al., ibid., 1986, 23, 1387; Y. Kurasawa, Y. Moritaki and T. Ebukuro, Chem. pharm. Bull., 1983, 31, 3897).

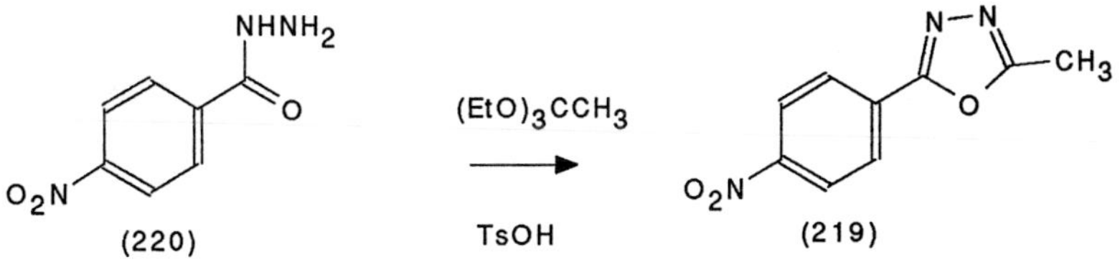

(220)

$(EtO)_3CCH_3$

TsOH

(219)

310

(ii) Ureas, *e.g.* 1,1'-carbonyldiimidazole (221) (E. Tiahanyi, M. Gal and P. Dvotsak, Heterocycles, 1983, 20, 571).

(iii) Isothiocyanates in the presence of N,N'-dicyclohexylcarbodiimide (S. Sunder, N.P. Peet and R.J. Barbuck, J. heterocyclic Chem., 1981, 18, 1601; A-M.M.E. Omar and O.M.A. Wafa, *ibid.*, 1984, 21, 1445).

(iv) Carbon disulphide under basic conditions (B. Kuchel, Monatsh, 1982, 113 793; B.N. Goswani, *et al.*, J. heterocyclic Chem., 1984, 21, 705).

(v) Aldehydes, followed by oxidative cyclisation with either iron (III) chloride (S.P. Hiremath, N.N. Gouder and M.G. Purohit, Indian. J. Chem., 1982, 213, 321) or lead (IV) oxide (R. Milcent and G. Barvier, J. heterocyclic Chem., 1983, 20, 77).

(vi) Carboxylic acids and phosphorus oxychloride (M.M. Dutta, B.N. Goswani and J.C.S. Kataky, J. heterocyclic Chem., 1986, 23 793).

(vi) Acyl halides or anhydrides followed by dehydration (B. Rigo and D. Couturier, J. heterocyclic Chem., 1986, 23, 253).

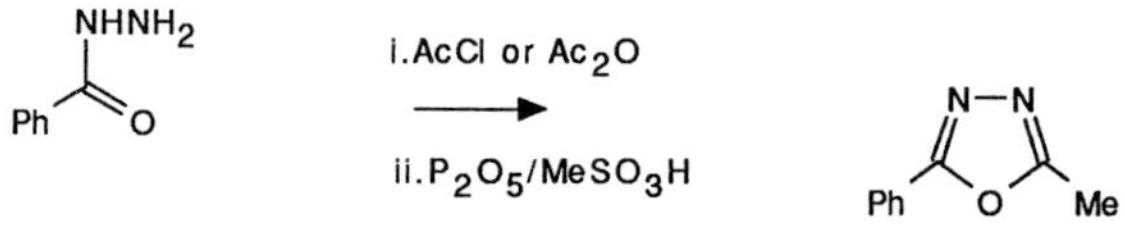

(viii) Ethyl chloroformate or chlorothioformate followed by thermolysis (I.T. Barnish, et al., J. heterocyclic Chem., 1986, 23, 417).

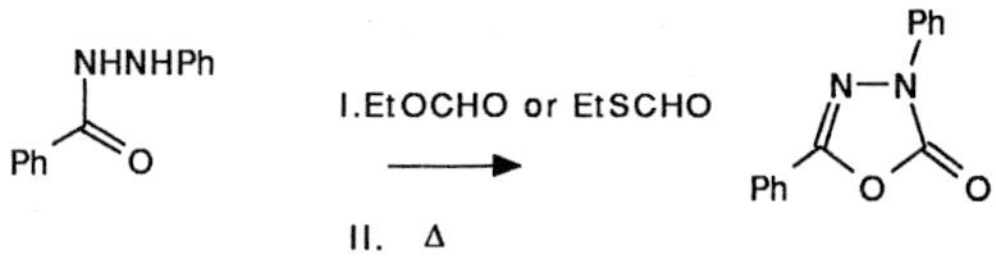

(ix) Phenyl isocyanate (I.T. Barnish et al., loc. cit.).

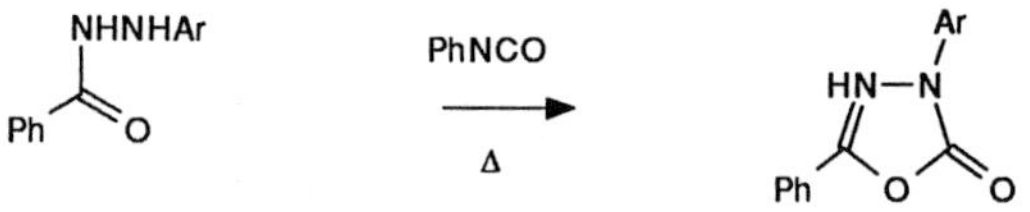

$$Ar = 2,4-Br_2-C_6H_3$$

(c) Oxidative methods

Oxidation of aldehyde semicarbazones (222) with lead (IV) oxide gives 2-amino-1,3,4-oxadiazoles (223) in excellent yield. A free radical mechanism has been proposed (R. Milcent and G. Barbier, J. heterocyclic Chem., 1983, 20, 77; T-H. Nguyen, R. Milcent and G. Barbier, ibid., 1985, 22, 1383).

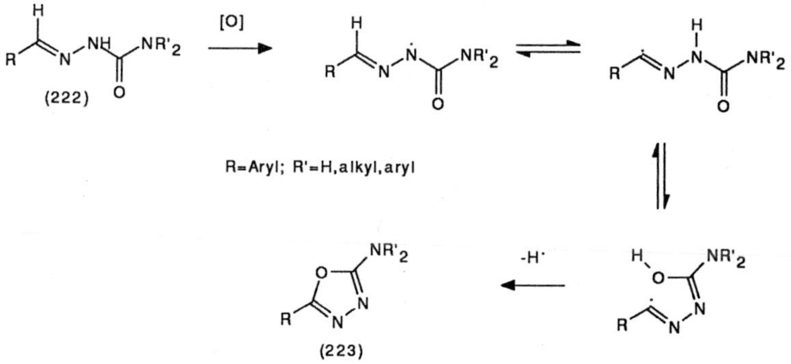

R=Aryl; R'=H,alkyl,aryl

Oxidation of benzoylthiosemicarbazides (224) with mercury (II) oxide gave a similar result (A.-M.M.E. Omar and O.M.A. Wafa, J. heterocyclic Chem., 1984, 21, 1415).

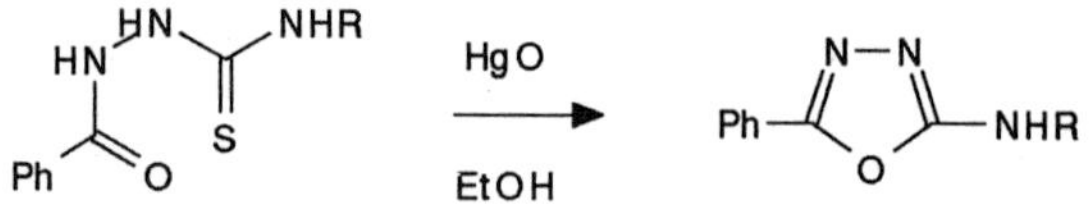

(224)

The thioaroyl semicarbazide (225) gives the disulphide (226) when treated with bromine. The latter when hydrolysed gives the oxadiazole (227) together with the thiadiazole (228) (F. Kurzer and K.M. Doyle, J. chem. Soc. Perkin 1, 1986, 1873).

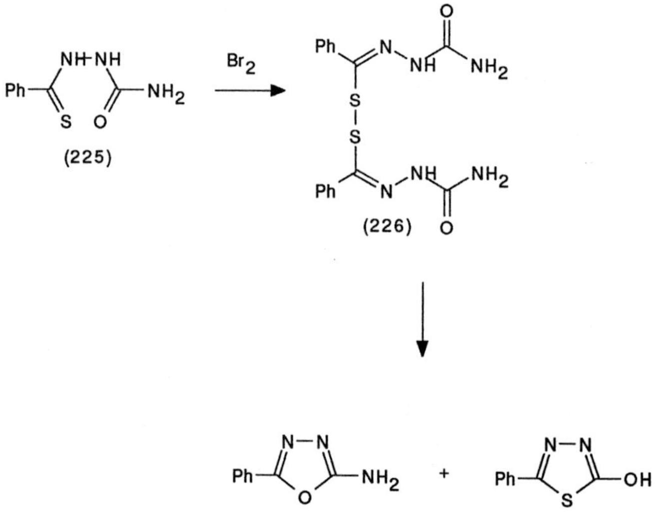

(225)   (226)

(227)   +   (228)

Oxidation of the hydrazonecarboxylic ester (229; R=t-Bu) with lead (IV) acetate gives a mixture of (230) and (231); iodosobenzene diacetate (IBDA) is a better oxidant, giving only (230) from (229; R=t-Bu) and oxidizing the ethyl ester (229; R=Et) with de-esterification (H.E. Baumgarten, D.-R. Huang and T.N. Rao, J. heterocyclic Chem., 1986, 23, 945).

(d) Other methods

Reduction of the carbonate esters (232) gives the 1,3,4-oxadiazol-2-ones (233) (A. Monge, *et al.*, J. heterocyclic Chem., 1984, <u>21</u>, 397).

(e) Reactions of 1,3,4-oxadiazoles

In common with 1,2,4-oxadiazoles (p.307)the 1,3,4-oxadiazole ring readily cleaves. Acid hydrolysis yields acyl hydrazides, making the 1,3,4-oxadiazole ring a useful "acid hydrazide synthon" in iodination (N.D. Heindel, *et al.*, J. heterocyclic Chem., 1985, <u>22</u>, 209).

1,3,4-Oxadiazoles (*e.g.* the 1,3,4-oxadiazol-2-one; 234) undergo ring-cleavage on treatment with an amine (Y. Saegusa, S. Harada and S. Nakamura, J. heterocyclic Chem., 1988, 25, 1337).

(234)

This reaction has been used in the synthesis of the 1,2,4-triazoles (237). The 1-imidoyl-2-acylhydrazines (236) formed by aminolysis of (235) can be re-cyclised to a 1,2,4-triazole (237). Hydrazinolysis of the ring proceeds similarly, leading to either a 1-(*N*-aminoimidoyl)-2-acylhydrazine (238) or a 1,2,4,5-tetrazine (239) (D.B. Reitz and M.J. Finkes, J. heterocyclic Chem., 1988, 26, 225; J. org. Chem., 1989, 54, 1760).

R=CF$_3$,C$_2$F$_5$,n-C$_3$F$_7$

R=CF$_3$,C$_2$F$_5$,n-C$_3$F$_7$

8. 1,2,5-Oxadiazoles

These have been reviewed by R.M. Paton in Comprehensive Heterocyclic Chemistry, Volume 6, Ch. 4.22, p. 393 and by A. Gasco and A.J. Boulton in Adv. heterocyclic Chem., 1981, 29, 251.

(a) Synthesis

The N-oxides of 1,2,5-oxadiazoles (furazans) are formed by dimerisation of nitrile oxides. Thus treatment of (240) with base in the absence of a 1,3-dipolarophile gives (241) via (242) (W.J. Middleton, J. org. Chem., 1984, 49, 919).

The anti-hypertensive agent (243) is prepared similarly (K. Schoenafinger, *et al.*, Chem. Abs., 1982, 97, 216189).

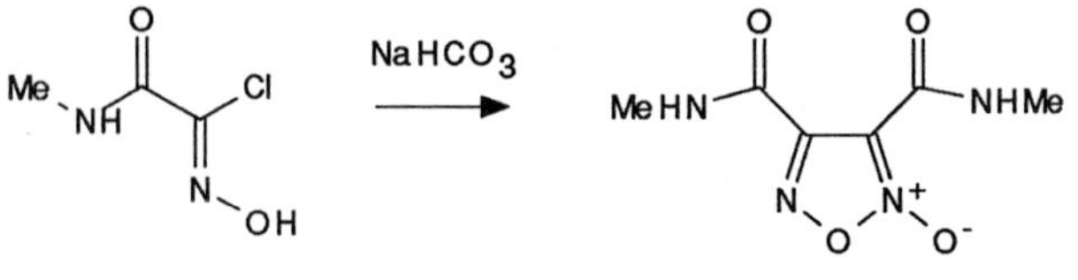

(243)

The dimerisation of nitrile oxides is discussed by P. Caramella and P. Grunanger in Padwa, Vol. 1, Ch. 3, p. 294. Benzofurazans(246) are formed when *o*-dinitrobenzenes (244) react with sodium azide, presumably *via* (245) (S.V. Eswaran and S.K. Sajadian, J. heterocyclic Chem., 1988, 25, 803).

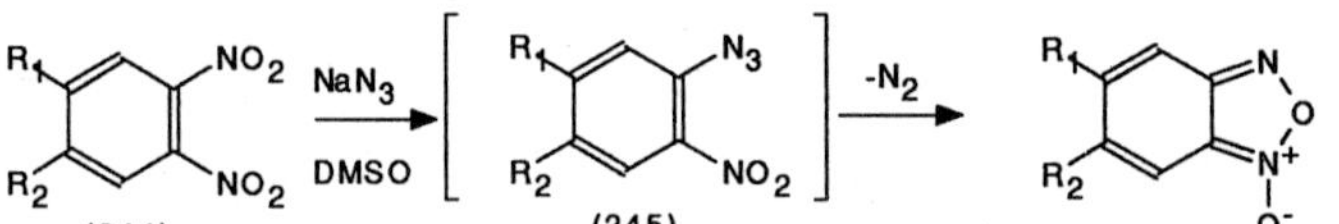

	R_1	R_2
a	OH	OH
b	OMe	OMe
c	OCH_2O	
d	OCH_2CH_2O	

The benzofuroxide (247) is formed by photolysis of (244b) in presence of butylamine; a reductive pathway is proposed *via* dinitroso- or nitrosohydroxylaminobenzene (J. Marquet, *et al.*, Tetrahedron, 1987, 43, 351).

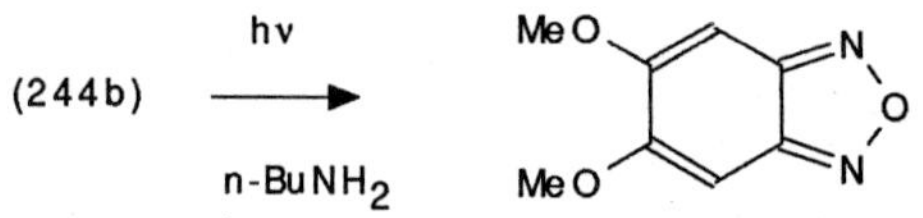

(247)

(b) Reactions

Variable temperature spectroscopy shows that the benzofurazan (248; R=H) is in rapid degenerate equilibrium with its isomer. A similar equilibrium does not occur with the nitro-derivative (248; R=NO₂) (S.V. Eswaran and S.K. Sajadian, J. heterocyclic Chem., 1988, 25, 803).

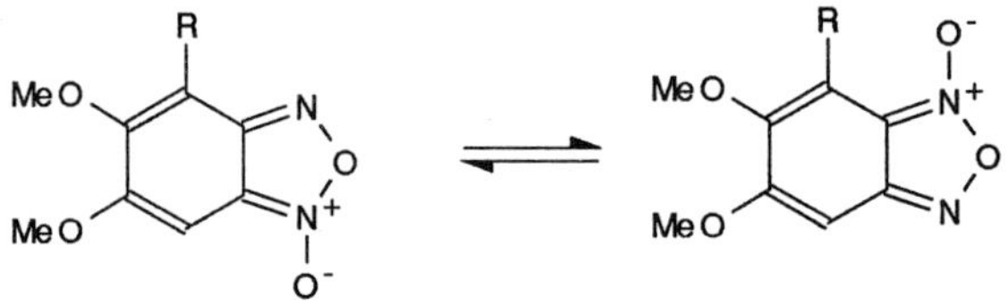

(248)

In contrast to 1,2,4-oxadiazoles, the 1,2,5-oxadiazole ring is relatively resistant to reductive cleavage, as illustrated by the reduction of either ring of the fused heterocycle (249) (S. Mataka, K. Takahishi and T. Imura, J. heterocyclic Chem., 1982, 19, 1481).

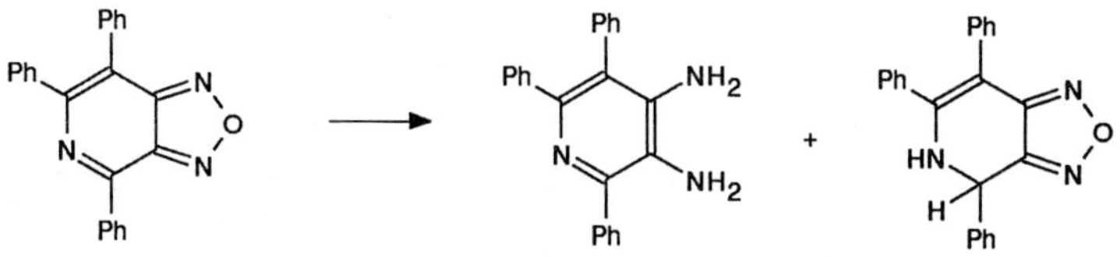

(249)

The 1,2,5-oxadiazole ring is cleaved when the benzofurazan (250) reacts with a 1,3-diketone to give a quinoxaline-1,4-dioxide (251) (the Beirut Reaction) (A. Atfah and J. Hill, J. chem. Soc. Perkin 1, 1989, 221).

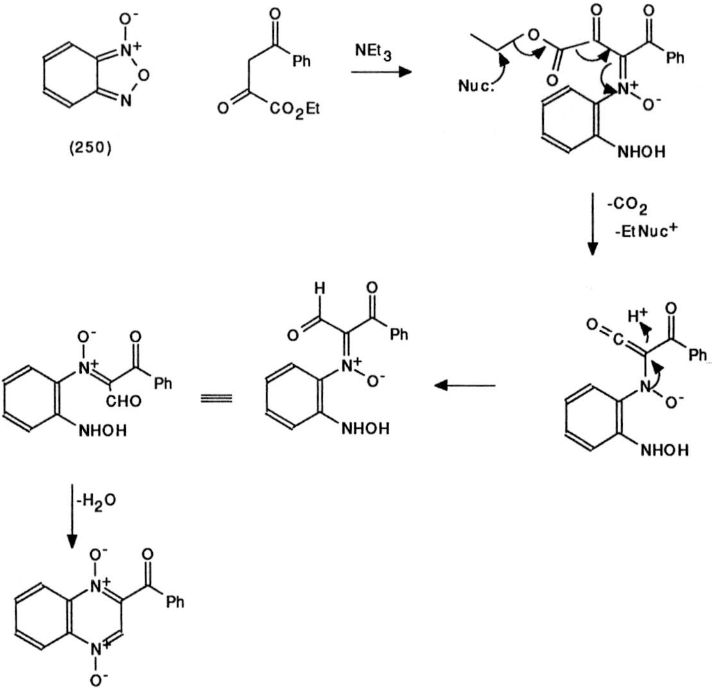

318

(250)

(251)

9. 1,2,3-Thiadiazoles

These have been discussed by E.W. Thomas in Comprehensive Heterocyclic Chemistry, Vol. 6, Ch. 4.24, p. 447.

(a) Synthesis and reactions

Cyclisation of the stereoisomeric hydrazones (252, 253) with thionyl chloride affords respectively the 1,2,3-thiadiazoles (254), (256); the structure of the final product is determined by the geometry of the hydrazone (O. Zimmer and H. Meier, Ber. ,1981, 114, 195).

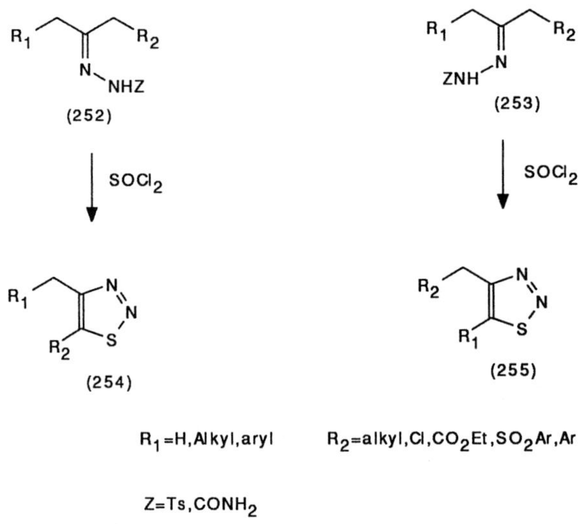

(252)

(253)

| SOCl$_2$ | SOCl$_2$ |

(254)

(255)

R$_1$=H,Alkyl,aryl   R$_2$=alkyl,Cl,CO$_2$Et,SO$_2$Ar,Ar

Z=Ts,CONH$_2$

Thionyl chloride with the hydrazone (256) gives (257) which can be cleaved to the salt (258) (W.V. Curran, et al., J. heterocyclic Chem., 1985, 22, 479).

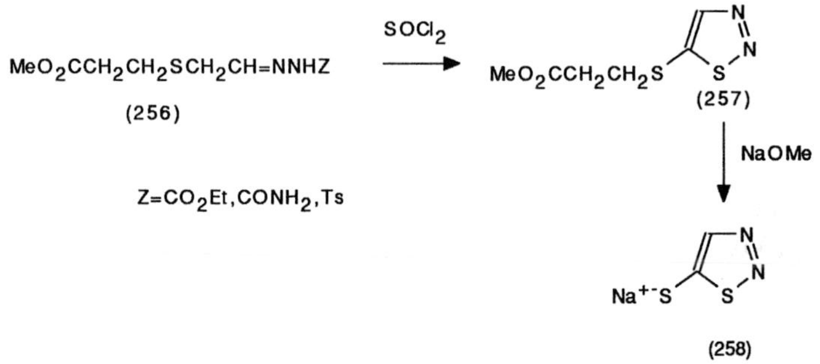

MeO$_2$CCH$_2$CH$_2$SCH$_2$CH=NNHZ

(256)

Z=CO$_2$Et,CONH$_2$,Ts

SOCl$_2$

MeO$_2$CCH$_2$CH$_2$S

(257)

NaOMe

Na$^+$$^-$S

(258)

Cyclisation of thiocarbazonate esters (259) with thionyl chloride gives the 1,2,3-thiadiazoles (260) which can be cleaved by base to the 1,2,3-thiadiazole-4-thiolates (261) (V.J. Lee, *et al.*, *ibid.*, 1988, 25, 1873).

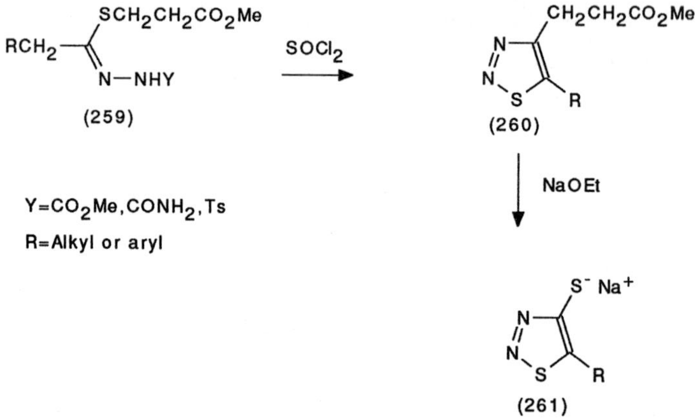

(259) → SOCl$_2$ → (260)

Y=CO$_2$Me, CONH$_2$, Ts
R=Alkyl or aryl

NaOEt

(261)

Diazomethane adds to acylisothiocyanates (262) to give the 1,2,3-thiadazoles (263); mesionic compounds 264) and (265) are formed respectively by alkylation with dimethyl sulphate and benzyl bromide (K. Masuda, *et al.*, J. chem. Soc. Perkin 1, 1981, 1591).

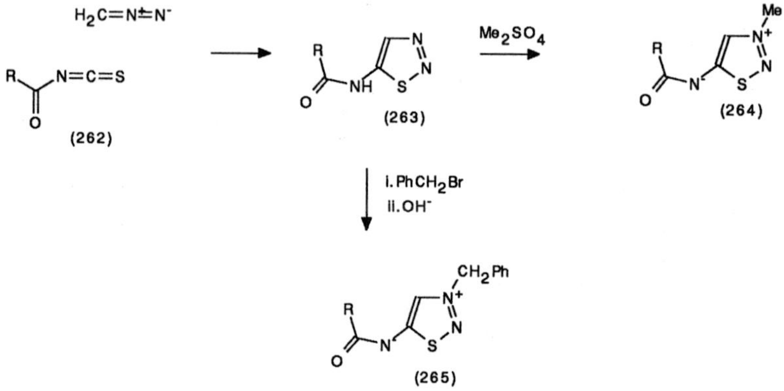

(262) → (263) → Me$_2$SO$_4$ → (264)

i. PhCH$_2$Br
ii. OH⁻

(265)

The reaction of diazoalkanes with isothiocyanates to form 1,2,3-thiadiazoles has been reviewed by M. Regitz and H. Heydt in Padwa, Vol. 1, Ch. 4, p. 493.

Mesoionic 1,2,3-thiadiazoles have been the subject of several investigations. Treatment of (266) with alkyl fluorosulphonates gives the alkylated heterocycles (267, a,b) which are readily deprotonated to give mesionic (268, a,b). Hydrolysis of (267 a) gives the amine (269) which on acylation gives (270) and on diazotisation gives (271) (K. Masuda, *et al.,* Chem. pharm. Bull, 1981, 29, 1743).

5-Phenylamino-1,2,3-thiadiazole (272) is alkylated at *N*-3 by dimethyl sulphate forming a salt which is readily deprotonated by ammonia to give (274), whose structure has been determined by X-ray crystallography. A similar compound (276) is obtained from dichlorocompound (275) and excess aniline; alkylamines give dehalogenation products (174) (V.A. Kozinskij, *et al.*, J. heterocyclic Chem., 1984, <u>21</u>, 1889).

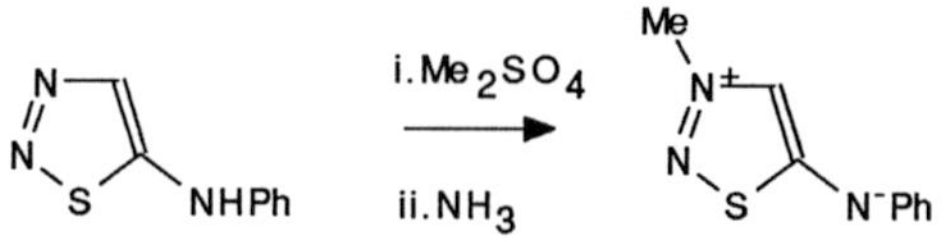

(272)   (273)

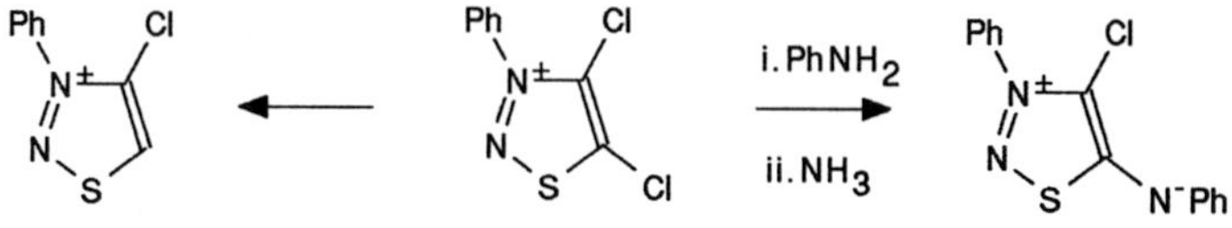

(274)   (275)   (276)

The salts (277) may be directly thiolated to give the mesionic heterocycles (278) in moderate yields; only low yields are obtained from the alkoxy compounds (280) and sodium hydrogen sulphide. The unsubstituted compound (277; R=H) gives predominantly the 5-thiolate. (281). The replacement of sulphur by oxygen may be effected by methylation of (281) followed by alkaline hydrolysis. (J. Adachi, *et al.*, Chem. pharm. Bull, 1983, <u>31</u>, 1746).

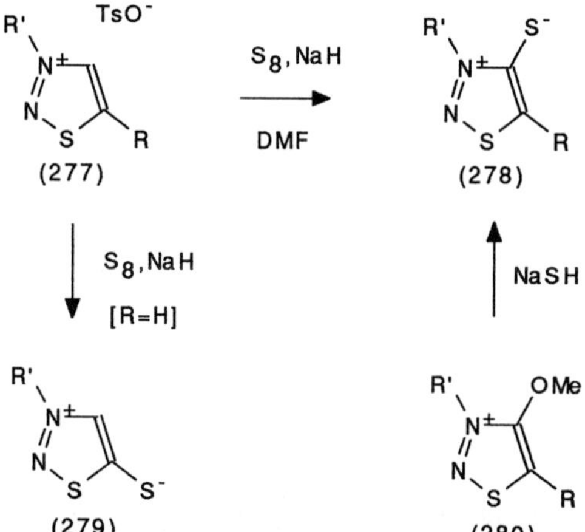

(277)   (278)

(279)   (280)

R = Me, Et

10. 1,2,4-Thiadiazoles

These have been reviewed by F. Kurzer (Adv. heterocyclic Chem., 1982, 32, 286). and by J.E. Franz and O.P. Dhingra in Comprehensive Heterocyclic Chemistry, Vol. 6, Ch. 4.25, p. 463.

(a) Synthesis

(b) Oxidative cyclisation

Oxidation of thioamides (283) with t-butyl hypochlorite gives 2,5-disubstituted-1,2,4-thiadiazoles (284). Two possible reaction pathways have been suggested (see below), and one of these, route B *via* a nitrile sulphide is supported by the isolation of nitriles as by-products (M.T.M. El-Wassimy, K.A. Jorgensen and S.O Laivesson, Tetrahedron, 1983, 39, 1729).

324

A similar pathway may be involved in the photo-oxidation of aryl thioamides (287) (M. Machida, K. Oda and Y. Kanaoka, Tetrahedron Letters, 1984, 409).

Ar=Ph,3- or 4-pyridinyl,2-furyl,2-thienyl

The unsymmetrically 2,5-disubstituted 1,2,4-thiadiazole (289) is prepared by oxidation of the adducts (288) of phenyl isothiocyanate and the guanidine (287) (F. Kurtzer, J. chem. Soc., Perkin 1, 1985, 311).

A formal oxidation also occurs in the formation, in good yield, of the 1,2,4-thiadiazoles (292) from the heterocycles (290), having a reactive methyl group, on treatment with thionyl chloride followed by an amidine (291). (A.H.M. Al-Shaar, et al., J. heterocyclic Chem., 1989), 26, 1819).

Ar		R	
2-pyridinyl		a	CH$_3$
2-quinoxalinyl		b	CCl$_3$
2-benzo-1,3-thiazolyl		c	NMe$_2$
2-benzo-1,3-oxazolyl		d	SMe
2-(1-methyl-5-nitropyrazolyl)			

(c) Cycloaddition of nitrile sulphides

Nitrile sulphides react by 1,3-dipolar addition with nitriles and with imines to give 1,2,4-thiadiazoles and 1,2,4-thiadiazolines respectively. The chemistry of nitrile sulphides has been reviewed (R.M. Paton, Chem. Soc. Rev., 1989, 18, 33).

Thermolysis of the 1,3,4-oxathiazoline (293) in refluxing mesitylene gives the nitrile sulphide (294) which is trapped by the nitrile (295) to give the 1,2,4-thiadiazole (296). Ethyl cyanoformate (295; R$_4$=CO$_2$Et) is an especially powerful 1,3-dipolarophile (R.M. Paton, et al., J. chem. Soc. Perkin 1, 1985, 1517).

R$_1$=Ph,4-X-C$_6$H$_4$[X=OMe,Cl]
R$_2$=CCl$_3$
R$_3$=H,CCl$_3$
R$_4$=Ph,CO$_2$Et

Polymers bearing "pendant" heterocycles have been prepared in this way. Co-polymerisation of styrene (297) with the 5-vinyl-1,3,4-oxathiazoline (298) gives a polymer of the form (299) which on heating with ethyl cyanoformate gives the polymer (300) (R.M. Mortimer, R.M. Paton and I. Stobie, Chem. Comm., 1983, 901).

The carbon-nitrogen bond of an aryl thiocyanate or a selenocyanate (301) is also an effective 1,3-dipolarophile (D.J. Grieg *et al.*, J. chem. Soc. Perkin 1, 1987, 607).

X=S or Se

R=Ph,4-Y-C$_6$H$_4$ [Y=Me,OMe,NO$_2$]

Ar=1-naphthyl,4-NO$_2$C$_6$H$_4$,2,4-(NO$_2$)$_2$C$_6$H$_3$,2-Cl-4,6-(NO$_2$)$_2$C$_6$H$_2$

1,2,4-Thiadiazolines can be prepared using nitrile sulphides. Thermolysis of 1,3,4-oxathiazolone (302) yields the nitrile sulphide (303) which with the imine (304) yields the thiadiazoline (305). The latter on further heating reverts to the nitrile sulphide which with ethyl cyanoformate yields the 1,2,4-thiadiazole (306) (R.O. Gould, *et al.*, J. chem. res. Synop., 1986, 156).

Ar=4-OMeC$_6$H$_4$

Ar'=4-X-C$_6$H$_4$[X=H,Cl,NO$_2$]

11. 1,2,5-Thiadiazoles

These have been reviewed by L.M. Weinstock and I. Shinkai in Comprehensive Heterocyclic Chemistry, Vol. 6, Ch. 4.26, p. 513.

(a) Synthesis

The reaction between dimethyl acetylenedicarboxylate and tetrasulphur tetranitride, gives dimethyl 1,2,5-thiadiazole-3,4-dicarboxylate (307) as the major product (67%) together with smaller amounts of (308), (309), and (310).

The formation of (307) may be rationalised by assuming an initial cycloaddition followed by ring-opening, ring-closure and subsequent cleavage (S.T.A.K. Daley and C.W. Rees, J. chem. Soc. Perkin 1, 1987, 203).

Trithiadiazepines such as (309) and (311) rearrange to 1,2,5-thiadiazoles when treated with either triphenyl phosphine or a strong base; the reaction in each case may involve an initial addition reaction followed by ring contraction (J.L. Morris and C.W. Rees, J. chem. Soc. Perkin 1, 1987, 217).

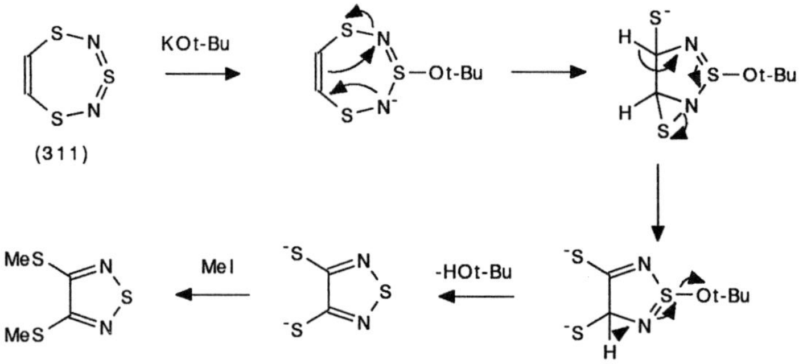

The nature of the substituents on the alkyne (312) influences the proportion of 1,2,5-thiadiazole (313) and trithiazapine (314) produced. Alkynes (312 (a) to (d)) gave mixtures of products; (312 (e)) gave (313(e)) as the major product, and (312 (f)) gave (313 (f)) as the only isolated product, but in low yield (20%). The best yield of 1,2,5-thiadiazole was (313 (g)) from the diacetal (312 (g)); (313 (g)) was readily hydrolysed to (313(f)) with dilute acid (P.J. Dunn and C.W. Rees, J. chem. Soc. Perkin 1, 1987, 1579, 1585).

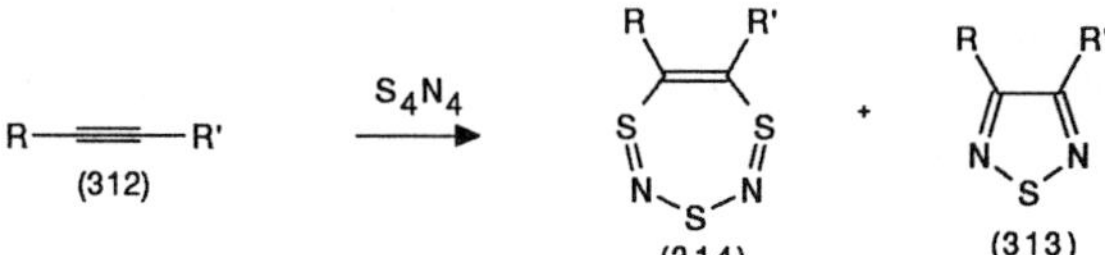

	R	R'	
a	CN	CN	mixture of (314) and (313)
b	CN	CF$_3$	mixture of (314) and (313)
c	CO$_2$Et	CHO	mixture of (314) and (313)
d	CO$_2$Et	SiMe$_3$	mixture of (314) and (313)
e	COMe	COMe	(313e) major product
f	CHO	CHO	(313f) only isolated product ; low yield
g	CH(OEt)$_2$	CH(OEt)$_2$	(313g) best yield ; gives (313f) on hydrolysis

The above reaction has been used in the synthesis of the piezochromic compound (315), which becomes an amorphous orange powder on grinding, but is converted to a yellow crystalline form when heated or washed (S. Mataka, *et al.*, J. heterocyclic Chem., 1989, 26, 215).

Ar=4-RO-C$_6$H$_4$[R=Me,Et,n-Bu]
Ar'=2-,3-,or4-MeO-C$_6$H$_4$;3,4-(MeO)$_2$C$_6$H$_3$ (315)

(b) Reactions

Reductive cleavage of the fused 1,2,5-thiadiazole (316) with sodium borohydride, lithium aluminium hydride or Raney nickel yields the diamine (317) (S. Makata, *et al.*, J. heterocyclic Chem., 1982, 19, 1481).

(316)   (317)

Cleavage with a Grignard reagent is a step in the synthesis of the macrocycle (319) from the 3,4-diaryl-1,2,5-thiadiazole (318) (T. Hatta, S. Mataka, and M. Tashiro, J. heterocyclic Chem., 1986, 23, 813).

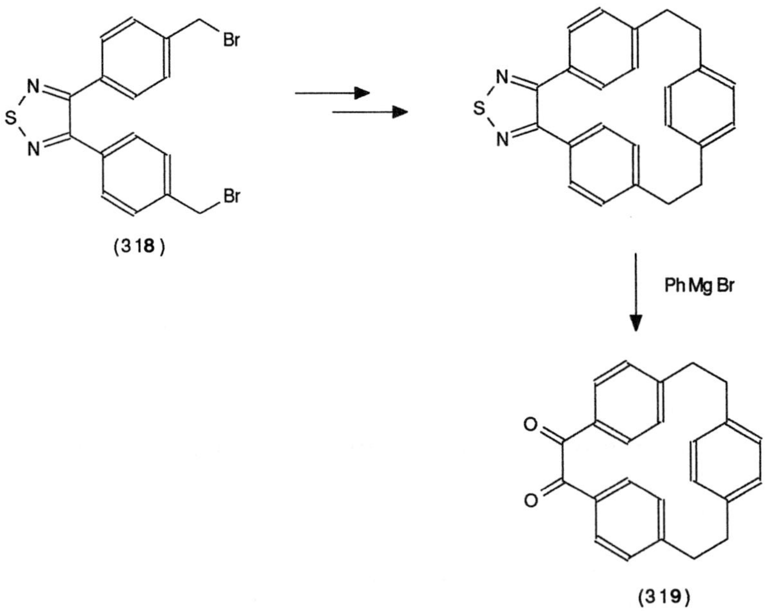

(318)

PhMgBr

(319)

12. 1,3,4-Thiadiazoles

These have been reviewed by G. Kornis in Comprehensive Heterocyclic Chemistry, Vol. 6, Ch. 4.27, p. 545.

(a) Synthesis

(i) Oxidative cyclisation

The oxidation of the thioaroylsemicarbazides (321) has been studied in some detail. (F. Kurzer and K.M. Doyle, J. chem. Soc., Perkin 1 1986, 1873) 1,3,4-Thiadiazoles (322) are by-products in the preparation of the thioarylsemicarbazides (321) due to further reaction with the thioaroylacetic acid (320).

Cautious oxidation of (321) gives the disulphide (323), which may be hydrolysed to a mixture of the thiadiazole (325) and the oxadiazole (324); under alkaline conditions, the oxadiazole predominates.

The imino analogues of (323) cannot be prepared as the thioaroylamidrazones give the 1,3,4-thiadiazoles (326) under the mildest oxidizing conditions. (F. Kurzer and K.M. Doyle, *loc. cit.*)

(326)

The thiosemicarbazone (328; R=Me) gives an excellent yield of 1,3,4-thiadiazole (329; R=Me) when treated with sulphur monochloride in acetic acid at room temperature; however (328;R = H) gives only a low yield of (329; R=H) in ethyl acetate at 40°C, and in refluxing acetic acid the triazole (327) is formed exclusively (R. Milcent and T.H. Nguyen, J. heterocyclic Chem., 1986, 23, 881).

Thiohydrazides can be cyclised by a variety of reagents. The arylthiohydrazides (331) are cyclised to the salts (332) by means of either a carboxylic anhydride - perchloric acid mixture of a nitrile-perchloric acid mixture (H. Mastalerz and M.S. Gibson, J. chem. Soc. Perkin 1, 1983, 245).

$$Ar = Ph, 2,4-Br_2C_6H_3 \ ; \ R = Me, Et.Ph$$

Acid hydrazides can also be cyclised to the 1,3,4-thiodiazolenone by ethyl chloroformate, ethyl chlorothioformate or phenyl isocyanate(I.T. Barnish, *et al.*, J.heterocyclic Chem., 1986, 23, 417).

(330)

(ii) Non-oxidative methods

Bithioureas are potential precursors of 1,3,4-thiadiazoles and the 1-acylbithiourea (333 a) forms the heterocycle (334a) when treated with base under phase transfer conditions; however, for reasons unknown, (333b to d) formed the pyrazoles (335b to d), and(333 e to f). However (333 a, b and f) form the related 1,3,4-thiadiazoline (334 a, b and f) in excellent yield when treated with tosyl chloride/triethylamine, probably *via* the intermediates (337).

	R_1	R_2	R_3				
a	Ph	H	Ph	d	Ph	Me	Ph
b	Ph	H	CH_2Ph	e	CH_2Ph	Me	Ph
c	Ph	Me	CH_2Ph	f	CH_2Ph	H	CH_2Ph
				g	$c-C_6H_{11}$	H	Ph

Moderate yields of 1,3,4-thiadiazoles are obtained when (333 a, b, d and f) are treated with iodomethane in the absence of base, the reaction being initiated by *N*- and/or *S*-methylation. (T. Okawara *et al.*, J. heterocyclic Chem., 1988, 25, 1071).

Isothiocyanates are precursors of 1,3,4-thiadiazoles. Phenyl isothiocyanate reacts with the acid hydrazide (338) to form the adduct (339) which undergoes acid-catalysed cyclisation to yield (340). The acidic protons α to the nitrile function can be further utilised, giving the aldol product (341) and the Japp-Klingemann products (342) and (343) (M.R.H. Elmoghayar, S.O. Abdalla and M.Y.A.-S. Nasr, J. heterocyclic Chem., 1984, 21, 781).

A closely related synthesis utilised a Japp-Klingeman reaction at an earlier stage. Aromatic diazonium salts couple with the isothiocyanate (344) to give the 1,3,4-thiadiazoles (346), presumably *via* (345). The alternative route *via* (347) also gives (346) without the isolation of (345) (A.O. Abdelhamid, J.M. Hassaneen and A.S. Shawali, J. heterocyclic Chem., 1983, 20, 719).

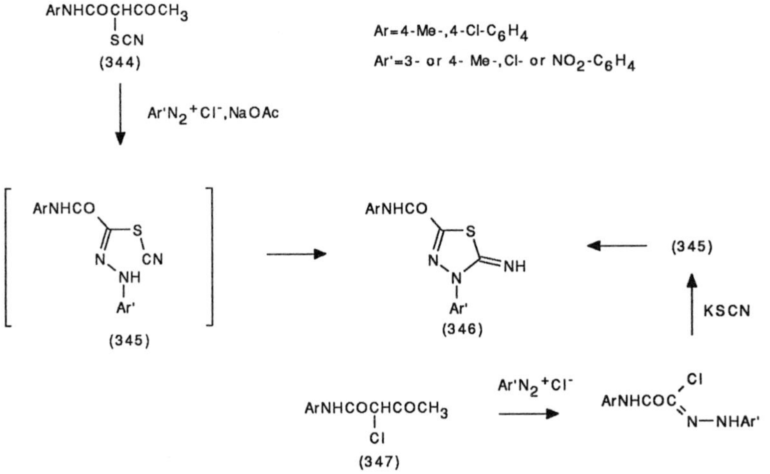

ArNHCOCHCOCH$_3$
|
SCN

(344)

Ar=4-Me-,4-Cl-C$_6$H$_4$
Ar'=3- or 4- Me-,Cl- or NO$_2$-C$_6$H$_4$

↓ Ar'N$_2$$^+Cl^-$,NaOAc

(345)

(346)

←——— (345)

↑ KSCN

ArNHCOCHCOCH$_3$
|
Cl

(347)

Ar'N$_2$$^+Cl^-$ ———→

The pyrrazolyl substituted 1,3,4-thiadiazole (348) have been prepared in an analogous manner (A.M. Farag, *et al.*, J. heterocyclic Chem., 1987, 24, 1341).

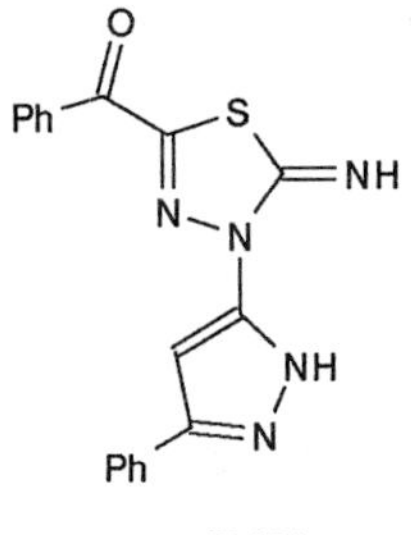

(348)

Hydrazidoyl bromides (349) yield 1,2,4-thiadiazoles (350) when treated with potassium isocyanate. The 2-iminofunction can be acylated and, *via* the *N*-nitroso derivative (351), hydrolysed (H.N. Hassaneen *et al.*,J. heterocyclic Chem., 1985, 22, 395).

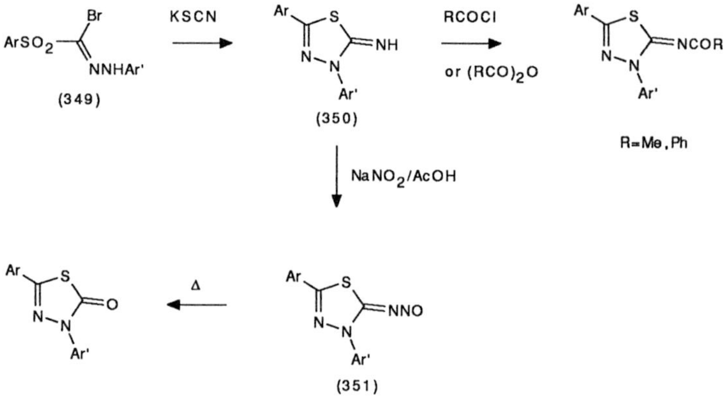

$Ar = Ph, 4-Me-C_6H_4$

$Ar' = 3-$ or $4-NO_2C_6H_4$

Similarly, the 3-nitro- and the 3-chloroformazans (352) react with potassium thiocyanate to give 2-arylazo-1,3,4-thiadiazoles (353). (A.O. Abdelhamid, *et al.*, J. heterocyclic Chem., 1985, 22, 813).

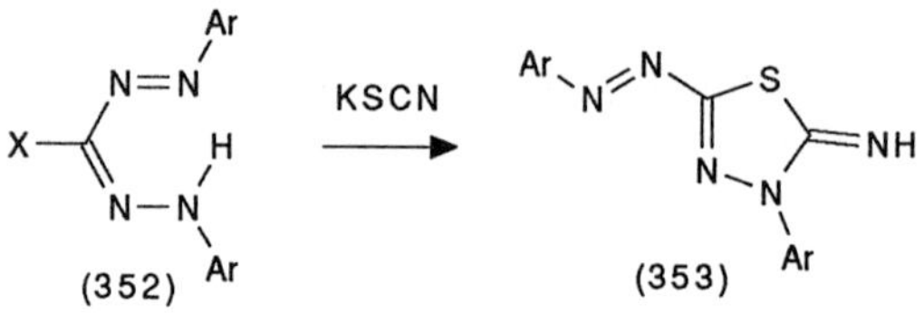

$X = Cl, NO_2$

$Ar = 4-Me-$ or $4-Cl-C_6H_4$

The organic isothiocyanate (355) is a precursor of 1,3,4-thiadiazoles. Hydrazine reacts with (355) to give the bithiourea (354), which cyclises under *acidic* condition to (359). The latter is obtained directly from (355) and either aminoguanidine (356a) or 1,2-diaminoguanidine (356b) possibly *via* the bis-adduct (357).

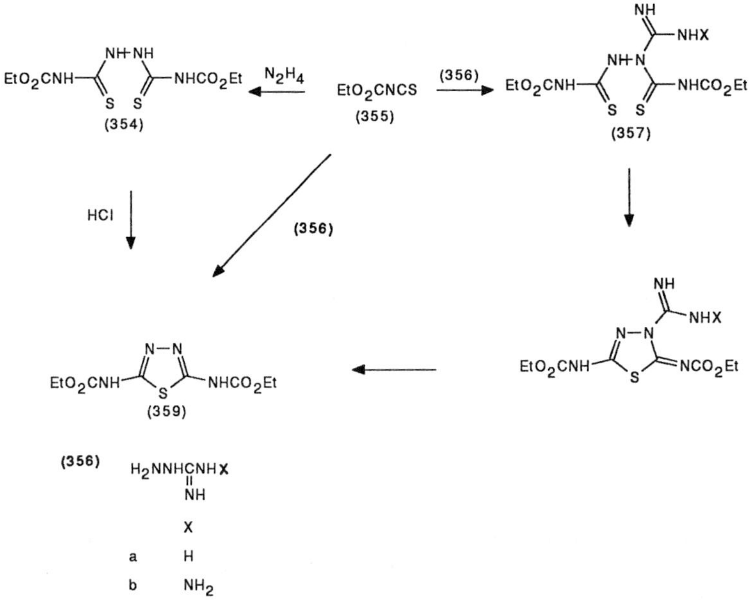

EtO₂CNH... (354) NH-NH ... S S ... NHCO₂Et

EtO_2CNH ... (354) ... N_2H_4

EtO_2CNCS (355)

(356)

EtO_2CNH ... (357) ... NHCO₂Et

HCl ↓

(356)

EtO_2CNH ... (359) ... N—N ... S ... NHCO₂Et

EtO_2CNH ... N—N ... S ... NCO₂Et ... NH—NHX

(356) H₂NNHCNHX
            ‖
            NH

	X
a	H
b	NH₂

The replacement of hydrazine by ethoxycarbonylhydrazine in the above reaction yields the 1,3,4-thiadiazole (360) (F. Kurzer and J.L. Secker, J. heterocyclic Chem., 1989, 26, 355).

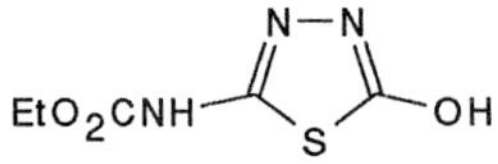

EtO_2CNH ... N—N ... S ... OH

(360)

(iii) Cycloaddition methods

The isothiocyanate ion participates regiospecifically in a 1,3-dipolar addition reaction with the nitrile imine (361) to form (362) (H.M. Hassaneen, *et al.*, J. heterocyclic Chem., 1987, 24, 577):

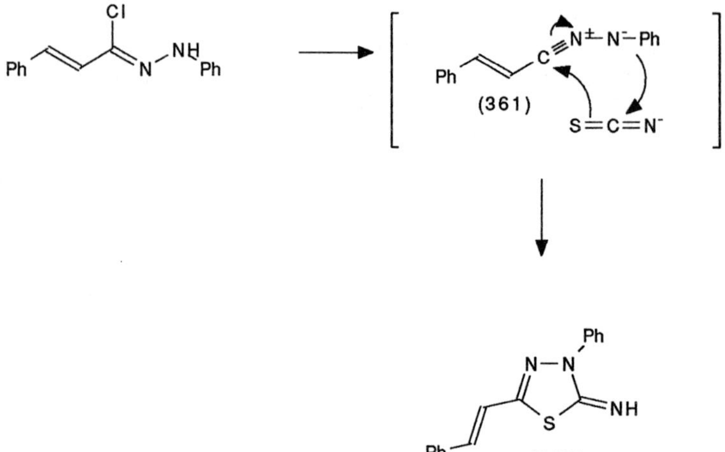

(361)

(362)

Nitrile imines also add to the C=S bond of carbon disulphide (W. Fliege, R. Graskey and R. Huisgen, Ber., 1984, <u>117</u>, 1194).

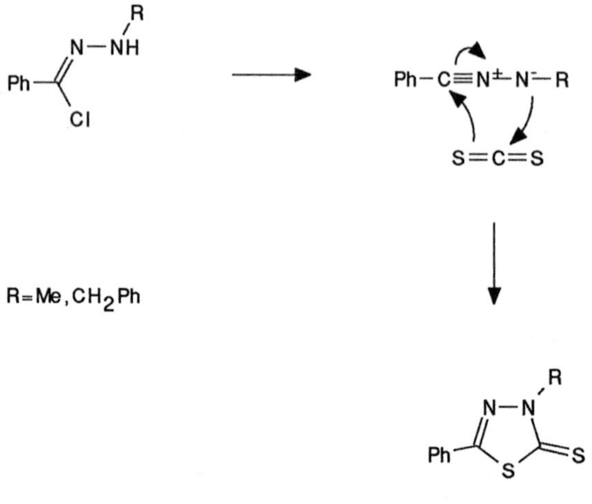

R = Me, CH$_2$Ph

The reaction between nitrile imines and the C=S bond is discussed in more detail by P. Caramella and P. Gninanger in Padwa, part A, Ch. 3, p. 360.

Two one-pot syntheses of 1,3,4-thiadiazoles are available. The 2-heteroaryl-1,3,4-thiadiazoles (364) are formed in moderate yields from heterocycles with acidic methyl groups on treatment with sulphur and an aroylhydrazide. A mechanism involving an intramolecular electrocyclisation of the nitrile imine intermediates (363) has been proposed (G. Mazzone, *et al.*, J. heterocyclic Chem., 1984, 21, 181):

Ar-CH$_3$ + H$_2$NNHAr' $\xrightarrow{\text{S}}$

(363)

$$\left[\text{Ar}-\text{CH}\!\!=\!\!\text{N}-\text{NH}-\text{C}\overset{\text{Ar'}}{\underset{\text{S}}{\diagup}} \right] \longrightarrow \left[\text{Ar}-\text{C}\!\!\equiv\!\!\text{N}^{\pm}-\text{N}^{-} \underset{\text{S}}{\diagdown}\text{Ar'} \right]$$

(364)

Ar=2- or 4-pyridinyl,2-quinolinyl

4-Acyl-2-(methylamino)-5-dialkyl-l3,4-thiadiazolines (366) are formed by reaction between methylthiosemicarbazide (365), an acyl chloride and a carbonyl compound. (E.V.P. Tao and G.S. Staten, J. heterocyclic Chem., 1984, 21 599):

MeNHCNHNH$_2$ + RCOCl + R$_1$COR$_2$ $\longrightarrow$
  ‖
  S

(365)

(366)

R=alkyl or aryl

R$_1$,R$_2$=H,Me,Et;-(CH$_2$)$_5$-

13. Selenadiazoles

These have been reviewed by I. Lalezari in Comprehensive Heterocyclic Chemistry, Vol. 6, Ch. 4. 20, p. 333.

342

(a) Synthesis and reactions

Selenadiazoles have been less studied than thiadiazoles, and many reactions and syntheses of the former parallel those of the latter.

1,2,3-Selenadiazoles are formed by the action of selenium dioxide on the hydroazones (367), (368); the E- and the Z- isomer each give a different product (O. Zummer and H. Meier, Ber.,1981, 114, 194).

R_1=H,Alkyl,aryl R_2=Alkyl,aryl,Cl,Ts

Z=Ts,CONH$_2$

Similarly, the semicarbazones (369) with selenium dioxide in acetic acid give 1,2,3-selenadiazoles. The ether (369a) exclusively (370a) whereas the sulphide (369b) and selenide (369c) give a mixture of (370), and (371) with the latter predominating. The reaction fails with (372a), whereas (372 b and c) give a single isomer (A. Shafiee, *et al.*, J. heterocyclic Chem., 1982, 19, 305).

X=(a) O , (b) S , (c) Se

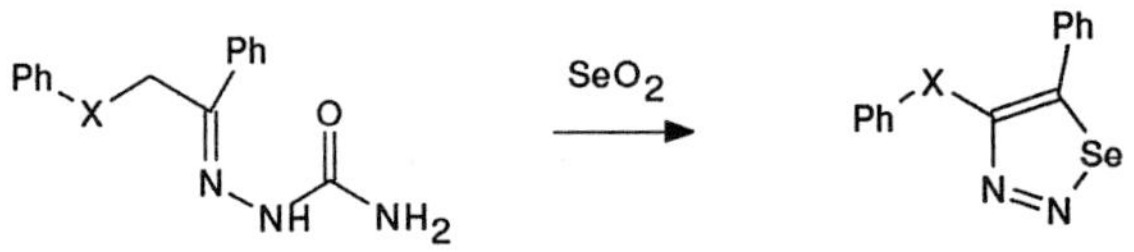

(372)

X=(a) O , (b) S , (c) Se

The heteroarylsemicarbazones (373) are cyclised similarly (A. Shafiee, M. Anaraki and A. Bazzaz, J. heterocyclic Chem., 1986, 23, 861).

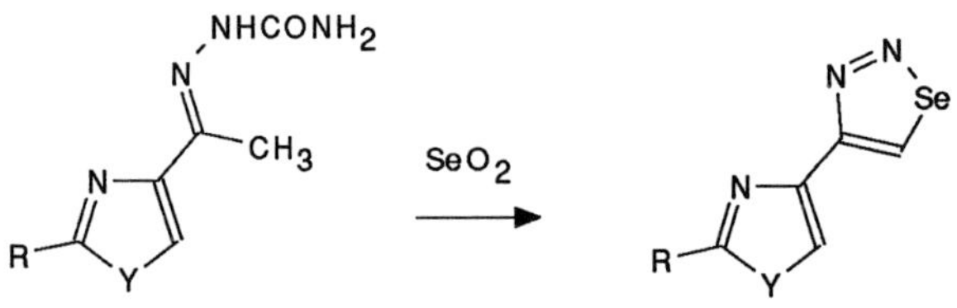

(373)

R=Me,Ph,4-X-C$_6$H$_4$ [X=Br,Cl,MeO]

Y=S,Se

1,2,3-Selenadiazoles (378) are formed from the selenocyanates (376), generated *in situ* either by a Japp-Klingeman reaction between (374) and the diazonium salts (375) or from the chloride (377) and potassium selenocyanate. The imine (378) with nitrous acid yields the *N*-nitrosimine, which on pyrolysis gives the oxo-derivative. When the *N*-aryl substituent bears an *ortho*-carboxy function, cyclisation occurs spontaneously, precluding isolation of the imine (A.O. Abdelhamid, H.M. Hassaneen and A.S. Shawali, J. heterocyclic Chem., 1983, 20, 719).

$Ar=2\text{-}CO_2R\text{-}C_6H_4$ [R=H,Me]

$Ar=4\text{-}CO_2R\text{-}C_6H_4$ [R=H,Me]

The 1,3,4-selenadiazole ring can be constructed by 1,3-dipolar addition. The nitrile imine (379) adds regioselectively to the selenocyanate ion to yield (380) (H.M. Hassaneen, *et al.*, J. heterocyclic Chem., 1987, 24, 577).

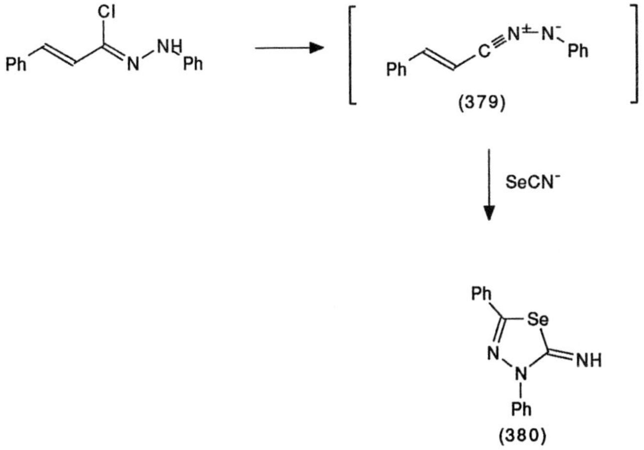

(379)

SeCN⁻

(380)

Cleavage of the 1,2,3-selenadiazole (381) yields the acetylenic selenide which can be recyclised with carbon disulphide to give the 1,3-thiaselenole-2-thione (382) (V.Z. Laisiev, M.L. Petrov and A.A. Petrov, Chem. Abs., 1982, 97, 6235; A Shafiee, M. Anarki and A. Bazzaz, J. heterocyclic Chem., 1986, 23, 861).

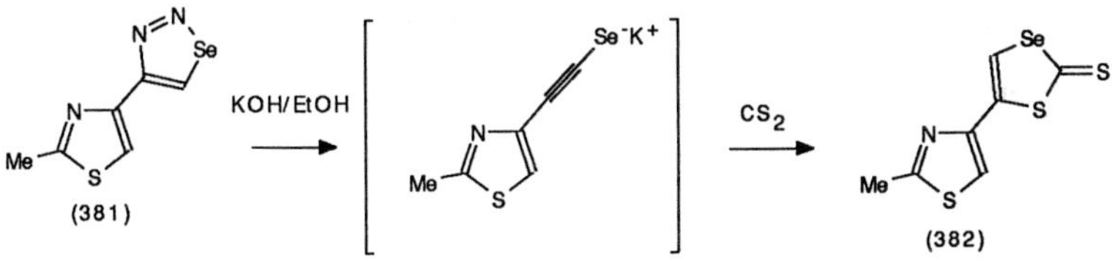

This cleavage reaction also forms the basis of a "one-pot" synthesis of the imide (383) (A. Shafiee and G. Fanaii, Synthesis, 1984, 512).

(383)

$$Ar=Ph,4-Y-C_6H_4[Y=Me,OMe]$$

$$R=Ph,Me,OEt,4-Me-C_6H_4$$

$$X=S,Se$$

14. Other five-membered rings with three heteroatoms

Five-membered rings containing three oxygen or sulphur atoms have been reviewed by G.W. Fisher and T. Zimmermann in Comprehensive Heterocyclic Chemistry, Vol. 6., Ch. 4.33, p. 851, and dioxazoles, oxathioazoles, and dithiazoles by M.P. Sammes, Ch. 4.34, p. 897.

(a) Oxathiazoles

1,3,4-Oxathioazol-2-ones (e.g. 385) decompose on heating to yield nitrile sulphides which can be trapped with 1,3-dipolarophiles. For example (385) on heating in xylene yields (387) *via* an intramolecular reaction of the nitrile sulphide (286) (P.A. Brownsort, R.M. Paton and A.G. Sutherland, J. chem. Soc. Perkin 1, 1989, 1679).

A similar reaction has been used in the synthesis of 1,2,4-thiadiazoles.

(b) 1,2,4-Dithiazoles

The dimethylaminoethyl thiocarbonates (388) react with chlorocarbonyl sulphenyl chloride (389) to give good yields of the 1,2,4-dithiazolidine-3,5-diones (390) (U. Slomczynska and G. Baranay, J. heterocyclic Chem., 1984, 21, 241).

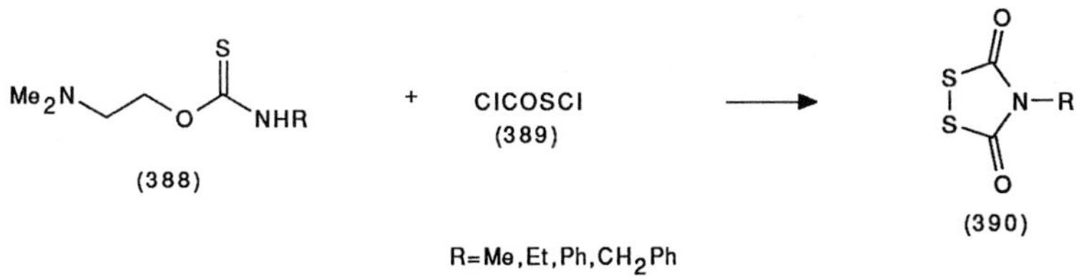

(388)   +   ClCOSCl
                              (389)

(390)

R=Me,Et,Ph,CH$_2$Ph

(c) 1,4,2-Dithiazoles

The 1,2,4-dithiazole-5-thione (393) is formed by the reaction of the thioamide (391) with trichloromethylsulphonyl chloride (392). It reacts both as a 1,3-dipole (formation of (394)) and as a 1,3-dipolarophile, with the azide (395) to yield (396) (D.J. Grieg, *et al.*, J. chem. Soc. Perkin 1, 1985, 1205).

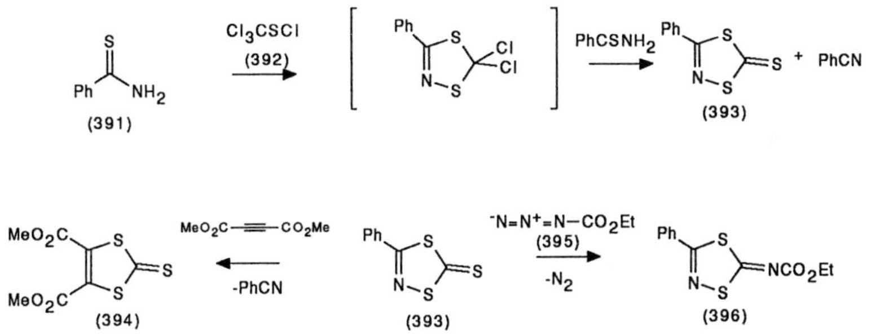

15. Heterocyclic rearrangements

These can be classified into three major types.

(a) Those involving ring opening and ring-closure, such as the Dimroth and Cornforth rearrangements.

(b) Those involving formation of a new ring at the side-chain, often known as the Boulton-Katritzky scheme.

(c) Photochemical rearrangements.

G. L'abbe (J. heterocyclic Chem., 1985, 21, 627) has reviewed all three types of rearrangement. The Boulton-Katritzky rearrangement has been the subject of a separate review (M. Ruccia, N. Vivona and D. Spinelli, Adv. heterocyclic Chem., 1981, 29, 142).

(a) The Dimroth rearrangement

This reaction, illustrated below by the equilibrium between 3-phenyl-4-amino-1,2,3-triazole (397) and 4-(aminophenyl)-1,2,3-triazole (398) is defined by Albert in a review on 4-aminotriazoles (A. Albert, Adv. heterocyclic Chem., 1986, 40, 129) as "rearrangement" from left to right and "regression" from right to left.

(397)   rearrangement / regression   (398)

L'abbe (*loc cit.*) describes this form of rearrangement as "ring-degenerate".

The position of equilibrium is determined by the following factors:

(a) The nature of the substituents. The bulkier substitutent favours the substitutent nitrogen, presumably due to steric factors.

(b) The reaction conditions. Basic conditions favour rearrangement to a 1-H triazole, due to formation of the salt (399).

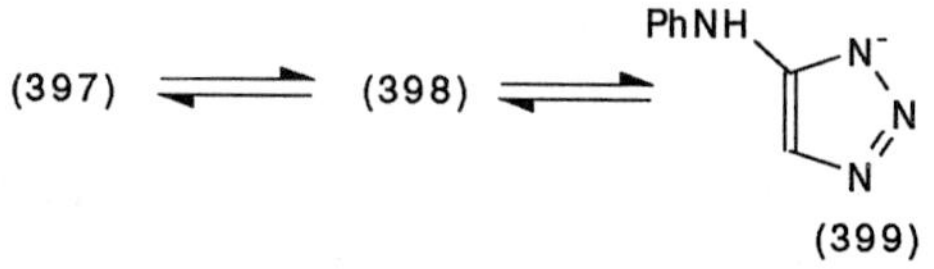

(397) ⇌ (398) ⇌ (399)

The pKa of the ring N-H of aminotriazoles is similar to that of phenol (A. Albert, *loc. cit*).

Judicious choice of reaction conditions can make this process of synthetic value. For example, heating the 4-(methylamino)triazole (400) (in pentanol; no reaction occurs in refluxing ethanol) gives the 3-methyltriazole (401); this enables 3-methyltriazoles to be prepared without the use of the dangerously explosive methyl azide (A. Albert, J. chem Soc. Perkin 1, 1981, 2344.)

A similar reaction occurs to "exchange" a methyl group and a benzyl group when (402) is heated in ethanol (A. Albert, *loc. cit.*).

The reaction may also be an unwanted complication. Arrangement occurs when (404) is heated with formamide, giving (405) rather than the expected 8-azapurine (403) (A. Albert, Adv. heterocyclic Chem., 1986, 39, 118).

The Dimroth rearrangement may occur subsequent to other reactions. For example cyclisation of the triazene (406) gives the 4-(arylamino) triazole (408) *via* the 3-aryltriazole (407) (K.M. Baines, T.W. Rourke, K. Vauaghan, J. org. Chem., 1981, 46, 856).

(406)   (407)   (408)

(i) The mechanism of the Dimroth rearrangement

This can be formulated as a cyclo-reversion followed by tautomerism and ring-closure.

Gilchrist has compared this with the commonly observed ring-opening reactions of the anions of 5-membered heterocycles, with the nitrogen lone pair taking the place of the electron pair on the carbanion in promoting ring opening; (T.L. Gilchrist, Adv. heterocyclic Chem., 1987, 41, 42).

A number of heterocycles undergo this ring-opening reaction. It is observed in the equilibrium between benzo-1,2,3-oxadiazole (409) and the diazaquinone (410) (R. Schultz and A. Schweigh, Angew Chem. intern. Edn., 1984, 23, 509) and between the 1-R-benzotriazole (411) and diazaquinoneimines (412) (C.L. Habraken *et al.*, J. org. Chem., 1984, 49, 2197).

The intermediate diazocompound has been observed in several cases. The diazoamide (414) can be isolated from the acid hydrolysis of the chlorotriazole (413) (P.H. Olesen, *et al.*, J. heterocyclic Chem., 1984, 21, 1693).

Parallels also exist with the ring-closure stage. Oxidation of the bis-sulphonylhydrazones (417) lead (IV), thallium (III) or mercury (II) acetate is believed to give a "diaza-imine" such as (418), which then undergoes ring closure in a manner parallel to the last stage of the Dimroth rearrangement (R.N. Butler, Chem. Reviews, 1984, 84, 249).

The exact nature of this process is unclear. The equilibrium between (419) and (420) is described as being due to a nucleophilic attack using a nitrogen lone pair, not an electrocyclic process (R.N. Butler, J.P. James, Chem. Comm., 1983, 627; R.N. Butler, et al.,J. Chem. Res. Synop., 1987, 332).

$Ar=4-NO_2C_6H_4$

Whereas the reacted cyclisation of (421) via the nitrilimine (422) is described as a 1,5-electrocyclisation (R.N. Butler, S.M. Johnston, Chem. Comm., 1981, 376; J. chem. Soc. Perkin 1, 1984, 2109).

(421)

(422)

E or Z

4-H-triazole (424) is formed by cyclisation of the intermediate azide (423) (R. Carrie, D. Danion, E. Ackermann, Angew Chem. intern. Edn., 1982, 21, 288); this is described (E.F.V. Scriven, K. Turnbull, Chem. Reviews, 1988, 88, 297) as a "cycloaddition", i.e. a 1,3-dipolar addition of an azide to an alkene, though this is geometrically impossible as a suprafacial process. A better description may be a 1,5-electrocyclisation analogous of the 1,6-electrocyclisations of hexa-1,3,5-trienes.

(423)

(424)

Closely related rearrangements occur in a number of 1,2,3-triazoles and related heterocycles. The iminomethyl compound (425) rearranges to the 4-amidino-1,2,3-triazole (426) and does not undergo the Dimroth Rearrangement to (427). (G. L'abbe and A. Vandendriersche, J. heterocyclic Chem., 1989, 26, 701).

(425)   (426)

(427)

354

A similar ring-closure at a nitrogen containing substituent occurs in the rearrangement of the diazocompound (428a) in benzene; the isolated norcaradiene (431) is produced *via* decomposition of the rearrangement product (429). As with the Dimroth Rearrangment the process is accelerated by electron-withdrawing substituents, and (428b) decomposes in benzene to the cycloheptatriene (430) without rearrangement (G. L'abbe and W. Dehaen, Tetrahedron, 1988, 44, 461).

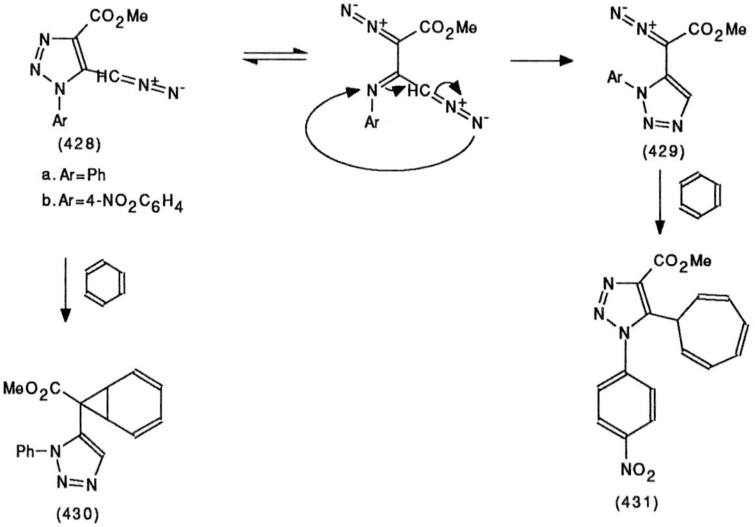

(428)

a. Ar=Ph

b. Ar=4-NO$_2$C$_6$H$_4$

(429)

(430)

(431)

An analogous process takes place in the rearrangement of the 5-azido-1,2,3-triazole (432) to the 5-diazoalkyltetrazole (433). (G. L'abbe, A. Vandendriessche and S. Tappet, Tetrahedron, 1988, 44, 3617).

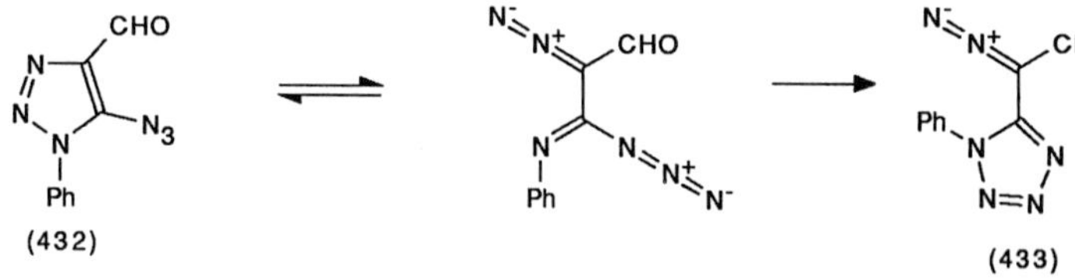

(432)

(433)

(ii) The Dimroth rearrangement in other heterocycles

Similar reactions to those described above occur in other heterocyclic systems. An apparently "innocent" reaction may involve a Dimroth rearrangement. It has been demonstrated by ^{15}N-labelling that acylation of 3-amino-1,2,4-thiadiazole (434) with an acyl chloride involves a Dimroth-type rearrangement where the ring-opening is initiated by nucleophilic attack by chloride ion on the initially formed ring-acylated compound. This may be described in L'abbes terms as a truly degenerate rearrangement as it can only be revealed using isotopic labelling.

R=OPh,CH$_2$Cl

$^{\bullet}$N= ^{15}N

By contrast acylation of (434) with methyl isocyanate gives the urea (435) without rearrangement, possibly due to the absence of a nucleophile to initiate ring opening of any ring acylated intermediate (A. Priess, W. Walek and S. Dietzel, J. pr. Chem., 1981, 323, 279).

(435)

356

A non-ring degenerate Dimroth rearrangement occurs when the chlorothiadiazole (436) is treated with hydrazine, forming the triazole (438). (G. L'abbe and E. Vanderstede, J. Heterocyclic Chem., 1989, 26, 1811). This rearrangement is readily reversed; acidification of (438) yielding (437). When (438) is treated with benzaldehyde, rearrangement to (439) occurs, whereas when (438) is methylated and then treated with benzaldehyde, the un-rearranged product (440) is formed.

A similar 1,2,3-thiadiazole to 1,2,3-triazole rearrangement takes place when the amine (441) is treated with base (E.F. Dankova, *et a.l*, Chem. Abs., 1986, 104, 68796; V.A. Bakulev, *et al., ibid*, 1988, 108, 150373).

(441)

The 1,2,3-thiadiazole (442; R=H) reacts with sodium azide to give the unrearranged product (443; R=H) but when the 4-substituent is electron-withdrawing (442; R=CO_2Et or COPh) the rearranged product (444; R=CO_2Et or COPh) is obtained. (G. L'abbe, *et al.*, Bull. Soc. chim. Belg., 1988, 97, 163; G. L'abbe, J.P. Kekerk, and M. Keketele, Tetrahedron Letters, 1982, 1103).

(442)   (443)   (444)

Diazo-substituted heterocycles are especially prone to rearrangement. When the tosylhydrazone (445) is treated with base, the intermediate (446) rearranges to the more stable (447) either by a concerted process (A) or ring-opening/ring-closure(B). The importance of the ester-function in stabilising the diazo group in (447) is shown by the clearn Forster reaction of (448) to give the un-rearranged product (449) (G. L'abbe, J.-P. Kekork and M. Deketele, Chem. Comm., 1983, 588).

(445)   (446)

Base
Route A

Route B
Ring-opening

(447)

Rearrangement of the isoxazolediazonium salt (450) yields the 1-hydroxytriazole (451) *via* a ring-opening and ring-closure sequence (G. L'abbe, F. Godta and S. Toppet, Tetrahedron Letters, 1983, 30, 3149).

A Dimroth rearrangement takes place when thiatrizole (452) reacts with 2-pyridinyl isothiocyanate (452) (which acts as a "masked" 1,3-dipole) giving (453). The latter then rearranges *via* bond-fission to form the related products (454) and (456). The intermediate (453) and (455) were identified by ^{13}C nmr spectroscopy. The basic nitrogen in 2-pyridinyl isothiacyanate may facilitate the ring-opening as shown in the formation of (454) from (453) (G. L'abbe, K. Allewaert and S. Toppet, J. heterocyclic Chem., 1988, 25, 1459).

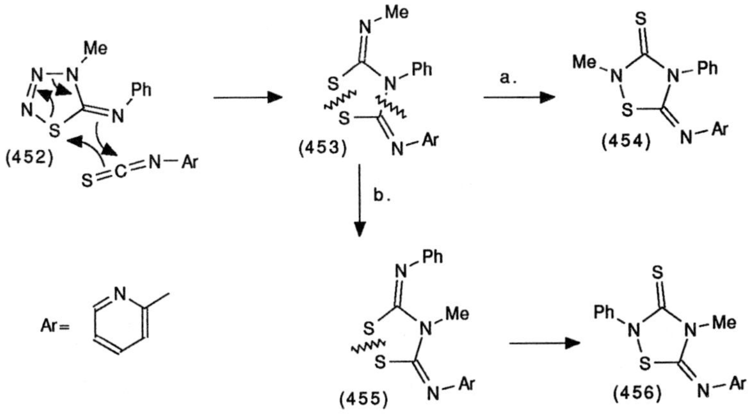
(452)
(453)
a.
b.
(454)
Ar=
(455)
(456)

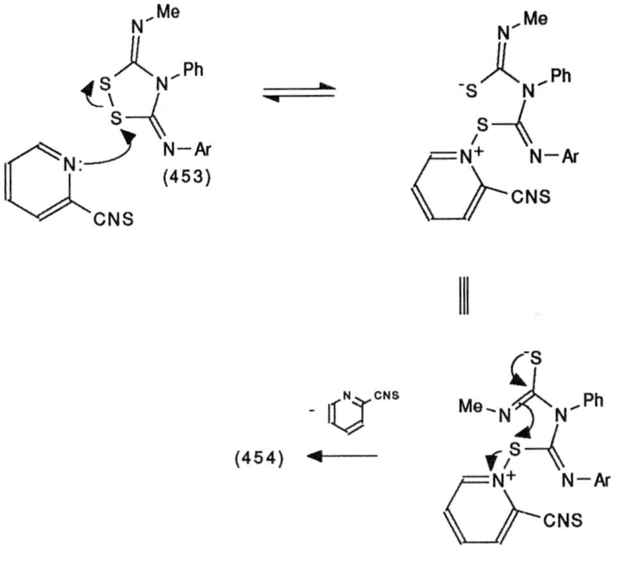
(453)
CNS
(454)

(b) The Boulton-Katritzky scheme

This can be generalised as shown below and hence this classification can be applied to 1,2,4- and 1,2,5-oxadiazoles.

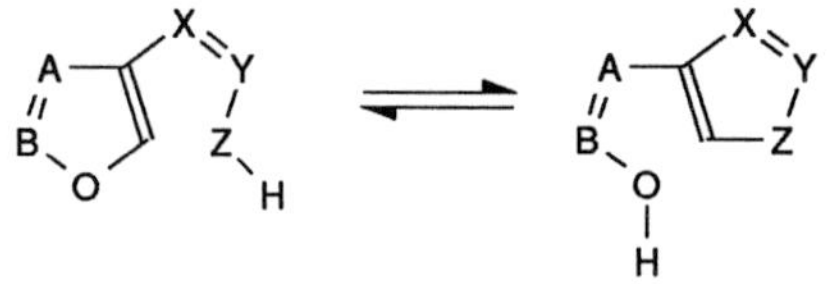

The mechanism of this process has been extensively studied. The conversion of the 1,2,4-oxadiazole (457) into the 2-H 1,2,3-triazole (458) is subject to general base catalysis (V. Frenna, *et al.*, J. chem. Soc. Perkin 2, 1981, 1325) and is best described as an S_Ni reaction with concerted N-O bond fission and N-N bond formation (for a detailed analysis, see V. Frenna, *et al., ibid*, 1984, 785, and references therein).

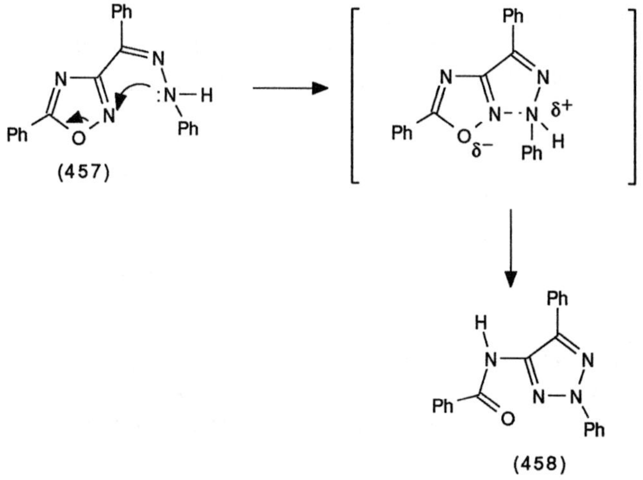

(457)

(458)

The stereochemistry of the side-chain is of great importance. Both the *E* (460) and the *Z* (461) oxime are formed when (459) reacts with hydroxylamine. With base only the Z-isomer (461) rearranges to the 1,2,5-oxadiazole (462).

Each of the isomeric oximes (463) when heated with copper yields the 1,2,5-oxadiazole (464). An E/Z interconversion is probably invovled (N. Vivona, G. Macaluso and V. Frenna, J. chem. Soc. Perkin 1, 1983, 483).

The oximes of 1,2,4-oxadiazoles can be rearranged. The reaction of (465) with hydroxylamine yields the E-oxime (466) together with the rearranged compound (467), presumably formed *via* the Z-oxime (468). The O-methyloximes (469) failed to rearrange irrespective of their configuratiOn (N. Vivona, *et al.*, J. heterocyclic Chem., 1985, 22, 97).

Steric factors may have a subtle influence on reactivity. When the 1,2,4-oxadiazole (470) is treated with 1-phenyl-1-methylhydrazine in acetic acid the hydrazone (471) is not isolable and immediately rearranges to the triazole (472), demethylation at N-2 being effected by the solvent.

In contrast, the hydrazone (473) is isolable and is only converted to (472) when treated with base or heated in acetic acid. Whilst the inductive effect of the methyl group in (471) may play a part in the increased reactivity of (471), it may be that the lower nucleophilicity of the nitrogen in (473) is due to conjugation of the nitrogen lone pair with the hydrazone π-bond. This interaction is reduced in (471) due to the bulk of the methyl group causing the lone pair to be twisted out of the plane of the π-band (N. Vivona et al., J. chem. Soc. Perkin 1, 1982, 165).

(473)

In a closely related reaction, the isoxazole (474) with excess 1-methyl-1-phenylhydrazine gives the 1,2,3-triazolylindole (476). Formation of the hydrazone of the initially formed triazole (475) is followed by a Fischer rearrangement.

(474)   (475)

(476)

In the absence of a demethylating agent such as acetic acid the preceding reaction yields the indole (478). A mechanism involving fission of the unstable intermediate (477) has been proposed (N. Vivona, et al., J. heterocyclic Chem., 1983, 20, 931).

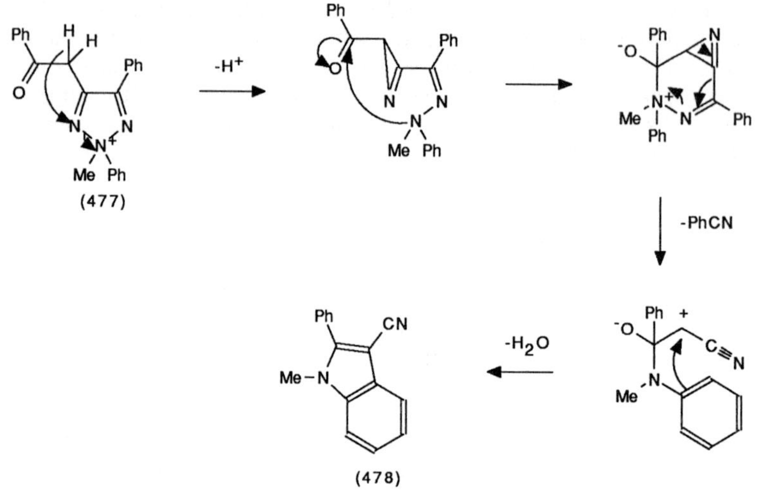

(477)

(478)

In a similar process 1,3,4-oxadiazoles have been converted into 1,2,4-triazolines. The acyl chloride (479) with methylhydrazine yields (482) *via* rearrangement of the intermediate hydrazide (480). Acid hydrolysis of the hydrazone (481) also yields (482) *via* (480) (R. Milcent, *et al.*, J. heterocyclic Chem., 1989, <u>26</u>, 231).

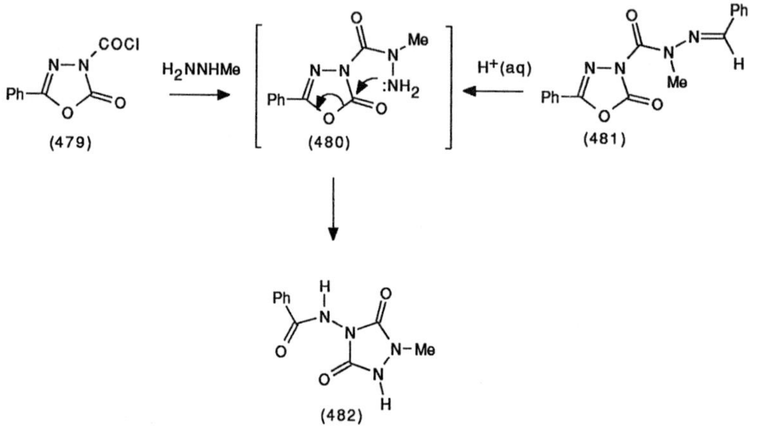

(479)   (480)   (481)

(482)

Some rearrangements are valuable in synthesis. The fused pyrrazoles (487)-(490) are formed in good yield from the 1,2,4-oxatriazoles (483)-(486). The reaction conditions are determined by the nucleophilicity of the arylamine, with the phenyl compound (483) rearranging simply on heating, whereas the pyrimidylamine (484) requires basic conditions (D. Korbonits, I. Kanzal-Szvoboda and K. Horvath, J. chem. Soc. Perkin 1, 1982, 759).

$R_1 = H, Et, CH_2Ph$

$R_2 = Ph, substituted\ Ph, CH_2Ph, Me, n-Pr$

The 1,2,4-oxadiazole azide (491) rearranges to the tetrazole (492), but this rapidly decomposes, yielding the acyl nitrile (493) as the isolated product (P. Choi, C.W. Rees and E.H. Smith, Tetrahedron Letters, 1982, 23, 121).

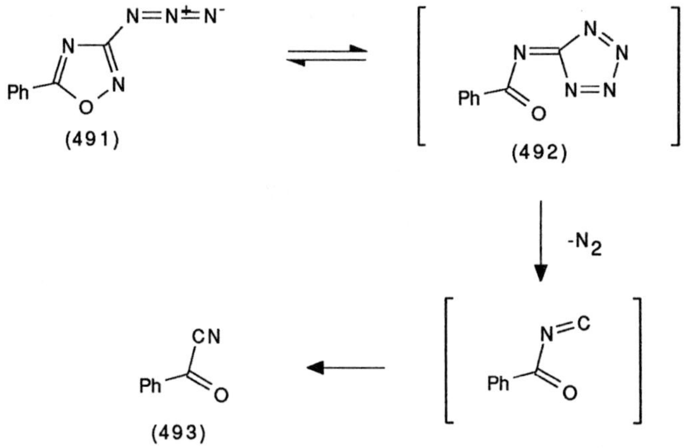
Ph
(491)
Ph
(492)
$-N_2$
Ph
CN
(493)
Ph

An aliphatic version of these rearrangements is possible. The amidoxime (494) when treated with a carboxylate ester under alkaline conditions gives the pyrrazolline (496). The latter presumably arises *via* the 1,2,4-oxadiazole (495).

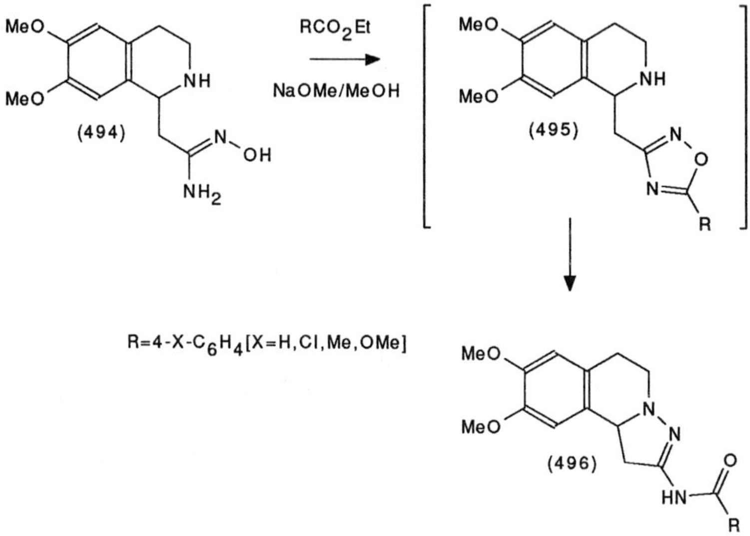
MeO
MeO
NH
(494)
OH
NH_2
RCO_2Et
NaOMe/MeOH
(495)
R
$R=4\text{-}X\text{-}C_6H_4$[X=H,Cl,Me,OMe]
(496)
HN
R

To test this hypothesis (496) was synthesised as below and shown to rearrange to (495) on heating; the unsaturated compound (497) is less reactive than (496) due to the lower nucleophilicity of the enamine nitrogen (I. Bata *et al.*, J. chem. Soc. Perkin 1, 1986, 9).

A closely related reaction is the formation of the pyrrazollone (500) from the *O*-acetyl amidoxime (498), *via* the non-isolated 1,2,4-oxathiazolines (499); in contrast, the unsaturated oxadiazolinones (501) are isolable (C-K. Kim, F. Devellis, C.A. Maggiulli, J. heterocyclic Chem., 1987, 24, 325);

To rationalise the occurrence of the rearrangement of a compound with either saturated or unsaturated side-chains, the modified scheme below has been suggested, where R_1 and R_2 represent either H or the components of a bond.

The rearrangemetns described above have all involved an N-O bond fission; fission of this weak bond is considered to be the driving force behind the rearrangement. A few examples of N-S bond fission are known. Reaction of the aroylisothiazolone (502) with hydroxylamine and with an arylhydrazine gives the 1,2,5-oxathiazoles (504) and the 1,2,3-thiadiazole (506) respectively *via* rearrangement of the intermediate oxime (503) and hydrazone (505) (A.Tsolomitis and C. Sandris, J. heterocyclic Chem., 1984, 21, 1679).

Ar=Ph,4-Cl-C$_6$H$_4$
Ar'=Ph,2,4-(NO$_2$)$_2$C$_6$H$_3$
R=Ph,CH$_2$Ph

A reaction believed to involve a Boulton-Katritzky rearrangement is the remarkable formation of the 1,2,4-thiadiazole (508) from the *S*-oxide (507). The proposed mechanism is given below. The interannular N-S separation in (507) is unusually small, and the rearrangement may be a consequence of this (P.J. Dunn and C.W. Rees, J. chem. Soc. Perkin 1, 1989, 2485).

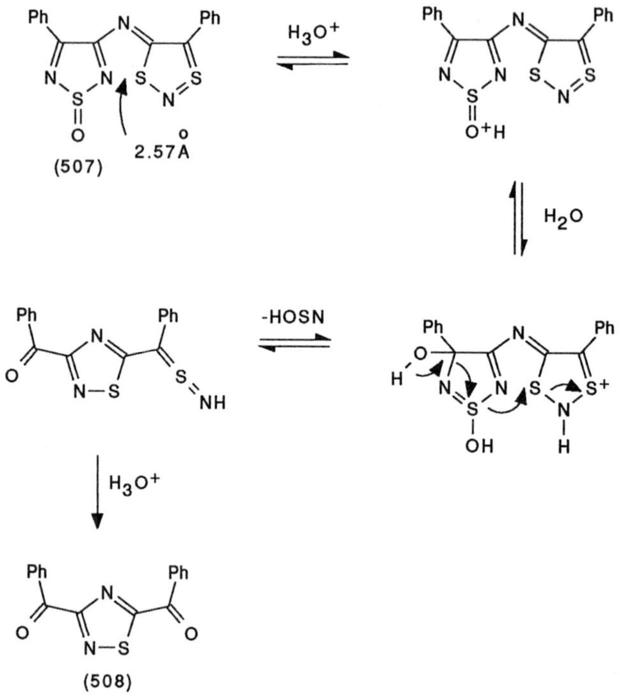

(507) 2.57Å

(508)

(c) Photochemical rearrangements

A parallel to the photoisomerism of isoxazoles to oxazoles has been observed. Irradiation of 1,2,4-oxadiazole (509) at 254 nm gives 1,3,4-oxadiazole (510) (S. Buscemi, *et al*, J. chem. Soc. Perkin 1, 1988, 1313; J. heterocyclic Chem., 1988, 25, 931).

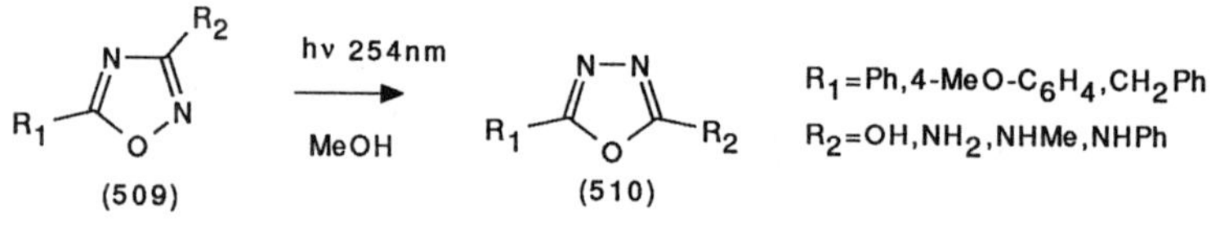

(509) $\xrightarrow[\text{MeOH}]{h\nu\ 254\text{nm}}$ (510)

R_1 = Ph, 4-MeO-C_6H_4, CH_2Ph

R_2 = OH, NH_2, NHMe, NHPh

The mechanism is thought to proceed *via* the oxo/imino tautomer (511) as compound (512) lacking a tautomerisable substituent at position 3 gives acyclic products (513) *via* O-N bond fission and reaction with the solvent. Further evidence comes from the clean rearrangement of the "fixed" compound (514).

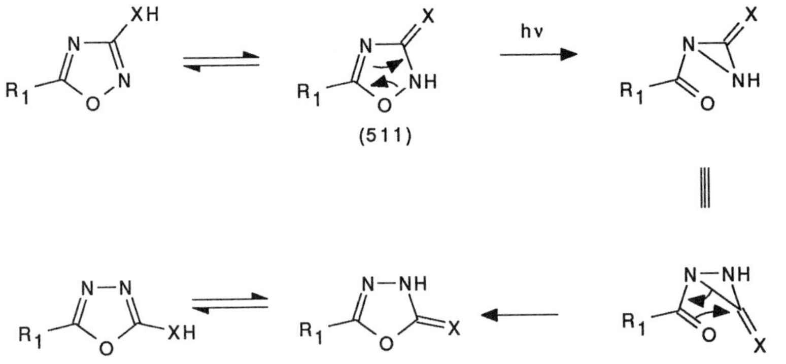

(511)

X = NR, O

R₁ as above

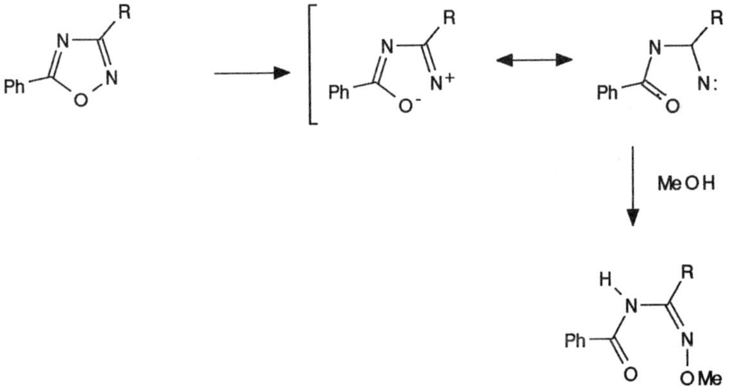

MeOH

(513)

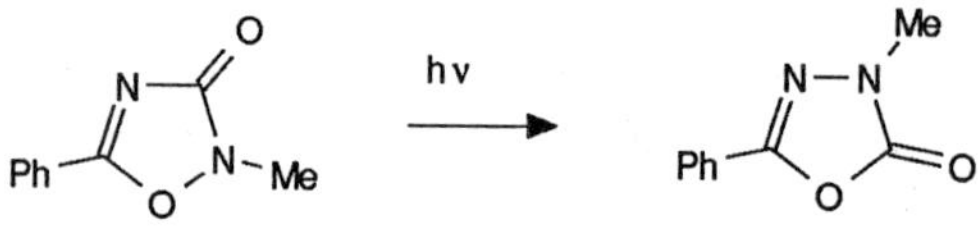

(514)

Investigation of compounds of the form (515) showed that rearrangement also occurs at the "side-chain", giving the fused heterocycles (516) in a photochemical Boulton-Katritzky rearrangement and ring-opened products as described above.

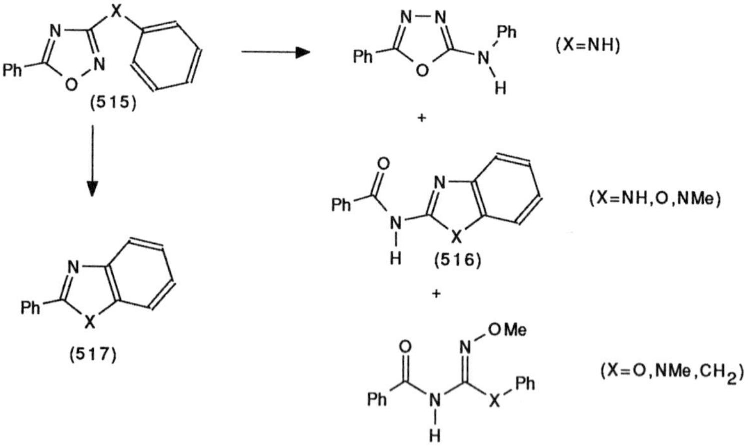

(515)

(X=NH)

+

(X=NH,O,NMe)

(516)

+

(X=O,NMe,CH$_2$)

(517)

The failure of the benzyl derivative (515, X=CH$_2$) to form the indole (517), and the successful rearrangement of the conjugated enamines (518) and (519) indicates that the ease of ring closure is enhanced by a cyclic delocalised transition state of the type (520) (S. Buscerni and N. Vivona, J. heterocyclic Chem., 1988, 25, 1551).

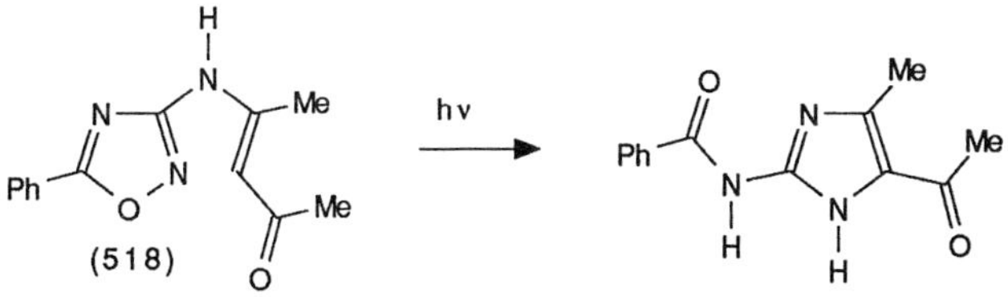

(518)

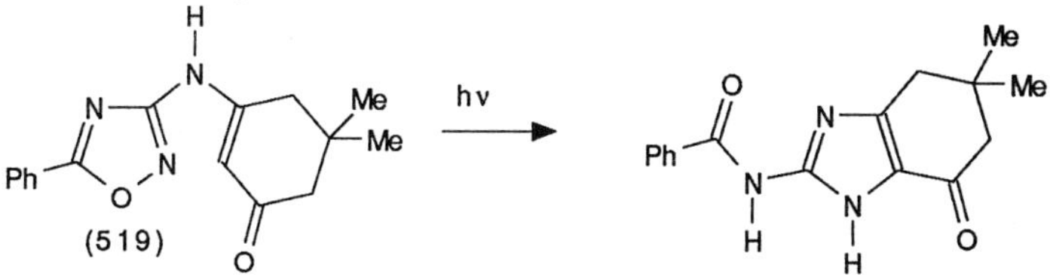

(519)

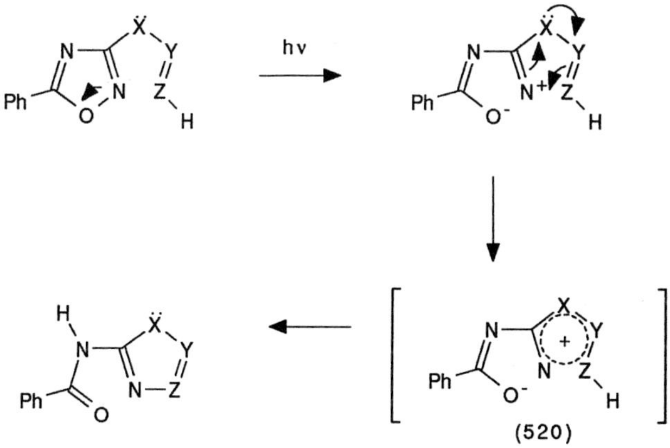

(520)

Chapter 19

FIVE-MEMBERED HETEROCYCLIC COMPOUNDS WITH FOUR HETEROATOMS IN THE RING

John H. Little

Introduction

Interest continues in the chemistry of this diverse range of compounds. Work on the established systems covered in this chapter of the Second Edition has been carried out, searching for new derivatives having commercial potential, most notably as pharmaceuticals. In addition, several new systems have been described and are included in this supplement either within an appropriate section dealing with related systems, or as an additional section, as in the case of the various ring systems containing phosphorus.

1. Tetrazoles

A general review of tetrazole chemistry has been published (R.N. Butler in "Comprehensive Heterocyclic Chemistry", ed. A.R. Katritzky and C.W. Rees, Pergamon, 1984. Vol 5, p.791). In addition, specific reviews are available: S.W. Schneller, *ibid.* Vol 5, p.847 (Tetrazoles with fused six-membered rings); R.P. Reddy *et al.*, Indian J. chem. Sci., 1987, 1, 45 (Steroidal Tetrazoles); J. Racek, Biochem. Clin. Bohemoslov., 1984, 13, 281 (Use of tetrazolium salts in clinical biochemistry); N.A. Klyuev, I.I. Grandberg and Y.V. Shurukhin, Khim. Geterotsikl. Soedin., 1985, 723 (Thermolysis and mass spectra of tetrazoles); H. Singh *et al.*, Heterocycles 1984, 22, 1821 (Tetrazole synthesis from hydrazines); G.I. Koldobskii,

376

V.A. Ostrovskii and V.S.Poplavskii, Khim. Geterotsikl. Soedin., 1981, 1299 (Advances in tetrazole chemistry).

a) *Synthesis*

The most frequently employed synthesis of tetrazoles involves the reaction of an azide with an appropriate substrate for cycloaddition. Established procedures have been modified to improve the efficiency of the reaction and to achieve greater selectivity. For example, the reaction of a nitrile with sodium azide is improved by the use of triethylamine hydrochloride as catalyst and *N*-methyl pyrrolidone as solvent (P.R. Bernstein and E.P. Vacek, Synthesis, 1987, 1133).

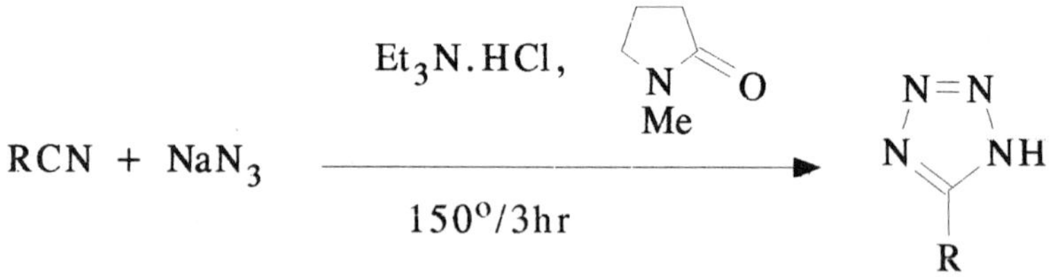

Among the novel substrates for azide addition are keten-*S,N*-acetals (1, R=alkyl, aryl). The initially formed imidoyl azide (2) cyclises to the tetrazole (R.T. Chakrasali, H. Ila and H. Junjappa, Synthesis, 1988, 453).

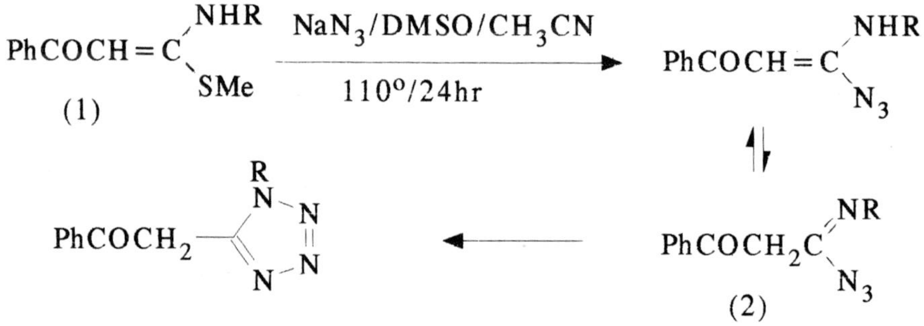

Aminoiminomethanesulphonic acids (3, R=Ph, R'=H; R=R'=Et) give the important 5-aminotetrazoles (4) (A.E. Miller *et al.*, Synth. Comm., 1990, 20, 217)

$$RN = C(NHR')SO_3H \quad + \quad NaN_3 \quad \longrightarrow$$

(3)

NHR'

N   NR

N=N

(4)

Cycloaddition of an alkyl azide to a nitrilium salt yields a tetrazolium salt (5) which gives a tetrazoline on reduction or alkylation. (R. Carboni and R. Carrié, Tetrahedron, 1984, 40, 4115).

$$MeC \equiv NMe \ \overset{+}{\ } \ \overset{-}{FSO_3} \quad + \quad MeN_3 \quad \longrightarrow$$

Me
          $\overline{FSO_3}$
MeN   +   NMe

N=N

(5)

Me   R

MeN   NMe

N=N

$\xleftarrow{\quad RX \quad}$ (5) $\xrightarrow{\quad redn. \quad}$

Me   H

MeN   NMe

N=N

1,5-Diaminotetrazole (6), a useful derivative for the synthesis of other 1,5-disubstituted tetrazoles has been prepared by the lead(II) oxide catalysed reaction of thiosemicarbazide with sodium azide in DMF (P.N. Gaponik and V.P. Karavai, Khim. Geterotsikl. Soedin, 1984, 1683).

$$NH_2NHCNH_2 \quad + \quad NaN_3 \quad \xrightarrow[100^\circ]{DMF/PbO}$$

with S double bonded to C above.

NH₂
          NH₂
N   N

N=N

(6)

1-Aminotetrazole derivatives are obtained in good yield by the reaction of a hydrazone with sodium azide and triethyl orthoformate in acetic acid (*idem. ibid.*, 1983, 841). For example, benzaldehyde hydrazone gives 1-amino-5-phenyltetrazole.

$$PhCH=NNH_2 \ + \ NaN_3 \xrightarrow[80°C/2.5hr]{(EtO)_3CH/AcOH}$$

A novel route to 1-vinyl tetrazole, investigated thoroughly for polymer applications, involves preparation of 1-(2'-hydroxyethyl)tetrazole (8) followed by dehydration (*idem. ibid.*, 1985, 1422). Formamidine and ethanolamine react together to give the N,N'-disubstituted formamidine (7) which with sodium azide gives the tetrazole (8).

$$HN=CHNH_2$$
$$+$$
$$2HO(CH_2)_2NH_2 \longrightarrow HO(CH_2)_2N=CHNH(CH_2)_2OH$$
$$(7)$$

$$NaN_3/ \ (EtO)_3CH/ \ AcOH$$

(8)

Trimethylsilyl azide is widely used as the azide component in tetrazole synthesis. An example of the use of this reagent is its reaction with an oxime ester or Reissert salt in the presence of trimethylsilyl trifluoromethanesulphonate ($Me_3SiOSO_2CF_3$) (K.Nishiyama and I. Miyata, Bull. chem. Soc. Japan, 1985, $\underline{58}$, 2419).

Trimethylsilyl azide, with tin(IV) chloride, reacts with acetals to give a *gem* -diazide which cyclises to a 1,5-disubstituted tetrazole, probably *via* a Schmidt rearrangement. Benzophenone gives the 1,5-diphenyl derivative (9), m.p. 145 °C, in 57% yield (S. Kirchmeyer, A. Mertens and G.A. Olah, Synthesis, 1983, 500).

$$Ph_2C(OMe)_2 + 2Me_3SiN_3 \xrightarrow{\text{SnCl}_4} Ph_2C(N_3)_2$$

(9)

Ketones themselves react similarly with trimethylsilyl azide in the presence of a Lewis acid to give tetrazoles (N. Nishiyama and A. Watanabe, Chem. Letters, 1984, 455). The mechanism of the thermal cyclisation of *gem*-diazides involved in such routes has been studied using deuterium labelling techniques (R.M. Moriarty, B.R. Bailey and I. Prakash, J. org. Chem., 1985, 50, 3710).

A high yield synthesis of 2-methyltetrazole involves initially the cyclocondensation of ethanedinitrile (cyanogen) with hydrazoic acid. Methylation of the sodium salt (10) followed by hydrolysis and decarboxylation gives 2-methyl tetrazole in 68% yield overall (D. Moderhach and A. Lembcke, Chem. Ztg., 1984, 108, 188).

Lithium trimethylsilyldiazomethane (11) is a new synthon for the preparation of tetrazoles (T. Aoyama and T. Shioiri, Chem. Pharm. Bull., 1982, 30, 3450). This reagent, prepared by lithiation of trimethylsilyl diazomethane, reacts smoothly with an ester to give a 2-substituted tetrazole (12) in good yield.

$$Me_3SiCHN_2 \xrightarrow{\text{LDA}} Me_3SiCLiN_2 \xrightarrow{\text{RCO}_2\text{Me}}$$

(11)

$$\begin{array}{c} \text{SiMe}_3 \\ \text{N} \quad \text{N} \\ | \quad \quad || \\ \text{N} - \text{N} \\ \text{RCOCH}_2 \end{array}$$

(12)

1-Aryltetrazoles have been prepared from aromatic aldoximes and ketoximes. These with thionyl chloride in carbon tetrachloride under mild conditions followed by treatment of the Beckmann rearrangement-type intermediate with hydrazoic acid give the tetrazole (R.N. Butler and D.A. O'Donohue, J. chem. Res. (S) 1983, 18). Acetophenone oxime gives 5-methyl-1-phenyltetrazole (13).

$$\begin{array}{c} \text{Ph} \\ \diagdown \\ \quad \quad \text{C} = \text{NOH} \\ \diagup \\ \text{CH}_3 \end{array} \xrightarrow[\text{CCl}_4]{\text{SOCl}_2} \begin{array}{c} \text{Ph} \\ \diagdown \\ \quad \quad \text{C} = \text{NOSOCl} \\ \diagup \\ \text{CH}_3 \end{array}$$

$$\Big\downarrow \; -\text{SO}_2$$

(13)

$$\begin{array}{c} \text{CH}_3 \\ \text{N} \quad \text{N}^{-\text{Ph}} \\ \diagdown \quad \diagup \\ \text{N} = \text{N} \end{array} \xleftarrow{\text{HN}_3} \begin{array}{c} \text{N} \\ \text{Ph} \overset{+}{\diagup} \quad \text{Cl}^- \\ \diagdown_{\text{C}} \\ \quad \text{CH}_3 \end{array}$$

Thionyl chloride is the cyclising agent in a synthesis of 2,3,5-triaryltetrazolium chlorides (15) from the formazan (14) (S. A. Belyakov, Zh. org. Khim., 1989, 25, 2252).

$$\text{ArN} = \text{NC(Ar')} = \text{NNHPh} \xrightarrow[\text{C}_6\text{H}_6]{\text{SOCl}_2}$$

(14)

$$\begin{array}{c} \text{Ar} \\ \text{N} \quad \text{N} \\ \diagdown \quad \overset{+}{\diagup} \quad \text{Cl}^- \\ \text{N} - \text{N} \\ \diagup \quad \diagdown \\ \text{Ph} \quad \quad \text{Ar'} \end{array}$$

(15)

Oxidation of triarylformazans using potassium permanganate in a two-phase system incorporating tetra-*n*-butylammonium bromide as catalyst gives 2,3,5-triaryltetrazolium salts, which themselves have been shown to act as efficient phase transfer catalysts (G. I. Koldobskii *et al.*, Khim. Geterotsikl. Soedin., 1985, 841).

The methoxyphenyl substituted formazan (16) from anisaldehyde can be converted into the tetrazolium salt (17) which when demethylated with boron tribromide and treated with sodium hydroxide gives the quinonoid mesion (18), a (2,3-diphenyltetrazol-5-ylium) phenolate (S. Araki, N. Aoyama and Y. Butsugan, Tetrahedron Letters, 1987, 28, 4289).

Tetrazole-5-thiol derivatives have been used extensively in the synthesis of cephalosporin and related antibiotic analogues. A synthesis of a 1-alkyltetrazole-5-thiol from the related thiosemicarbazide using butyl nitrite and potassium *t*-butoxide has been described (USP 4,515,958. Chem. Abs. 1985, 103, 54082).

$$\underset{\text{MeNHCNHNH}_2}{\overset{\overset{\displaystyle S}{\|}}{}} \quad \xrightarrow[\text{tBuOK}]{\text{BuONO}} \quad \text{(1-methyl-1H-1,2,3,4-tetrazole-5-thiol)}$$

b) Spectroscopic and other studies

The tautomeric equilibrium in tetrazole and in its 5-substituted derivatives
has been extensively studied.

1-*H* Tetrazole ⇌ 2-*H* Tetrazole

The dipole moment of tetrazole is 4.88D suggesting a 1-*H* tautomer
content of 78% (H. Lumbroso *et al.*, J. mol. Struct., 1982. *82*, 283).
For 5-(p-substitutedphenyl)tetrazoles the preference for the 1-*H* form is
evident from dipole moment measurements in solution, but the tautomeric
equilibrium is affected by the nature of the substituent on the benzene ring
- electron-withdrawing groups displace the equilibrium towards the 2-*H*
form (R. N. Butler *et al.*, J. heterocyclic Chem., 1980, 17, 1374; J.
chem. Soc. Perkin 2, 1984, 721).

In the gas phase, the 2-*H* tautomer is energetically favoured as predicted
by *ab initio* MO calculations (A. P. Mazurek and R. Osman, J. phys.
Chem., 1985, 89, 460), and this result is supported by photo-electron
spectroscopy of tetrazole and 1- and 2-methyltetrazole (M. H. Palmer, I.
Simpson and J. R. Wheeler, Z. Naturforsch., 1981, 36A, 1246).

Tetrazole tautomerism in the solid state has been investigated using ^{13}C-
nmr spectroscopy of solids using the cross polarisation - magic angle
spinning technique (R. Fauré, E. J. Vincent and J. Elguero, Heterocycles,
1983, 20, 1713).

Data on the ^{13}C-nmr spectra of substituted tetrazoles in solution is available. The composition of a mixture of isomeric vinyl tetrazoles (19 and 20, R= H, Me, vinyl) has been determined using ^{13}C - ^{13}C coupling constants (L. B. Krivdin *et al.*, Zh. org. Khim., 1985, 21, 1138)

The ^{13}C-nmr spectra of 5-aryltetrazoles have been widely discussed (P. J. Kothari *et al.*, J. heterocyclic Chem., 1980, 17, 637; I. Goljer, J. Svetlik and I. Hrusovsky, Monatsch., 1983, 114, 65; R. N. Butler *et al.*, J. chem. Soc., Perkin 2, 1984, 721). The nicotinic acid analogue, 5-(3-pyridyl)-tetrazole, has been examined by ^{1}H and by ^{13}C nmr-spectroscopy in (CD$_3$)$_2$SO/D$_2$O solution. In the tautomeric equilibria (21$\rightleftharpoons$22$\rightleftharpoons$23) the 1-*H* tetrazole (21) is favoured (R. N. Butler and V. C. Garvin, J. chem. Res. (S), 1982, 122).

The ^{13}C and ^{1}H-nmr chemical shifts for 1-(substitutedphenyl) tetrazoles have been related to Hammett and Taft values (V. P. Karavai, P. N. Gaponik and O. A. Ivashkevich, Magn. Res. Chem., 1989, 27, 611). Tetrazoles have been examined by ^{15}N-nmr spectroscopy and further information obtained on their tautomeric equilibria. The results confirm the findings from ^{1}H- and ^{13}C-nmr spectroscopy that tetrazole exists in solution predominantly in the 1-*H* tautomeric form. A large chemical shift difference of around 100ppm is observed between the two types of N atoms in 1-*H* tetrazole. The nitrogen atoms at positions 1 and 4 are pyrrole-like and their resonance position in the spectrum is at lower field than that of the nitrogens at positions 2 and 3 which are pyridine-like (G. A. Webb *et al.*, Bull. Pol. Acad. Sci., Chem., 1985, 33, 375; 1987, 35, 301; Bull. chem. Soc. Japan, 1986, 59, 3125; Org. Magn. Res., 1984, 22, 215; Magn. Res. Chem., 1985, 23, 166; J. H. Nelson *et al.*, Magn. Res. Chem., 1986, 24, 984; D. S. Wofford, D. M. Forkey and J. G. Russell, J. org. Chem., 1982, 47, 5132).

The absorption and the fluorescence spectra of a series of 5-aryltetrazoles have been recorded (I. Gryczynski *et al.*, Z. Naturforsch., 1982, 37A, 1259). Infrared spectra of tetrazole and its *N*-deuterio analogue have been analysed in detail (P. N. Gaponik *et al.*, Zh. Prikl. Spektrosk., 1990, 53, 323; Chem Abs., 1990, 113, 161468) and infrared spectroscopic data on *N*-alkyltetrazoles and their Cu(II) complexes published (*idem.*, Spectrochim Acta, 1987, 43A, 349).

Analysis of the mass spectral fragmentation of several tetrazole derivatives has appeared. For 1-methyl- and 2-methyltetrazoles significant differences in fragmentation behaviour are noted (G. I. Koldobskii *et al.*, Zh. org. Khim., 1984, 20, 2458). Benzimidazoles (24) are believed to be among the fragmentation products of 1,5-diaryltetrazoles, possibly arising by rearrangement of a nitrene (Y. V. Shirukhin *et al.*, Khim. Geterotsikl. Soedin, 1988, 925).

(24)

c) Thermolysis and Photolysis

Thermolysis of tetrazole derivatives is often related to fragmentation in their mass spectrum as exemplified by the behaviour of 1-aryl-5-methyltetrazoles which give rise to benzimidazoles and carbodiimides *via* nitrene intermediates (*idem. ibid.*, 1984, 1422).

A kinetic analysis of the thermal decomposition of a number of tetrazole derivatives is available (S. V. Vyazovkin, A. I. Lesnikovich and V. Lyutsko, Thermochem. Acta., 1990, 165, 17).

Heterocyclic transformations are frequently observed in thermolysis reactions. For example, *N*-benzoyltetrazoles yield oxadiazoles (G. I. Koldobskii *et al.*, Zh. org. Khim., 1988, 24, 1550).

Oxadiazoles are produced on thermolysis of 5-aryl-2-carbazoyltetrazoles (R. Milcent and G. Barbier, J. heterocyclic Chem., 1987, 24, 1233). 1-(*o*-Nitrophenyl)-5-phenyltetrazole (27), when heated in boiling bromobenzene, eliminates nitrogen and carbon dioxide, and *via* a series of electrocyclic ring opening and closing steps yields the fused triazole (28). (C. W. Rees, D. F. Pipe and P. G. Houghton, J. chem. Soc. Perkin 1, 1985, 1471).

Photolysis of diaryltetrazoles gives rise to nitrile imines, yet trapping with an added 1,3-dipolarophile has rarely been achieved. However, photolysis of a 1-aryl-5-phenyltetrazole (29) in the presence of methyl methacrylate yields the dihydropyrazole (30). (C. Csongar, P. Weinberg and H. Slezak, J. prakt. Chem., 1988, <u>330</u>, 629).

Photolysis of 1-(2,2-disubstitutedalk-1-enyl)-5-phenyltetrazoles (31), however, proceeds *via* an imidoyl nitrene (32) to yield a 4-*H* imidazole (33). (C.W.Rees, C.J.Moody and R.G.Young, J. chem. Soc. Perkin 1, 1987, 1389; 1991, 335).

1-Vinyltetrazole (34) loses nitrogen on photolysis to yield imidazole (E. Sato, Y. Kanaoka and A. Padwa, J. org. Chem., 1982, <u>47</u>, 4256).

$$\text{(34)}$$

Tetrazolin-5-one derivatives have been photolysed and the interesting heterocyclic products identified. The 1-allyl-4-t-butylcompound (35) eliminates nitrogen to give the diaziridinone (36) in high yield (L. Quast and U. Nahr, Ber., 1983, 116, 3427).

$$\text{(35)} \qquad \text{(36)}$$

The phenylsubstituted derivatives (37; R=H, Me, Ph), however, give a benzimidazolone (38) in excellent yield (*idem ibid.*, 1985, 118, 526).

$$\text{(37)} \qquad \text{(38)}$$

The corresponding 5-thione behaves differently, the major photolysis product being a carbodiimide.

$$PhN = C = NMe$$

Photolysis of the 5-iminoanalogues (Quast, Nahr and A. Fuss, Ber., 1985, 118, 2164) gives rise to a number of products including, for the diphenyl compound, the unusual aminobenzodiazepine (39). Trisiminomethane diradicals (40) are postulated as intermediates.

d) Other reactions

Methylation of 5-substituted tetrazolates with iodomethane gives a mixture of the 1- and the 2-methyl product with the proportions dependent on the nature of the 5-substituent, both electronic and steric effects playing a part (R. J. Spear, Austr. J. Chem., 1984, 37, 2453). Increased steric bulk *and* substituent electronegativity favour substitution at the 2-position.

$$(R = H, Me, Me_2CH, Me_3C, Ph, NH_2, NO_2, SMe, SO_2Me, Cl, CF_3)$$

Selective N-2-alkylation is achieved with an alcohol (41, R=*t*-Bu, *i*-Pr, cyclohexyl) in 80-96% sulphuric acid at room temperature (A. O. Koren and P. N. Gaponik, Khim. Geterotsikl. Soedin, 1990, 1643).

$$R' = H, Me, Ph, CF_3$$

The components of a mixture of a 1- and a 2-substituted tetrazole may be separated by a variety of techniques. One method involves the formation and then separation of the copper (II) chloride complexes. Such complexes may be used for the isolation and purification of tetrazoles. These and other complexes have been described (P. N. Gaponik, Dokl. Akad. Nauk USSR, 1984, 28, 543; Zh. Obshch. Khim., 1985, 55, 516; G. I. Koldobskii et al., ibid. 1988, 58, 825).

Chromatographic separation of 1-methyl- and 2-methyl- 5-aminotetrazole has been achieved (V. N. Andreev et al., Zh. Prikl. Khim., 1982, 55, 2103). Dichloromethane preferentially elutes the 2-methyl isomer from an activated carbon adsorbent.

Methylation of 1,5-disubstituted tetrazoles gives a mixture of 1,3,5- and 1,4,5-trisubstituted tetrazolium salts, which can be separated by fractional crystallisation (H. Quast, L. Bieber and G. Meichsner, Ann., 1987, 469).

A regioselective alkylation of 5-aryltetrazole triethylammonium salts (42) with methyl vinyl ketone has been observed. Conjugate addition at the 2-position occurs (G. I. Koldobskii et al., Zh. org. Khim., 1989, 25, 2182).

Direct substitution at the 5-position of tetrazoles can be achieved. For example, 1-alkyltetrazoles (43, R=Me, Et) react with iodine as shown below to give the 5-iododerivative (44) in 55-75% yield (P. N. Gaponik, Y. V. Grigoriev and A. O. Koren, Khim. Geterotsikl. Soedin, 1988, 1699).

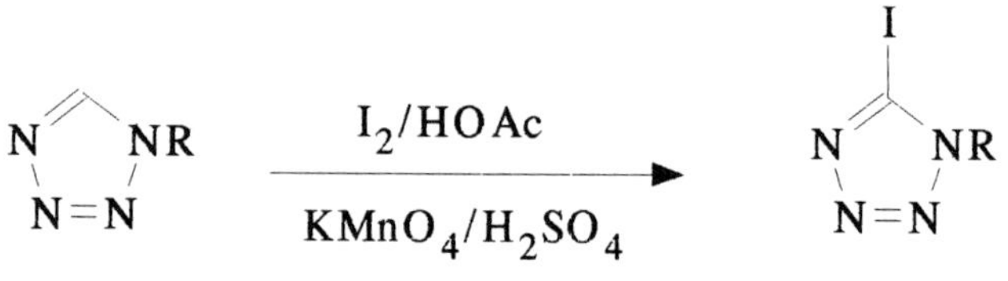

1-Methyltetrazole with dimethylamine hydrochloride and 38% aqueous formaldehyde gives the 5-(dimethylaminomethyl)-1-methyltetrazole (45). The 2-methyltetrazole does not react under similar conditions (Gaponik and V. P. Karavai, *ibid.*, 1985, 564).

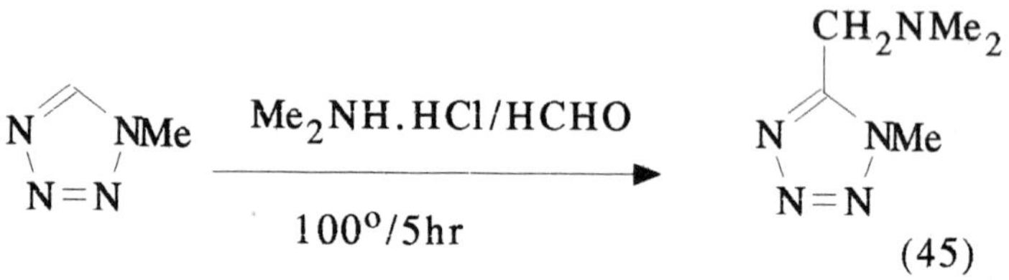

N-(α-Lithioalkyl)tetrazoles (e.g. 46) are prepared from the alkyltetrazole with t-butyl lithium.

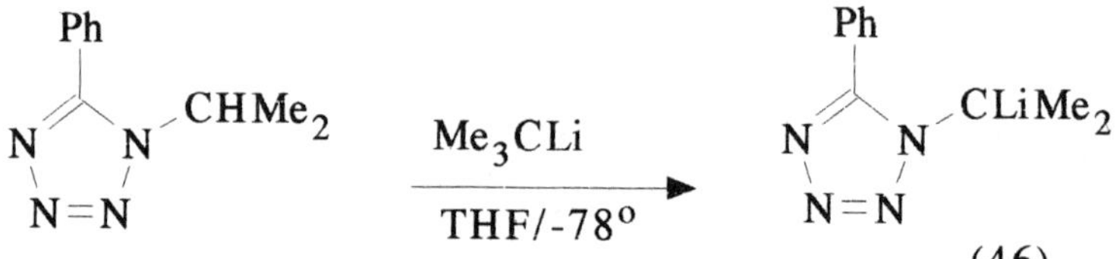

Such lithioderivatives react readily with electrophiles such as D_2O, Me_3SiCl, Me_2SO_4, ArCN, ArCHO etc to give the corresponding modified side chain (C. J. Moody, C. W. Rees and R. G. Young, Synlett., 1990, 413; J. chem. Soc. Perkin 1, 1991, 323).

Tetrazolium-*N*-phenacylides (47 and 48) are stable solids generated when the tetrazolium salts react with potassium carbonate. They react with electrophiles at the carbanionic centre (D. Moderhach and A. Lembcke, J. chem. Soc. Perkin 1, 1986, 1157).

(47)   (48)

5-Aminotetrazoles with thionyl chloride give the *N*-sulphinylamine derivative (49) which can undergo cycloaddition reactions with, for example, 2,3-dimethylbuta-1,3-diene giving a Diels-Alder adduct (50) (R. N. Butler, G. A. O'Halloran and L. A. Burke, J. chem. Soc. Perkin 2, 1989,1855).

(49)   (50)

With excess thionyl chloride, 1-alkyl-5-aminotetrazoles undergo a ring interconversion leading to 3-azido-1,2,4-thiadiazoles (51). An intermediate 5-*N*-sulphinylaminotetrazole is again involved (Butler, O'Halloran and D. A. O'Donaghue, Chem. Comm., 1986, 800; J. chem. Res.,(S), 1988, 188).

(51)

Hypobromite oxidation of 5-(*N*-alkylamino)tetrazoles is the last stage in an efficient conversion of aldehydes into isocyanides (G. Hoefle and B. Lange, Org. Synth., 1983, <u>61</u>, 14).

The mesionic tetrazolium-5-olate (52), with trifluoromethanesulphonic anhydride gives the dimeric salt (53) which can undergo nucleophilic substitution by a carbanion such as that derived from either malononitrile or ethyl cyanoacetate to give a new mesionic species (S. Araki, J. Mizuya and Y. Butsugan, J. chem. Soc. Perkin 1, 1985, 2439).

Mesionic olates can be converted into the corresponding thiolates using Lawesson's reagent (Araki, Butsugan and T. Goto, Bull. chem. Soc. Japan, 1988, <u>61</u>, 2977). Such thiolates (e.g. 54) with carbanions undergo a cycloaddition/elimination to give a thiadiazoline (55).

An air stable mesionic fulvalene analogue (57) is obtained when 5-ethoxy-1,3-diphenyl-1,2,3,4-tetrazolium tetrafluoroborate (56) reacts with sodium cyclopentadienide (Araki and Butsugan, Chem. Comm., 1983, 789).

e) Applications

(i) Pharmaceuticals

Tetrazole derivatives are potential therapeutic agents in a wide range of categories. The tetrazole ring as a mimic of the carboxylic acid group has formed the basis of many of the investigations. For example, analogues (58 and 59) of the anti-inflammatory compounds ibuprofen and flurbiprofen have been prepared (P. Valent *et al.,* Arch. Pharm., 1983, 316, 752).

(58)   (59)

Other tetrazoles with potential as anti-inflammatory agents have been described (K. Faber and T. Kappe, J. heterocyclic Chem., 1984, 21, 1881; K. Pande *et al.,* Pharmacology, 1987, 35, 333).
  The triiodoallylderivative (60) has potent antifungal activity (M. Koyama *et al.,* J. med. Chem., 1987, 30, 552).

(60)

A selection from the many references dealing with pharmaceutical applications follows:
ß-Lactam antibiotics: F. R. Atherton and R. W. Lambert, Tetrahedron, 1983, 39, 2599; A. Andrus *et al.,* J. Amer. chem. Soc., 1984, 106, 1808; D. W. Anderson, M. M. Campbell and M. Malik, Tetrahedron Letters, 1990, 31, 1755.
Leucotriene receptor antagonists: P. R. Bernstein and E. P. Vacek, Synthesis, 1987, 1133; D. M. Gapinski *et al.,* J. med. Chem., 1988 ,31, 172; H. Fu-Chih *et al., ibid.,* 1990, 33, 1194.

Anti-allergic agents: N. Peet *et al.*, J. med. Chem., 1986, <u>29</u>, 2403; J. heterocyclic Chem., 1987, <u>24</u>, 223 and 1531; 1989, <u>26</u>, 97.

Anti-diabetic agents: K. L. Kees, D. H. Prozialech and R. S. Cheesman, J. med. Chem., 1989, <u>32</u>, 11.

Anti-hypercholinesterolemic agents: P. J. Brown *et al.*, J. med. Chem., 1989, <u>32</u>, 2038.

(ii) Herbicides

1-Aryl-5-aryloxytetrazoles (61) are herbicidal (BP 2,203,739; Chem. Abs., 1989, <u>110</u>, 114842).

(iii) Polymers

Vinyl tetrazole derivatives are potential monomers for polymer production. 5-Vinyltetrazole and 5-amino-1-vinyltetrazole readily polymerise by radical initiation and the ease of polymerisation of various 5-substituted-1-vinyltetrazoles has been compared (V. N. Kizhnyaev *et al.*, Vysokomol Soedin., 1986, <u>28A</u>, 765; 1989, <u>31A</u>, 2490). 5-Vinyltetrazole can be co-polymerised with styrene (V. A. Kruglova, V. V. Annenkov and S. R. Buzilova, *ibid.*, 1986, <u>28A</u>, 257).

The 5-ureidotetrazole (62, R=alkyl) may be used as a blowing agent in polycarbonate polymerisation (USP 4,871,861; Chem. Abs., 1990, <u>112</u>, 118829).

(iv) Oligonucleotide synthesis

Tetrazole derivatives have been used as condensing agents in the search for increased efficiency in the phosphotriester approach to oligonucleotide synthesis. 1-Arylsulphonyltetrazoles such as the mesityl derivative (63) are effective for this purpose (J. Stawinski, T. Hozumi and S. A. Narang, Canad. J. Chem., 1976, 54, 670).

(63)

Other tetrazoles used in this process include the phosphoditetrazolide (64) (T. M. Cao, S. E. Bingham and M. J. Sung, Tetrahedron Letters, 1983, 24, 1019).

(64)

2. Oxatriazoles

The chemistry of oxatriazoles, together with that of the thiatriazoles (next section), has been reviewed (A. Holm in "Comprehensive Heterocyclic Chemistry" ed. A. R. Katritsky and C. W. Rees, Pergamon, 1984, Vol 6, p 579).

2,3-Dihydro-1,2,3,5-oxatriazole derivatives may be prepared by the reaction of an aldoxime with a diazadicarboxylate (G. A. Lee, USP 4,268,674; Chem. Abs., 1981, 95, 62228).

$$MeO_2CN = NCO_2Me$$

$$+$$

$$\xrightarrow[\text{NaOCl/} -15°]{\text{CH}_2\text{Cl}_2/\text{Et}_3\text{N}}$$

$$PhCH = NOH$$

[structure: ring with CO_2Me on N, O, N–CO$_2$Me, N=, Ph]

1,2,3,4-Oxatriazoles are normally mesionic species and such compounds are of interest as potential therapeutic agents, most notably anti-hypertensives (M. Lund et al., J. med. Chem., 1982, 25, 1503). Attention has been concentrated on the 3-aryl derivatives which may be prepared from an isonitrile and trinitromethane (nitroform) (J. H. Boyer, T. Moran and T. P. Pillai, Chem. Comm., 1983, 1388). With p-nitrophenyl isocyanide the reaction also yields an oxadiazole derivative (65) and p-nitroaniline.

[structure: NC-phenyl-NO$_2$] $\quad + \quad CH(NO_2)_3 \quad \xrightarrow[25°C/14hr]{\text{hexane/Et}_2\text{O}}$

[structure: NH$_2$-phenyl-NO$_2$] $\quad + \quad$ [structure: O_2N–phenyl–ring with N–N, N, +, O, O–] (42%) $\quad + \quad$ [structure: O_2N–phenyl–ring N–N, +, O, O–] (65)

With p-tolyl isocyanide no oxadiazole is formed but nitration of the benzene ring occurs.

A 3-aryl derivative is formed by the thermal cyclisation of the hydrazone (66) (M. N. Martynova *et al.*, Khim Geterotsikl. Soedin., 1981, 1682).

Acid hydrolysis of such 3-aryl substituted mesionic compounds involves the initial formation of carbon dioxide and an aryl azide. The latter then undergoes further hydrolysis (Z. Said and J. Tillett, J. chem. Soc. Perkin 2, 1982, 701).

3-*t*-Alkyl mesionic derivatives are anti-thrombotic agents (Ger. Pat. DE 3,911,688; Chem Abs., 1991, 114, 102011). These are produced by nitrosation of a *t*-alkyl semicarbazide followed by cyclisation of the resultant nitroso compound with acetic acid.

The mesionic 5-amino-3-aryl compounds (67) show anti-tumour activity (USP. 4,329,355; Chem. Abs., 1982, 97, 61018).

A study of the ^{1}H, ^{13}C, ^{14}N and ^{15}N nmr spectra of the 3-phenyl-1,2,3,4-oxatriazolium-5-thiolates (68 and 69) together with those of the analogous thiatriazolium olates (70 and 71) has been reported. (J. Jazwinski, L. Stafaniak and G. A. Webb, Mag. Res. Chem., 1988, 26, 1012). The chemical shift of N-2 can be used to distinguish the O- from the S-system.

(68)   S⁻

(69)   SEt   BF₄⁻

(70)   O⁻

(71)   OEt   BF₄⁻

3. Thiatriazoles

In the Second Edition only the 1,2,3,4-thiatriazoles were considered, and although these are the most commonly encountered compounds in this class, several syntheses of 1,2,3,5-thiatriazoles have now been reported.

As stated in the previous section, thiatriazole chemistry has been reviewed alongside the oxatriazoles in the "Comprehensive Heterocyclic Chemistry" series (Vol 6, p 579).

a) 1,2,3,4-Thiatriazoles

(i) Synthesis

Modifications to established synthetic methods have been used to prepare 1,2,3,4-thiatriazole derivatives showing commercial potential in a variety of fields, for example photographic emulsions (Jap. Pat. 62,270,949, Chem. Abs.. 1988, 109, 64177: Ger. Pat. DE 3,307,506, Chem Abs.,

1985, 102, 157,904), pharmaceuticals (Czech. Pat. 189,370, Chem Abs., 1982, 92, 69050) and explosives (H. Graubaum and D. Martin, Z. Chem., 1985, 25, 136).

Piperidinium salts of thioacids (72, R=alkyl, aryl) react with 2-chloro-1-methylpyridinium iodide to give the thioesters (73) which with sodium azide give the 5-substituted thiatriazole (74). This route is particularly valuable for the preparation of 5-alkylderivatives which are comparatively rare due to the difficulty of obtaining the precursors (thiacyl chlorides and hydrazides) required for older methods (S. Ikeda *et al.*, Synthesis, 1990, 415).

The formation of ethyl thiatriazolyldiazoacetate (76) by the reaction of 4-carboethoxy-5-chloro-1,2,3-thiadiazole (75) with sodium azide presumably involves rearrangement of the intermediate 5-azide (G. L'Abbé, M. Deketele and J. P. Dekerk, Tetrahedron Letters, 1982, 23, 1103).

The complexing ability of the 5-thiolate is reflected in a cycloaddition within a palladium complex (W. Fehlhammer and W. Beck, Z. Naturforsch., 1983, 38B, 546).

$$(Ph_3P)_2Pd(N_3)_2 \quad + \quad CS_2 \quad \longrightarrow \quad (Ph_3P)_2Pd \left[S - \underset{N-N}{\overset{S-N}{\underset{\|}{\diagdown}}} \right]_2$$

Similar behaviour has been observed within Cu(II) complexes (G. LaMonica *et al.*, J. organometallic Chem., 1984, <u>273</u>, 263).

ii) Properties and Reactions

Photolysis of the 5-phenyliminothiatriazoline (77) gives the carbodiimide (78) (H. Quast and U. Nahr, Ber., 1985, <u>118</u>, 526).

$$PhN = \underset{Me}{\overset{S-N}{\underset{N-N}{\diagdown}}} \quad \overset{h\nu}{\longrightarrow} \quad PhN = C = NMe$$

(77)   (78)

5-Phenyliminothiatriazolines (e.g. 77) react with isothiocyanates to yield thiadiazolines and dithiazolidines. The reaction involves cycloaddition and elimination *via* fused ring adducts incorporating hypervalent sulphur. A reaction scheme is given below (G. L'Abbé *et al.*, Tetrahedron, 1990, <u>46</u>, 1281; J. heterocyclic Chem., 1990, <u>27</u>, 199; Bull. Soc. chim. Belg., 1990, <u>99</u>, 391; 1989, <u>98</u>, 879).

Similarly, cycloaddition of an isothiocyanate to the imine in the opposite sense gives the thiadiazolidine thione (79).

(79)

The thermal decomposition of 5-phenyl-1,2,3,4-thiatriazole has been re-examined using a flash-pyrolysis technique, and contrary to previous reports dinitrogen sulphide, N_2S, is a product (C. Wentrup *et al.*, J. org. Chem., 1986, <u>51</u>, 1908).

5-Picryliminothiatriazolidines, prepared from picryl isothiocyanate and an alkyl azide, are particularly stable members of this class (G. L'Abbé, P. Brems and E. Albrecht, J. heterocyclic Chem., 1990, <u>27</u>, 1059).

Methylation of a 5-arylamino-1,2,3,4-thiatriazole using trimethyloxonium fluoroborate yields a mixture of the *N*-methylimine (80) and the mesionic compound (81). Diazomethane methylates the side-chain secondary amino group (G. L'Abbé, K. Buelans and L. Bastin, Bull Soc. chim. Belg., 1991,<u>100</u>, 25).

(80)

(81)

5-amino-1,2,3,4-thiatriazole undergoes a number of ring transformation reactions.

With isothiocyanates a thioureidothiadiazoline thione (82) is formed (H. Graubaum, J. Prakt. Chem., 1989, 331, 115).

(82)

With chloroformates, chlorothioformates and aryl cyanates, thiadiazoles are formed (Graubaum, H. Seeboth and P. Zalupsky, Monatsch., 1989, 120, 997).

With the thiacyl chloride (83, Ar=Ph,p-tol) the product is an aryloxy-dithiazolimine (84) (Graubaum et al., J. Prakt. Chem., 1990, 332, 208).

Condensation of 5-amino-1,2,3,4-thiatriazole with pentane-2,4-dione in a strong acid medium gives mainly the pyrimidine derivative (85) (V. A. Chuiguk and K. G. Nazarenko, Khim. Geterotsikl. Soedin., 1984, 269).

There have been a number of studies on the 1,2,3,4-thiatriazole-5-thiolate system. Quantum mechanical calculations show preference for the thiolate structure (86) rather than the thione structure (87) (M. Conti, D. W. Franco and M. Trsic, Inorg. chim. Acta., 1986, 113, 71).

Photochemical decomposition of the anion and acid forms gives sulphur, nitrogen and the thiocyanate anion (M. L. Manzano, E. Tfouni and D. W. Franco, Polyhedron, 1986, 5, 2119). Oxidation of the 5-thiolate ion (86) using permanganate in alkaline solution has been investigated over a wide range of oxidant concentrations. With a high permanganate concentration the ion undergoes a 26-electron reaction yielding sulphate, carbon dioxide and nitrite as final products, whereas with only a moderate excess of permanganate a 17-electron reaction takes place yielding sulphate, carbon dioxide and nitrogen. These two different stoichiometries arise from the existence of competitive mechanisms at different relative concentrations of thiolate and oxidant (W. L. Polito et al., Talanta, 1981 28, 867).

(b) 1,2,3,5-Thiatriazoles

Three routes to dihydro-1,2,3,5-thiatriazole-1,1-dioxides have been described (M. Knollmuller and P. Kosma, Monatsch., 1985, 116, 1321).

(i) The reaction of a benzamidrazole (88, R=H, Ph, $PhCH_2$) with sulphuryl fluoride leading to the Δ^3 derivative (89).

$$RNHC(Ph) = NNHPh + SOF_2 \longrightarrow$$

(88)

(89)

(ii) The reaction of *N'*-acylsulphamoyl hydrazide (90) with phosphorus pentachloride followed by cyclisation gives the Δ^3 derivative.

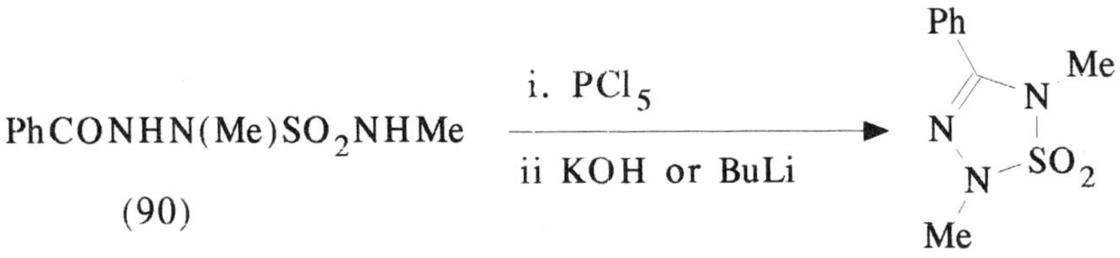
PhCONHN(Me)SO2NHMe
i. PCl5
ii KOH or BuLi
Ph
Me
N
N
SO2
N
Me

(90)

(iii) The reaction of the formamidine (91) with a hydrazine yields the Δ^4 derivative (92).

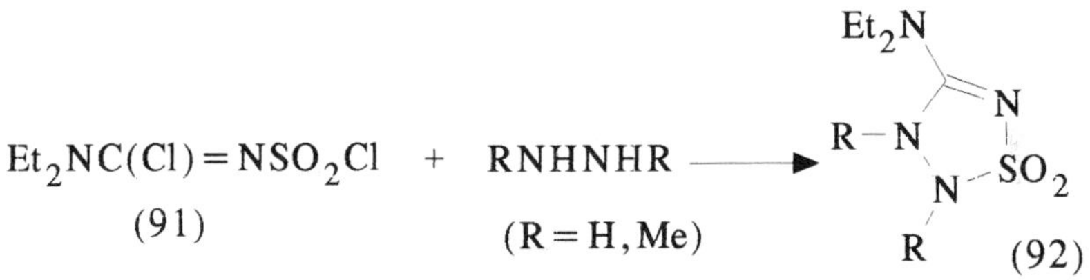
Et2NC(Cl)=NSO2Cl + RNHNHR
(R=H,Me)
Et2N
N
R-N
SO2
N
R

(91) (92)

Cycloaddition of heteroaryl *N*-sulphonylimines such as the tetrazole (93, $R=CH_3$, $PhCH_2$) with a nitrile imine gives the *S*-oxide (94) (R. N. Butler and G. A. O'Halloran, Chem. Ind., 1986, 750).

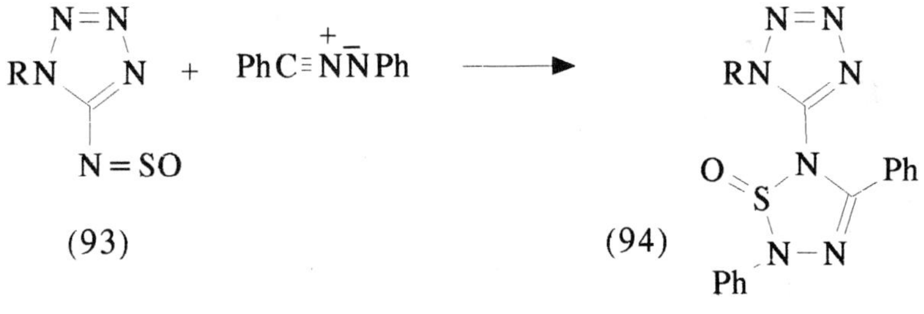
N=N
RN
N
N=SO
+
PhC≡NNPh
N=N
RN
N
O
N
Ph
S
N-N
Ph

(93) (94)

1,2,3,5-Thiatriazoles are products of the reaction of 2-alkyl-1,2,5-thiadiazol-3-one-1-oxides (95) with phenylhydrazine (S. Karady *et al.*, Tetrahedron Letters, 1985, 26, 6155).

(95)

Mesionic derivatives, described as anhydrothiatetrazolium hydroxides, are obtained from the aryl substituted thiourea (G. Felding and A. Holm, Acta. chem. Scand., 1988, B42, 63).

4. Dioxathiazoles and Oxathiadiazoles

The relative instability of many of these compounds means they are encountered only rarely. However, 4-substituted-1,2,3,5-oxathiadiazole-2-oxide derivatives (96, R=alkyl, aryl, heteroaryl etc.) have been described in a series of patents as potential anti-hyperglycemics for the treatment of diabetes (see T. R. Alessi *et al.*, Chem. Abs. 1990, 113, 23925, 40693/4/5; 1991, 114, 122379, 143424).

Synthesis involves the preparation and cyclisation of an amidoxime.

$$RCH_2Cl \longrightarrow RCH_2CN \xrightarrow[NaOMe]{NH_2OH.HCl} RCH_2C\overset{NH_2}{\underset{NOH}{\diagdown}}$$

$$SOCl_2/py \ /$$
$$CH_2Cl_2/0°$$

(96)

Cyclisation of an aryl amidoxime using thionyl chloride also provides a route to the 4,5-diaryl substituted derivative (97) (C. W. Rees, D. F. Pipe and P. G. Houghton, J. chem. Soc. Perkin 1, 1985, 1471). The oxathiadiazole (97) readily loses sulphur dioxide with rearrangement to give the carbodiimide (98).

Oxidation of 2-hydroxybenzaldehyde thiosemicarbazone with hydrogen peroxide in an ammoniacal medium catalysed by iron (III) salts gives 4-amino-2,2,5-trioxo-1,5,3,2-oxathiadiazole (99). The reaction has been described for use as a sensitive analytical method for iron (III) based on kinetic fluorimetric measurements in which the reaction is followed by monitoring the appearance rate of the fluorescence of the oxidation product (99). A calibration graph which is linear in the range 10-60 ng ml^{-1} iron (III) can be obtained. Similar catalytic effects of manganese (II) and zinc

(II) salts have been described (M. Valcarcel *et al.*, Anal. chim. Acta., 1984, 157, 333; Talanta, 1983, 30, 107; Analyst, 1983, 108, 85).

$$\text{(99)}$$

5. Dithiadiazoles and Dioxadiazoles

There has been notable interest in both 1,2,3,5- and 1,2,3,4-dithiadiazolium salts and their reduction to radical species.

a) 1,2,3,5-dithiadiazoles

A generally applicable route to 1,2,3,5-dithiadiazolium salts (102) involves the reaction of trithiazyl trichloride, usually formulated as the cyclic species (100) with a nitrile. Examples include (102, R=CF$_3$; G. M. Sheldrick *et al.*, Angew. Chem., 1984, 96, 1001) and (102, R=CCl$_3$ or Ph; L. N. Markovski *et al.*, Tetrahedron Letters, 1982, 23, 761).

The intermediate in the reaction is the dithiatriazine (101) which can be isolated under mild conditions but undergoes ring contraction on heating (A. Apblett and T. Chivers, Chem. Comm., 1989, 96; Chivers *et al.*, Inorg Chem., 1986, 25, 2119; 1989, 28, 4554; Phosphorus Sulfur, 1989, 41, 439).

(100)   (101)   (102)

$$R = Et_2N, \ Me_2N, \ (Me_2CH)_2N, \ Cl_3C, \ Me_3C.$$

If an amidine such as benzamidine (103) is used in place of the nitrile a dithiadiazolium salt is again formed but the major product is the thiatriazine (104) (P. J. Hayes *et al.*, J. Amer. chem. Soc., 1985, 107, 1346).

(103)   (104)

An earlier synthetic method based on sulphur dichloride has been modified. A 4-phenyl-1,2,3,5-dithiadiazolium chloride has been prepared by passing chlorine through a heated suspension of ammonium chloride in sulphur dichloride and benzonitrile (A. J. Banister, N. R. M. Smith and R. G. Hey, J. chem. Soc. Perkin 1, 1983, 1181).

Tris(trimethylsilyl)benzamidine reacts with sulphur dichloride to give 4-phenyl-1,2,3,5-dithiadiazolium chloride in 60% yield.

Benzamidoxime rects similarly, but perhaps the simplest route to this salt involves the reaction of benzamidine hydrochloride with sulphur monochloride in the presence of DBU (C. W. Rees *et al.*, J. chem. Soc. Perkin 1, 1989, 2495).

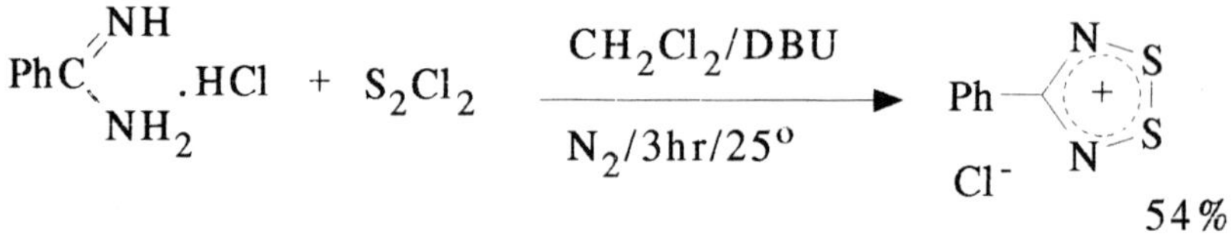

The crystal structure of this salt (as an adduct with toluene of crystallisation) has been described (A. Hazell and R. G. Hazell, Acta. Crystallogr., 1988, C44, 1807).

Sulphur dichloride reacts with F_2SNCN or di(trimethylsilyl)carbodiimide to give 4-chloro-1,2,3,5-dithiadiazolium chloride. The corresponding bromide salt (105) is similarly prepared (G. M. Sheldrick *et al.*, Ber., 1983, 116, 416).

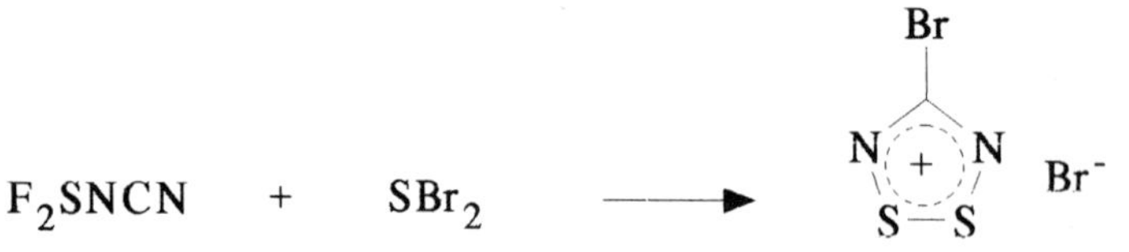

(105)

Reduction of 1,2,3,5-dithiadiazolium salts using powdered sodium in benzene or THF gives the corresponding radical (106) (L. N. Markovski *et al.*, Tetrahedron Letters, 1982, 23, 761).

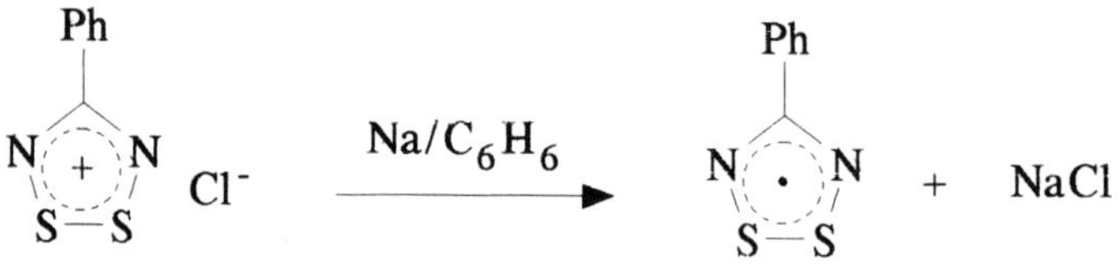

(106)

The radicals can be isolated as stable species under controlled conditions but dimerise in solution at low temperature. For esr-spectra see J. Passmore, A. J. Banister et al., (J. chem. Soc. Dalton, 1986, 1465) and for gas phase HeI photoelectron spectroscopy see R. T. Oakley et al., (J. Amer. chem. Soc., 1989, 111, 1180, 6147).

The dimeric 4-phenyldithiadiazole from (106) on reaction with atomic nitrogen generated in a cool direct current plasma undergoes an unusual insertion of atomic nitrogen into the S-S bond to produce a 1,3,2,4,6-dithiatriazine dimer.

In the crystal of the 4-methyl-1,2,3,5-dithiadiazole dimer (107) the rings overlap in parallel planes and are linked by a perpendicular S-S bond (A. J. Banister et al., J. chem. Soc. Dalton, 1989, 1705).

(107)

b) *1,3,2,4-Dithiadiazoles*

1,3,2,4-Dithiadiazolium salts are prepared by the reaction of dithionitronium hexafluoroarsenate (V) with a nitrile (J. Passmore et al., Chem Comm., 1983, 807; 1986, 140).

The crystal structure of this salt has been described (J. Passmore et al., J. chem. Soc. Dalton, 1985, 1405) and its reduction by silver powder in acetonitrile to the stable radical (108) achieved (idem. Chem. Comm., 1983, 807).

$$(108)$$

The 4-iodo and 4-trifluoromethyl salts, when reduced using triphenyl-antimony in liquid sulphur dioxide with tetramethylammonium chloride give the corresponding radicals initially, but isomerisation yields the 1,2,3,5-dithiadiazolium radical.

Thiazyl trichloride, $NSCl_3$, can be generated from $(NSCl)_3$ and SO_2Cl_2 and although unstable acts as a useful *in situ* reagent for heterocyclic synthesis (A. Apblett and T. Chivers, Chem. Comm., 1987, 1889). Thus addition of thioacetamide to a solution of $(NSCl)_3$ in SO_2Cl_2 gives 5-methyl-1,3,2,4-dithiadiazolium chloride (109) as an orange solid, m.p. 199 °C.

$$(109)$$

Under similar conditions thiobenzamide gives the 5-phenyl analogue (42% yield) but the major product is 3,5-diphenyl-1,2,4-thiadiazole (110), 51%) (A. Apblett and T. Chivers, Canad. J. Chem., 1990, <u>68</u>, 650).

$$PhC\overset{S}{\underset{NH_2}{\diagdown}} + (NSCl)_3 + SO_2Cl_2 \longrightarrow \overset{Ph}{\underset{S-N}{\underset{\oplus}{N}}}\overset{}{S} \ Cl^- + \overset{Ph}{\underset{S-N}{\diagup}}\overset{N}{\diagdown}Ph$$

$$(110)$$

The 1,3,2,4-dithiadiazolidin-5-one (112) is prepared by addition of ClSCOCl to di(methylamino)sulphone (111) in triethylamine (A. H. Cowley. S. K. Mehotra and H. W. Roesky, Inorg. Chem., 1983, 22, 2095).

$$(CH_3NH)_2SO + ClSCOCl \overset{Et_3N}{\longrightarrow} \overset{Me}{\underset{O}{\overset{S-N}{\diagdown}}}\overset{}{\underset{N}{\diagdown}}SO_2$$

$$(111) \qquad\qquad\qquad (112)\ \ Me$$

Tetrasulphurtetranitride reacts with the tin compound (113) to give 5,5-dimethyl-1,3 λ^4,2,4,5-dithiadiazastannole (114) which with fluorophosgene gives the dithiadiazole (115) as yellow crystals, m.p. 40.5 °C (H. W. Roesky and M. Witt, Inorg. Synth., 1989, 25, 49).

$$S_4N_4 + (Me_3Sn)_3N \longrightarrow Me_2Sn\overset{N=S}{\underset{S-N}{\diagdown}}\overset{COF_2}{\longrightarrow} O=\overset{N=S}{\underset{S-N}{\diagdown}}$$

$$(113) \qquad\qquad (114) \qquad\qquad (115)$$

c) Dioxadiazoles

The 1,2,3,4-dioxadiazole (116) has been claimed to be the unstable minor product of the solid phase decomposition of benzophenone oxime (M. P. Prashad and A. P. Bhaduri, Ind. J. Chem., 1980, 19B, 1074).

$$\begin{array}{c} Ph \\ \diagdown \\ \diagup \\ Ph \end{array} C = NOH \quad \xrightarrow[\text{30 days}]{\text{30°C}} \quad \begin{array}{c} Ph \quad Ph \\ \diagdown\diagup \\ N \qquad O \\ \| \qquad / \\ N - O \end{array} \quad < 1\%$$

(116)

6. Phosphorus – containing heterocycles

Five-membered rings containing four heteroatoms of which one or more is phosphorus are now well known and although not featured in the Second Edition are included here as much work has been reported since its publication. A review covering heterocyclic systems containing the P-N-N linkage considers some earlier work in this field (J. P. Majoral, Synthesis, 1978, 577).

a) Triazaphospholes

(i) 1,2,3,4-Triazaphospholes

Synthesis of these compounds involves cycloaddition of an azide to either a phosphaalkyne or a phosphaalkene derivative.

$$Me_3CC{\equiv}P \; + \; RN_3 \quad \longrightarrow \quad \begin{array}{c} R \diagdown N \diagdown N \\ \quad N \qquad N \\ \quad \diagdown \quad \diagup \\ \quad P = \diagup \\ \qquad CMe_3 \end{array}$$

(117)

The reaction of 2,2-dimethylpropylidynylphosphine with an organic azide (or hydrazoic acid) to give a 3-substituted 5-t-butyl-1,2,3,4-triazaphosphole (117) has been reported (W. Roesch and M. Regitz,

Angew. Chem. intern. Edn., 1984, 23, 900; K. Lam, Y. Yeung and R. Carrié, Chem. Comm., 1984, 1643).

Other hindered phosphaalkynes have been prepared by the sodium hydroxide-promoted elimination reaction of the phosphalkene (118) obtained from an acid chloride and tris(trimethylsilyl)phosphine (M. Regitz *et al.*, J. organometallic Chem., 1986, 306, 39; Synthesis, 1986, 31).

$$P(SiMe_3)_3 + RCOCl \longrightarrow (Me_3Si)_2PCOR$$

$$\Big\downarrow \begin{array}{c} 1,3\text{-}SiMe_3 \\ shift \end{array}$$

$$P{\equiv}CR \xleftarrow{\quad NaOH \quad} Me_3SiP{=}C\begin{array}{c} OSiMe_3 \\ R \end{array}$$

$$(118)$$

(R = isopropyl, neopentyl, 1-methylcyclohexyl, 1-methylcyclopentyl)

Two flash pyrolysis methods have been reported for the generation of the unstable methylidynephosphine (HC≡P), characterised by [3+2] cycloaddition reactions with dipoles such as methyl azide.

2,2-dimethylpropylidynylphosphine eliminates isobutene, and dichloromethylphosphine eliminates hydrogen chloride (M. Regitz *et al.*, J. organometallic Chem., 1988, 338, 329).

$$(H_3C)_3CC{\equiv}P \xrightarrow[-(CH_3)_2C=CH_2]{950^\circ\ C} HC{\equiv}P \xleftarrow[-2HCl]{1100^\circ C} H_3CPCl_2$$

Ethylidynephosphine (CH$_3$C≡P) is also unstable but can be similarly generated by flash pyrolysis of dichloroethylphosphine, eliminating hydrogen chloride which is neutralised with triethylamine (Regitz *et al.*, Phosphorus Sulfur, 1989, 46, 31).

1-Chloro-2-phenyl-2-trimethylsilyl-1-phosphaethene (119) reacts with an azide to give a 5-phenyl-1,2,3,4-triazaphosphole (G. Märkl, I. Trötsch-Schaller and W. Hoelzl, Tetrahedron Letters, 1988, 29, 785).

$$ClP{=}C\underset{SiMe_3}{\overset{Ph}{}} \quad + \quad RN_3 \longrightarrow$$

(119)

The bis(trimethylsilyl)phosphaalkene (120) with phenyl azide forms a dihydrotriazaphosphole but this is unstable and loses trimethylsilyl chloride to give the phenyl substituted heterocycle (121) (L. K. Yeung, Y. C. Yeung and R. Carrié, Chem. Comm., 1984, 1640).

$$(Me_3Si)_2C{=}PCl \quad + \quad PhN_3 \longrightarrow$$

(120)

(121)

A stable triarylphosphaalkyne with the bulky mesityl group on phosphorus adds phenyl azide to give 4-mesityl-1,5,5-triphenyl-4,5-dihydro-1,2,3,4-triazaphosphole (122) (F. Bickelhaupt et al., Tetrahedron, 1984, 40, 991).

$$Ph_2C{=}PAr \quad + \quad PhN_3 \longrightarrow$$

$$Ar = 2,4,6(Me)_3C_6H_2$$

(122)

P-Phenyl-C-aminophosphaalkenes undergo a similar cycloaddition with phenyl azide to yield a triazaphospholine (R. Carrié et al., New J. Chem., 1989, 13, 891).

$$Me_2NCH=PPh \ + \ PhN_3 \xrightarrow{-20^\circ}$$

(structure: five-membered ring with N–N at top, N–Ph, P–Ph, and NMe$_2$ substituents)

ii) 1,2,4,3-Triazaphospholes

A convenient synthesis of this ring system is the reaction of an amidrazone hydrochloride with a phosphorus triamide (J. Barrans and Y. Charbonnel, Tetrahedron, 1976, 32, 2039).

$$\begin{array}{c} R \quad NH_2 \\ C \\ \| \\ N \\ NHR' \end{array} + \ P(NEt_2)_3 \xrightarrow[\text{reflux}]{C_6H_6} \text{(triazaphosphole ring)}$$

A similar procedure leads to 5-phenyl-1,2,4,3-triazaphosphole 2-H and 4-H tautomers (J. Barrans and L. Lopez, Chem.Comm., 1984, 183).

$$\begin{array}{c} PhC(NH_2)=NN=CHPh \\ + \quad (123) \\ \\ P(NMe_2)_3 \end{array} \xrightarrow[\text{reflux}]{PhMe} \text{(2-H ring)} + \text{(4-H ring)}$$

The amidrazone imine (123) reacts with the chlorophosphine (124) to give initially the third (1-H-) tautomer (125) which slowly isomerises in solution to the 2-H-form.

$$\begin{array}{c} ClP(NMe_2)_2 \\ (124) \\ + \\ (123) \end{array} \xrightarrow[\text{0}^\circ C]{C_6H_6/Et_3N} \text{(125)} \longrightarrow \text{(2-H ring)}$$

N-Substituted analogues of these tautomers, the 1,5-, 2,5- and 4,5-disubstituted-1,2,4,3-triazaphospholes, have been studied by nmr-spectroscopy (^{1}H, ^{13}C, ^{31}P and ^{15}N) techniques (M. Haddad *et al.*, J. chem. Res.,(S), 1989, 250). Significant differences in chemical shift and coupling constant values from those observed for 1,2,4-triazole and tetrazole analogues are noted. Thus the ^{13}C-nmr chemical shift of the ring carbon in 1,2,4,3-triazaphospholes is deshielded significantly (typically 20-30 ppm) compared to that of these homologues, an effect ascribed to differences in ring conjugation.

1,2,4,3-Triazaphospholes have been used as ligands for a variety of platinum and palladium complexes (A. Schmidpeter *et al.*, J. organometallic Chem., 1983, 256, 375; Inorg. chim. Acta., 1984, 85, 1018; Inorg. Chem., 1988, 27, 2612).

Dimethyl acetylenedicarboxylate can replace the nitrile unit from the 4,5-position of the 1,2,4,3-triazaphosphole ring to give a 1,2,3-diazaphosphole by a cycloaddition-cycloreversion mechanism (A. Schmidpeter and H. Klehr, Z. Naturforsch., 1983, 38B, 1484).

Reactions at phosphorus in 1,2,4,3-triazaphospholes have been observed, leading to spirocyclic phosphorus compounds. For example:
(i) With 1,2-diketones (J. Barrans *et al.*, Tetrahedron Letters, 1984, 25, 5521).

(ii) With 1,4-diaza-1,3-dienes (*idem. ibid.*, 1986, 27, 2971).

(iii) With the dimethyl azodicarboxylate (126) an oxidative [4+1] cycloaddition at phosphorus occurs (H. Tautz and A. Schmidpeter, Ber., 1981, 114, 825).

1,2-Addition reactions of amines result in amination at phosphorus (J. Barrans and L. Lopez, Canad. J. Chem., 1986, 64, 1725). Thus 2,5-dimethyl-1,2,4,3-triazaphosphole reacts with diethylamine to give 2,5-dimethyl-3-(diethylamino)-3,4-dihydro-1,2,4,3-triazaphosphole (127).

Flash photolysis of various cyclic organophosphorus compounds including 1,2,4,3-triazaphospholes has been undertaken and the formation of transient species such as MeP: and PhP: detected (J. Barrans, L. Lopez and O. S. Diallo, Phosphorus Sulfur, 1990, 49, 363).

b) Diazadiphospholanes

1,4,2,3-Diazadiphospholan-5-one derivatives are prepared from an appropriately substituted urea with a dichlorophosphine (N. Weferling, R. Schmutzler and W .S. Sheldrick, Ann., 1982, 167; R. Vogt and R. Schmutzler, Z. Naturforsch., 1989, 44B, 690).

$(Me_3SiNMe)_2CO$

+

$MePCl_2$

$\longrightarrow$ $(MePClNMe)_2CO$ $\xrightarrow{-Cl_2}$

5-Thione derivatives are prepared similarly (H. Riffel et al., Z. anorg. allg. Chem., 1984, 508, 61).

Cyclisation of methylene bis(dichlorophosphine) with a hydrazine gives a 1,2,3,5-diazadiphospholane (128) which with triethylamine eliminates hydrogen chloride to give the diazadiphosphole (129) (A. Schmidpeter, C. Leyh and K. Karaghiosoff, Angew. Chem., 1985, 97, 127).

$CH_2(PCl_2)_2$

+

$RNHNH_2$

R = Me, Ph

$\xrightarrow{\hspace{1cm}}$

(128)

$\xrightarrow{Et_3N}$

(129)

The diazadiphosphole (129) undergoes addition reactions with HX (X=Cl, OMe, NEt_2) to reverse this final step and yield appropriately P- substituted diazadiphospholanes. A modification of the synthesis leads to tetraalkyl derivatives (D. J. Brauer et al., Ber., 1986, 119, 2767).

$CH_2[P(Cl)CMe_3]_2$

+

$MeNHNHMe$

$\xrightarrow{\hspace{1cm}}$

1,3,2,4-Diazadiphospholan-5-ones are also known and are reported to show herbicidal activity (R. Chen *et al.*, Chem. Abs., 1990, 112, 235438; 179217).

c) Oxadiazaphospholes

1,3,4,2-Oxadiazaphospholines are prepared by reaction of phosphorus pentachloride with an acyl hydrazide such as the trifluoroacetyl compound (130) (K. Tanaka *et al.*, Chem Letters, 1983, 507).

$$F_3CCONHNHPh + PCl_5 \longrightarrow$$

(130)

Such compounds break down easily on thermolysis to yield the nitrile imine ($F_3CC\equiv N^+N^-Ph$) which can be trapped by 1,3-dipolar cycloaddition reactions with alkenes (K. Tanaka *et al.*, Bull. chem. Soc. Japan, 1984, 57, 2689; J. heterocyclic Chem., 1985, 22, 565).

The preparations and the reactions of other 1,3,4,2-oxadiazaphospholines have been reported (M. M. Yusupov and A. Razhabov, Zh. Obshch. Khim., 1989, 59, 1895; 1990, 60, 1042; 1985, 55, 748. T. N. Dudchenko, S. K. Tupchienko and A. D. Sinistra, *ibid.*, 1989, 59, 1500. L. S. Khaikin, O. E. Grikina and L. V. Vilkov, J. molec. Struct., 1982, 82, 115. G. Y. Gadzhiev *et al.*, Azerb. Khim. Zh., 1981, 41; Chem. Abs., 1982, 97, 6407. A. Razhabov *et al.*, Zh. obshch. Khim., 1982, 52, 2209).

(131)

2-Chloro-5-methyl-3-phenyl-1,3,4,2-oxadiazaphospholine (131) has been used as a condensation agent in polymerisation reactions leading to polyamides (H. Kimura, I. Kawakami and F. Suzuki, Chem. Abs., 1987, <u>106</u>, 85160).

The synthesis of a 1,3,5,2-oxadiazaphospholine has been achieved by the cycloaddition of nitrosobenzene with a (methylidenamino)phosphine (J. W. Bats, W. Reid and M. Fulde, Helv., 1989, <u>72</u>, 969).

$$Ph_2C=NPPh_2 \ + \ PhN=O \ \longrightarrow$$

d) Thiadiazaphospholes

1,3,4,2-Thiadiazaphospholines are prepared by routes similar to those used for the oxygen analogues; for example, from thiobenzhydrazide and phosphorus trichloride (A. Schmidpeter *et al.*, Phosphorus Sulfur, 1982, <u>14</u>, 49).

$$PhCSNHNH_2 \ + \ PCl_3 \ \longrightarrow$$

A synthon for heterocycles with a P-S bond is the dithiadiphosphetane (132). With the hydrazine derivative (133) a 1,3,4,2-thiadiazaphospholine (134) is obtained (S. O. Lawesson *et al.*, Bull. Soc. chim. France, 1985, 62).

$$+ \ CH_3CONHNHCO_2Et \ \longrightarrow$$

(132)   (133)   (134)

With hydroxamoyl chlorides, a mild efficient synthesis of the 1,3,5,2-oxathiazaphospholine is achieved, a five membered ring compound with five different ring atoms (A. A. El-Barbary, R. Shabana and S. O. Lawesson, Phosphorus Sulfur, 1985, 21, 375).

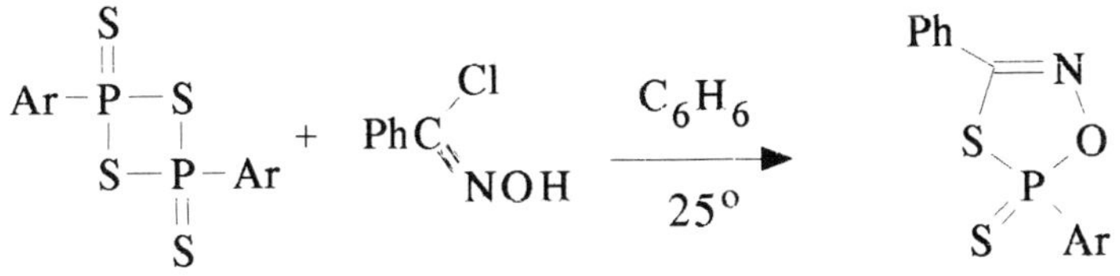

e) Tetraphospholanes

Tetraphenyltetraphospholane (135) has been observed as a by-product of the reaction of PhPHNa with dichloromethane (K. Langhans and O. Stelzer, Ber., 1987, 120, 1707).

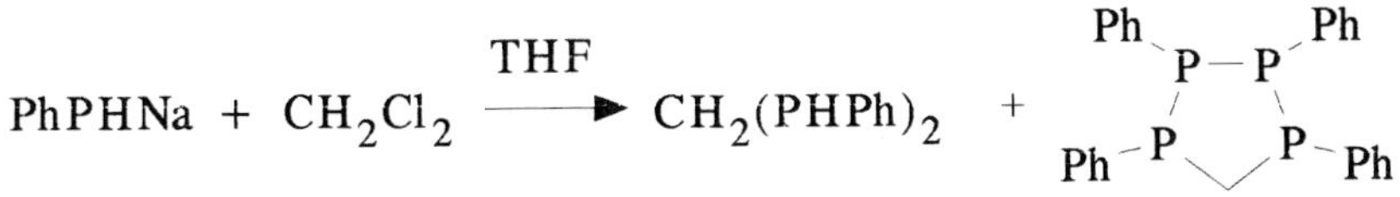

(135)

A cyclic phosphaurea, tetra t-butyltetraphospholanone (137) results from the reaction of the phosphine (136) with phosgene (R. Appel and W. Paulen, Ber., 1983, 116, 109).

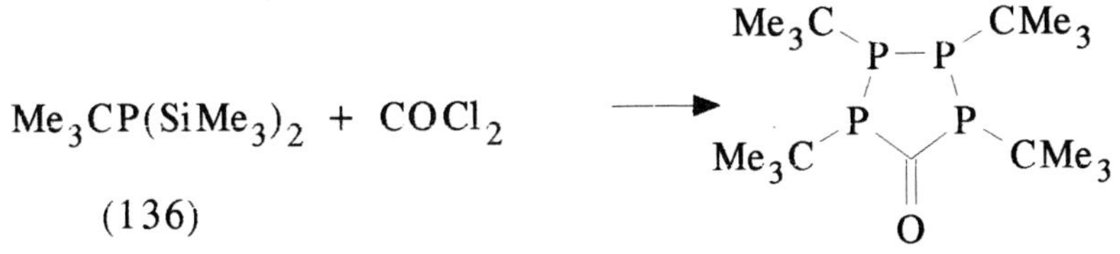

$$Me_3CP(SiMe_3)_2 + COCl_2 \longrightarrow$$

(136)

(137)

Guide to the Index

This index is constructed in a similar manner to the volume indexes of the first edition of the Chemistry of Carbon Compounds. However, to make the index easier to use, more descriptive entries have been made for the commonly occurring individual, and groups of chemicals.

The indexes cover primarily the chemical compounds mentioned in the text, and also include reactions and techniques, where named, and some sources of chemical compounds such as plant and animal species, oils, etc.

Chemical compounds have been indexed alphabetically under the names used by authors, editing being restricted to ensuring uniformity of entries under the same heading. In view of the alternative nomenclature that can often be used, a limited amount of cross-referencing has been done where it is considered to be helpful, but attention is particularly drawn to Convention 2 below.

For this and the succeeding volumes, the indexing conventions listed below have been adopted.

1. *Alphabetisation*

(a) The following prefixes have not been counted for alphabetising:

n-	*o-*	*as-*	*meso-*	D	C
sec-	*m-*	*sym-*	*cis-*	DL	*O-*
tert-	*p-*	*gem-*	*trans-*	L	*N-*
	vic-				*S-*
		lin-			*Bz-*
					Py-

Some prefixes and numbering have been omitted in the index, where they do not usefully contribute to the reference.

(b) The following prefixes have been alphabetised:

Allo	Epi	Neo
Anti	Hetero	Nor
Cyclo	Homo	Pseudo
	Iso	

(c) A letter by letter alphabetical sequence is followed for entries, firstly for the main entry, followed by the descriptive entry. The only exception to this sequence is the placing of plural entries in front of the corresponding individual entries to prevent these being overlooked by a strict alphabetical sequence which could lead to a considerable separation of plural from individual entries. Thus "butanes" will come before *n*-butane, "butenes" before 1-butene, and 2-butene, etc.

2. Cross references

In view of the many alternative trivial and systematic names for chemical compounds, the indexes should be searched under any alternative names which may be indicated in the main body of the text. Only a limited amount of cross-referencing has been carried out, where it is considered that it would be helpful to the user.

3. Esters

In the case of lower alcohols esters are indexed only under the acid, e.g. propionic methyl ester, not methyl propionate. Ethyl is normally omitted e.g. acetic ester.

4. Derivatives

Simple derivatives are not normally indexed if they follow in the same short section of the text.

5. Collective and plural entries

In place of "– derivatives" or "– compounds" the plural entry has normally been used. Plural entries have occasionally been used where compiunds of the same name but differing numbering appear in the same section of the text.

6. Main entries

The main entry of the more common individual compounds is indicated by heavy type. Multiple entries, such as headings and sub-headings over several pages are shown by "–", e.g., 67–74, 137–139, etc.

Index

452